Cardamom

Medicinal and Aromatic Plants – Industrial Profiles

Individual volumes in this series provide both industry and academia with in-depth coverage of one major medicinal or aromatic plant of industrial importance.

Edited by Dr Roland Hardman

(Continued)

Cardamom
The genus *Elettaria*

Edited by

P.N. Ravindran

Indian Institute of Spices Research,
Kozhikode (Calicut), Kerala, India

and

K.J. Madhusoodanan

Indian Cardamom Research Institute,
Myladumpara, Kerala, India

CRC Press
Taylor & Francis Group
Boca Raton London New York

CRC Press is an imprint of the
Taylor & Francis Group, an **informa** business

A TAYLOR & FRANCIS BOOK

CRC Press
Taylor & Francis Group
6000 Broken Sound Parkway NW, Suite 300
Boca Raton, FL 33487-2742

First issued in paperback 2019

© 2002 by Taylor & Francis Group, LLC
CRC Press is an imprint of Taylor & Francis Group, an Informa business

Typeset in Garamond by
Integra Software Services Pvt. Ltd, Pondicherry, India

No claim to original U.S. Government works

ISBN-13: 978-0-415-28493-6 (hbk)
ISBN-13: 978-0-367-39574-2 (pbk)

British Library Cataloguing in Publication Data
A catalogue record for this book is available from the British Library

Library of Congress Cataloging in Publication Data
A catalog record for this book has been requested

Visit the Taylor & Francis Web site at
http://www.taylorandfrancis.com

and the CRC Press Web site at
http://www.crcpress.com

Dedicated to
Prof. K.L. Chadha
National Professor (Horticulture) &
Former Deputy Director General (Hort.), ICAR

A visionary, whose untiring efforts led to the documentation of horticultural research accomplishments in India;

An administrator par excellence;

A person of rare distinction, a pathfinder, whose footsteps will remain imprinted in the portals of Indian Horticulture.

Contents

Contributors

Dr K. Nirmal Babu
Senior Scientist (Biotechnology)
Indian Institute of Spices Research
P. Box No. 1701, Calicut-673 012
Kerala, India

Dr A.K. Biswas
Scientist (Botany)
ICRI Regional Station
Gangtok-735 102
Sikkim, India

Dr S.S. Chandrasekar
Scientist (Entomology)
ICRI Regional Research Station
Donigal, Sakleshpur-573 134
Hassan Dist., Karnataka, India

Dr C.K. George
Advisor
Peermade Development Society
P. Box No. 11, Peermade-685 531
Idukki Dist., Kerala, India

Dr B. Gopakumar
Senior Scientist (Entomology)
Indian Cardamom Research Institute
Regional Station
Thadiyankudisai-624 212
Tamil Nadu

Dr V.S. Korikanthimath
Principal Scientist
IISR Cardamom Research Centre
Madikeri-571 201, Appangala
Kodagu, Karnataka, India
[P.a: Director, ICAR Research Complex
    for Goa, Ela, Old Goa, 403 402
    Goa, India]

Dr V. Krishnakumar
Senior Scientist (Agronomy)
Indian Cardamom Research Institute
Myladumpara, Kailasanadu-685 553
Idukki Dist., Kerala, India
[P.a: Senior Scientist
    Central Plantation Crops Research
    Institute
    Kasaragod-670 124, Kerala, India]

Dr N. Krishnamurthy
Deputy Director & Head
Division of Food Processing
Central Food Technological
    Research Institute
Mysore-570 013
Karnataka State, India

Dr K. Pradip Kumar
Scientist (Botany)
Indian Cardamom Research Institute
Myladumpara
Kailasanadu-685 553
Idukki Dist., Kerala, India

Dr M.S. Madan
Senior Scientist (Economics)
Indian Institute of Spices Research
P. Box No. 1701
Calicut-673 012
Kerala, India

Dr K.J. Madhusoodanan
Project Officer
Indian Cardamom Research Institute
Myladumpara, Kailasanadu-685 553
Idukki Dist., Kerala, India

Dr S.N. Potty
Director
Indian Cardamom Research Institute
Myladumpara
Kailasanadu-685 553
Idukki Dist., Kerala
India

Dr V.V. Radhakrishnan
Scientist (Botany)
Indian Cardamom Research Institute
Myladumpara
Kailasanadu-685 553
Idukki Dist., Kerala, India

Dr P. Rajeev
Scientist (Extension)
Indian Institute of Spices Research
P. Box No. 1701, Calicut-673 012
Kerala, India

Dr P.N. Ravindran
Principal Scientist & Project Coordinator
Indian Institute of Spices Research
P. Box No. 1701, Calicut-673 012
Kerala, India
[P.a: Coordinating Director
   Centre for Medicinal Plants Research
   Arya Vaidya Sala, Kottakkal-676 503
   Kerala, India]

Dr S.R. Sampathu
(Scientist E)
Division of Food Processing
Central Food Technological Research
   Institute
Mysore-570013
Karnataka State, India

Dr M. Shylaja
Lecturer (Sel. Gr.)
Department of Botany
Providence Women's College
Calicut-673 009, Kerala, India

Dr R. Suseela Bhai
Scientist (Plant Pathology)
Indian Cardamom Research Institute
Myladumpara
Kailasanadu-685 553
Idukki Dist., Kerala, India
[P.a: Senior Scientist
   Indian Institute of Spices Research
   P. Box No. 1701, Calicut-673 012
   Kerala, India]

Dr Joseph Thomas
Deputy Director (Res)
Indian Cardamom Research Institute
Myladumpara
Kailasanadu-685 553
Idukki Dist., Kerala, India

Dr S. Varadarasan
Senior Scientist (Entomology)
Indian Cardamom Research Institute
Myladumpara
Kailasanadu-685 553
Idukki Dist., Kerala, India

Dr M.N. Venugopal
Principal Scientist (Pathology)
IISR Cardamom Research Centre
Madikeri-571 201, Appangala
Kodagu, Karnataka, India

Dr K.K. Vijayan
Professor
Department of Chemistry
Calicut University
Calicut-673 635
Kerala, India

Dr T. John Zachariah
Senior Scientist (Biochemistry)
Indian Institute of Spices Research
P. Box No. 1701
Calicut-673 012
Kerala, India

Preface to the series

There is increasing interest in industry, academia and the health sciences in medicinal and aromatic plants. In passing from plant production to the eventual product used by the public, many sciences are involved. This series brings together information which is currently scattered through an ever increasing number of journals. Each volume gives an in-depth look at one plant genus, about which an area specialist has assembled information ranging from the production of the plant to market trends and quality control.

Many industries are involved such as forestry, agriculture, chemical, food, flavour, beverage, pharmaceutical, cosmetic and fragrance. The plant's raw materials are roots, rhizomes, bulbs, leaves, stems, barks, wood, flowers, fruits and seeds. These yield gums, resins, essential (volatile) oils, fixed oils, waxes, juices, extracts and spices for medicinal and aromatic purposes. All these commodities are traded worldwide. A dealer's market report for an item may say 'Drought in the country of origin has forced up prices'.

Natural products do not mean safe products and an account of this has to be taken by the above industries, which are subject to regulation. For example, a number of plants which are approved for use in medicine must not be used in cosmetic products.

The assessment of safe-to-use starts with the harvested plant material which has to comply with an official monograph. This may require absence of, or prescribed limits of, radioactive material, heavy metals, aflatoxin, pesticide residue, as well as the required level of active principle. This analytical control is costly and tends to exclude small batches of plant material. Large scale contracted mechanised cultivation with designated seed or plantlets is now preferable.

Today, plant selection is not only for the yield of active principle, but for the plant's ability to overcome disease, climatic stress and the hazards caused by mankind. Such methods as *in vitro* fertilization, meristem cultures and somatic embryogenesis are used. The transfer of sections of DNA is giving rise to controversy in the case of some end-uses of the plant material.

Some suppliers of plant raw material are now able to certify that they are supplying organically-farmed medicinal plants, herbs and spices. The European Union directive (CVO/EU No 2092/91) details the specifications for the *obligatory* quality controls to be carried out at all stages of production and processing of organic products.

Fascinating plant folklore and ethnopharmacology leads to medicinal potential. Examples are the muscle relaxants based on the arrow poison, curare, from species of *Chondrodendron*, and the anti-malarials derived from species of *Cinchona* and *Artemisia*. The methods of detection of pharmacological activity have become increasingly reliable and specific, frequently involving enzymes in bioassays and avoiding the use of laboratory animals. By using bioassay-linked fractionation of crude plant juices or extracts, compounds can be specifically targeted which, for example, inhibit blood platelet aggregation, or have anti-tumour, or anti-viral, or any other required activity. With the assistance of robotic devices, all the members of a genus may be readily screened. However, the plant material must be *fully* authenticated by a specialist.

The medicinal traditions of ancient civilisations such as those of China and India have a large armamentaria of plants in their pharmacopoeias which are used throughout South-East Asia. A similar situation exists in Africa and South America. Thus, a very high percentage of the World's population relies on medicinal and aromatic plants for their medicine. Western medicine is also responding. Already in Germany all medical practitioners have to pass an examination in phytotherapy before being allowed to practise. It is noticeable that throughout Europe and the USA, medical, pharmacy and health related schools are increasingly offering training in phytotherapy.

Multinational pharmaceutical companies have become less enamoured of the single compound magic bullet cure. The high costs of such ventures and the endless competition from "me too" compounds from rival companies often discourage the attempt. Independent phyto-medicine companies have been very strong in Germany. However, by the end of 1995, eleven (almost all) had been acquired by the multinational pharmaceutical firms, acknowledging the lay public's growing demand for phytomedicines in the Western World.

The business of dietary supplements in the Western World has expanded from the health store to the pharmacy. Alternative medicine includes plant-based products. Appropriate measures to ensure the quality, safety and efficacy of these either already exist or are being answered by greater legislative control by such bodies as the Food and Drug Administration of the USA and the recently created European Agency for the Evaluation of Medicinal Products, based in London.

In the USA, the Dietary Supplement and Health Education Act of 1994 recognised the class of phytotherapeutic agents derived from medicinal and aromatic plants. Furthermore, under public pressure, the US Congress set up an Office of Alternative Medicine and this office in 1994 assisted the filing of several Investigational New Drug (IND) applications, required for clinical trials of some Chinese herbal preparations. The significance of these applications was that each Chinese preparation involved several plants and yet was handled as a *single* IND. A demonstration of the contribution to efficacy, of *each* ingredient of *each* plant, was not required. This was a major step forward towards more sensible regulations in regard to phytomedicines.

My thanks are due to the staff of Taylor & Francis publishers who have made this series possible and especially to the volume editors and their chapter contributors for the authoritative information.

Roland Hardman

Preface

Among the crops used by humankind, the history of spices is perhaps the most adventurous, and most romantic. In the misty distant past, when the primitive man was roaming around the forests in search of food and shelter, he might have tested and tasted many plants, and might have selected those that were aromatic and spicy as of special value, and used them to propitiate his primitive gods to save him from the raging storm, thunder and rain. Out of the darkness of those distant past blossomed the early human civilizations. In all civilizations the aromatic plants were given special status, and many were probably used as offerings to gods. Gradually man might have started using them for curing various illnesses and in course of time spices and aromatic plants had acquired magical associations about their properties.

From the dawn of human civilization spices were sought after as eagerly as that of gold and precious stones. Discovery of the spice land was one of the aims of most circumnavigations and great explorations that the period of Renaissance had witnessed. One such navigational exploration in search of the famed land of spices, reached the Malabar Coast on May 20, 1498. Vasco da Gama discovered the sea route to India. Subsequently the Portuguese established trade relations with the Malabar Coast and was trading on pepper, cardamom, ginger and cinnamon.

Cardamom, otherwise known as Malabar cardamom, true cardamom or small cardamom, is often qualified as the queen of spices, because of its very pleasant, mild aroma and taste, and it was next to pepper in importance. It was the Arab traders who took cardamom from the Malabar Coast and traded with ancient and medieval Greece and Rome, and from there this spice had spread to other Mediterranean and West European countries. Though we are not yet sure whether the "cardamomum" known to ancient Greece and Rome is the same as the present cardamom.

The Western Ghat forests of the Malabar Coast (the present Kerala) is the centre of origin and diversity for cardamom. It might have been nature's design, that the king and queen of spices (black pepper and cardamom) originated in the same Western Ghat forests. Cardamom was collected mainly by tribals as a forest produce. Later during the close of 19th century, cardamom plantations came up in Western Ghat forest areas and also in Ceylon (Sri Lanka). The credit for starting organized cultivation of cardamom goes to the British planters. Later cardamom was taken to Guatemala from Ceylon, and that country by 1970s became the largest commercial producer of cardamom, eclipsing the monopoly of India.

Research and development work in cardamom started only in the last quarter of the last century. Though much information has been generated during the past decades, no effort has so far been made to collect and collate the information in one place. This volume is the first attempt in this direction. I have tried to bring together the best talents available in the field of cardamom research to author the various chapters.

The present volume contains sixteen chapters, fourteen on cardamom, one each on large cardamom and false cardamoms. All aspects of the cardamom crop have been covered in this volume, and each chapter is written by experts in the respective fields. The research work on cardamom is concentrated in India, and little is being done elsewhere. Hence, the authors of various chapters of this volume are from India only, from various agencies involved in cardamom research and development. This volume is visualized both as a textbook and reference work for scientists and students of horticulture, plantation crops, botany and related fields. It is hoped that this book will continue to serve as the main source book and reference volume on cardamom at least for the next one decade.

I am thankful to Taylor & Francis publishers for the confidence bestowed on me. There can be lapses, for which I am solely responsible. I hope that this book will be helpful in kindling interest on this crop in the minds of readers and will act as a catalyst for more research into the many problems besetting this wonderful spice.

P.N. Ravindran
December 20, 2000

Acknowledgements

The senior editor (PNR) would like to express his deep gratitude to each one of the contributors for agreeing to author the various chapters of this monograph.

I have received a lot of help and support from the project staff of the Biotechnology laboratory of IISR, and I am grateful to them. I have a special word of thanks to Ms Minoo Divakaran and Ms Geetha S. Pillai for their sincere help in the preparation of this volume. I am thankful to Dr K.V. Peter, Former Director of the Indian Institute of Spices Research for his constant encouragement and support. I am also thankful to Drs M.N. Venugopal, R. Ramakrishnan Nair and K. Nirmal Babu for supplying unpublished materials for inclusion in this monograph.

I have received immense help from Mr K. Jayarajan and Ms M. Seema of the IISR Computer Cell in the preparation of this volume, and I would like to express my gratitude to them. I am thankful to Mr A. Sudhakaran for the cover page design and for the drawings incorporated in the chapters on Botany and crop improvement and Processing. I am also thankful to the Indian Council of Agricultural Research for giving permission to edit this volume.

In this effort of collecting and collating the information on cardamom, I have made use of many published material by the cardamom workers. I am sure the contributions of most cardamom research workers find a place in this book, and that their studies and findings will live long through the pages of this volume. I am thankful to all of them, whose research and writings gave us the present insight and knowledge base on the queen of spices, Cardamom.

Finally I express my deep appreciation to Taylor & Francis publishers for providing this great opportunity of producing this volume.

P.N. Ravindran

1 Introduction

P.N. Ravindran

Cardamom is a large perennial, herbaceous rhizomatous monocot, belonging to the family Zingiberaceae. It is a native of the moist evergreen forests of the Western Ghats of southern India, which incidentally is also the centre of origin and diversity for black pepper (*Piper nigrum* L). The cardamom of commerce is the dried ripe fruit (capsules) of cardamom plant. This is often referred as the "Queen of Spices" because of its very pleasant aroma and taste, and is highly valued from ancient times. It is grown extensively in the hilly regions of South India at elevations of 800–1300 m as an under crop in forest lands. Cardamom is also grown in Sri Lanka, Papua New Guinea, Tanzania and in Guatemala. It is grown on a commercial scale in Guatemala, which incidentally is also now the largest producer of cardamom.

Cardamom belongs to the genus *Elettaria*, and species *cardamomum* (Maton). The genus name is derived from the Tamil root *Elettari*, meaning cardamom seeds. The genus consists of about six species (Mabberly, 1987). Only *E. cardamomum* Maton occurs in India, and this is the only economically important species. The closely related *E. ensal* (Gaertn.) Abeywick. (*E. major* Thaiw.), a native of Sri Lanka, is a much larger and sturdier plant; and is known as the Sri Lankan (Ceylon) wild cardamom, and its taste and flavour are far inferior to the true cardamom. The Malaysian species, *E. longituba* (Ridl.) Holtt., is a large perennial herb, its flowering panicles sometimes reaching a length of over 3 m (Holttum, 1950). Its flowers appear singly, and it seems that the cincinnus stops flowering as soon as a fruit is formed. The fruit is large, and is not used. Sakai and Nagamasu (2000) listed seven species from Borneo (Indonesia). The related genera are *Elettariopsis* and *Cyphostigma*, both genera occur in Malaysia–Indonesia region.

1 HISTORY

Cardamom is known to be in use in India from ancient times. It is known as *Ela* in Sanskrit and references to this can be found in ancient Sanskrit texts. *Taitreya Samhita*, which belongs to the later Vedic period (ca. 3000 BC), contains mention of cardamom among the ingredients to be poured in the sacrificial fire on the occasion of a marriage ceremony (Mahindru, 1982). The ancient Indian Ayurvedic texts, *Charaka Samhita* and *Susrutha Samhita*, written in the post-epic period (1400–600 BC) also mention cardamom on many occasions; although it is not clear whether the *ela* mentioned in those texts is cardamom or large (Nepal) cardamom, it is the latter that occurs in the northeastern region of India.

Both Babylonians and Assyrians were quite well informed about many plants of medicinal importance. Assyrian doctors and chemists were known to use many herbs, and among the 200 or so plants known to them cardamom, cumin, dill, fennel, *Origanum*, thyme, saffron and sesame have been identified (Parry, 1969). It was mentioned that the ancient king of Babylon, Merodach-Baladan II (721–702 BC) grew cardamom among other herbs in his garden. However, in the ancient Egyptian texts there was no mention of the use of cardamom. Probably at that time cardamom might have been reaching the ancient Babylonia and Asseria through the land routes. We come across references of cardamom in ancient Greece and Rome. In those days in Greece and Rome, spices were symbols of luxury, and they occupied a proud place in social ceremonies and functions. Cardamom, cassia, cinnamon and sweet marjoram were among the ingredients of their perfumes, while anise, basil, fennel leaves, coriander and garlic were among their aphrodisiacs (Parry, 1969). Dioscorides (40–90 AD), the Greek physician and the author of the famous *Materia Medica*, mentioned cardamom, cinnamon, ginger, pepper and turmeric and many herbs as useful medicines. Cardamom has been recommended as an aid in digestion especially after heavy meals, and this was probably one of the reasons for importing large quantities of cardamom from India to Greece and Rome. It was one of the most popular oriental spices in Roman cuisine. Because of its importance, cardamom was listed as an item liable to duty at Alexandria in AD 176.

Sitting far away from that distant past one cannot be sure of the exact identity of the plants that the ancients used. This is true of cardamom as well, as there seems to have some confusion existing in the literature on cardamom. Linschoten in his Journal of Indian Travels (1596) describes two forms of cardamoms used in South India; and he called them Lesser and Greater cardamom. Does it mean that the Nepal (large) cardamom was reaching South India 4000 years ago? Dymock, referring to the introduction of cardamom into Europe writes "when they were first introduced into Europe is doubtful, as their identity with the *Amomum* and *Cardamomum* of the Greeks and Romans cannot be proved". Linschoten writes about lesser cardamom as "it most growth in Calicut and Cannanore, places on the coast of Malabar". Paludanus, contemporary of Linschoten, wrote that according to Avicenna, there are two kinds of cardamoms, the Greater and the Lesser, and continues to add that cardamom was unknown to Greeks such as Galen, and Dioscorides. Galen in his *Seventh Book of Simples* said, "cardamom is not so hot as Nasturium or water cresses", "but pleasanter of savour and smell with some small bitterness" and the properties indicated do not agree with that of the Indian cardamom. Dioscorides, in his *First Book* commented on the cardamom brought from Armenia and Bosphorus and added that "we must choose that which is full, and tough in breaking, sharp and bitter of taste, and the smell there of, cause the heaviness in a man's head" (Watt, 1872). Evidentially Dioscorides was writing not about the Indian cardamom, but some other distinctly different material. Such evidences led Paludanus (Watt, 1872) to conclude that the *Amomum* and *Cardamomum* of the ancient Greeks were not the spices of India.

In general, the references about the use of cardamom in ancient and early centuries of the Christian era and even in the middle ages are much scanty compared to spices like black pepper, cinnamon or cassia. Even Auboyar in his classical work on daily life in ancient India from 200 BC–700 AD, gave only passing references about cardamom (Mahindru, 1982).

Arabs were the major traders of Indian spices and they were successful in hoodwinking the Mediterranean merchants by keeping the sources of the spices a secret.

Cardamom was no exception, and even historians like Pliny thought that cardamom was produced in Arabia. This situation continued till the discovery of the sea route and the landing of Portuguese in the West Coast of India. This event was the beginning of the end of the Arabian monopoly on spices trade, which they have enjoyed for so long. The Portuguese started collecting and exporting pepper, ginger and cardamom directly to Europe.

The European colonizers were more interested in the procurement of black pepper and as a result pepper as well as ginger cultivation and production picked up considerably during the sixteenth–eighteenth century. Cardamom was considered a minor forest produce. Only in the beginning of the nineteenth century plantations were established for cardamom cultivation, that too as a secondary crop in coffee plantations. But its cultivation spread rapidly in the hilly terrains of Western Ghats and the portion south of the Palghat gap came to be known as Cardamom Hills.

The earliest writings on cardamom growing in India were those of the officers working for the British East India Company. The most important among such writings was that of Ludlow, an Assistant Conservator of forests. The other sources were Pharmacographia, Madras Manual and Rices Manual. Watt (1872) also described briefly the cardamom cultivation in South India. This system of cardamom collection from naturally growing plants continued till 1803 at least, but in later years the demand became too large, and large-scale organized cultivation was started in India and Ceylon (Ridley, 1912).

Previously in the native states of Travancore and Cochin, cardamom was a monopoly of the respective governments. The *Raja* (King) of Travancore made it compulsory that all the produce shall be sold to his official, who forwarded it to the main depot in Alleppey, then the most important port of the state of Travancore. At Alleppey, cardamom was sold by auction. The main buyers were the Muslim merchants, and the best quality (designated as Alleppey Green) was reserved for export. In the forest lands owned by the British Government cardamom was a miscellaneous produce, while in Coorg, forest lands were leased out to private individuals for cardamom cultivation. It was noted by Clighorn, the Conservator of forests of Madras Presidency, that the spread of coffee led to the eclipse of cardamom in many areas of the "Malabar mountains" (Watt, 1872).

In the Madras Manual there were references about how cardamom was grown. It is mentioned there that "in the hills (of Travancore) the cardamom grows spontaneously in the deep shade of the forests: it resembles some what the turmeric and ginger plants, but grows to a height of 6 to 10 feet, and throws out the long shoots which bear the cardamom pods". The type of cardamom management is clear from the following passage:

> The owners of the gardens, early in the season come up from the low country east of the Ghats, cut the brushwood and burn the creepers and otherwise clear the soil for the growth of the plants as soon as the rains fall. They come back to gather the cardamom when they ripen, about October or November (Watt, 1872).

From the writings of British officials we also learn that a process of bleaching of cardamom was carried out in Karnataka, and for carrying out this, cardamom was transported to a place in Dharwar (Haveri) for bleaching with the aid of water from a particular well which is supposed to have the power of bleaching and improving the flavour of dried cardamom fruit (Watt, 1872). Mollison (1900) describes an elaborate method of bleaching cardamom using soapnut-water.

2   THE PRESENT SCENARIO

The cardamom producing areas are given in Fig. 1.1. Today, its production is concentrated mainly in India and Guatemala. Cardamom was introduced in Guatemala only in early 1920s from Sri Lanka or India with the help of a New York broker and was planted in the vicinity of Coban in the Department at Alta Verapaz (Lawrence, 1978). After the second world war cardamom production in Guatemala expanded considerably mainly because of the shortage for cardamom and the high prize prevailing at that time, and soon became the largest cardamom producer in the world. The natives do not relish the taste of cardamom, and so almost the entire production is exported. Currently, Guatemala produces around 13,000–14,000 tons of cardamom annually.

In India the area under cardamom has come down over the last one decade from 1,05,000 ha during 1987–88 to 69,820 ha during 1997–98, while production has gone up from 3200 tons during 1987–88 to 9290 tons in 1999–2000. During the period, the productivity has increased from almost 47 kg/ha to 173 kg/ha (Table 1.1). Cardamom cultivation is located in three states; Kerala, Karnataka and Tamil Nadu. Kerala has 59 per cent of area and contributes 70 per cent of production; Karnataka 34 per cent of area and 23 per cent production, while Tamil Nadu has 7 per cent area as well as production. Most of the cardamom growing areas in Kerala is located in the districts of Idukki, Palakkad and Wynad. In Karnataka, cardamom is grown in Coorg, Chikmagalur and Hassan districts and to a lesser extent in North Kanara district. In Tamil Nadu, cardamom cultivation is located in certain localities in Pulney and Kodai hills (Table 1.2). In India it is a small holders' crop, almost 40,000 holdings covering an area of 80,000 ha of cardamom (George and John, 1998).

The cardamom growing regions of South India lies within 8° and 30° N latitudes and 75° and 78° 30′ E longitudes. Cardamom growing areas are located at elevations ranging from 800–1300 m above mean sea level (msl) and these areas lie on both sides – the windward and leeward – of the Western Ghats which acts as a climatic barrier of the monsoon trade winds, thereby determining the spatial distribution of rainfall. The rainfall pattern differs among the cardamom growing regions located in Kerala, Tamil Nadu and Karnataka (Nair *et al.*, 1991).

The productivity increase in recent years is due to the use of high-yielding varieties and better agro-production technology. However the export of cardamom has touched

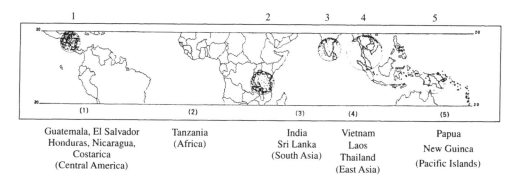

Figure 1.1 Cardamom growing countries.

Table 1.1 Area, production and productivity of cardamom in India

Year	Area (ha)	Growth index	Production (mt)	Growth index	Productivity (kg/ha)	Growth index
1985–1986	100,000	109.31	4700	148.26	47.00	135.64
1986–1987	105,000	114.78	3800	119.87	38.00	109.67
1987–1988	105,000	114.78	3200	100.95	30.48	87.96
1988–1989	81,113	88.67	4250	134.07	40.48	116.83
1989–1990	81,113	88.67	3100	97.79	38.22	110.30
1990–1991	81,554	89.15	4750	149.84	58.24	168.08
1991–1992	81,845	89.47	5000	157.73	61.09	176.31
1992–1993	82,392	90.06	4250	134.07	51.58	148.86
1993–1994	82,960	90.69	6600	208.20	79.56	229.61
1994–1995	83,651	91.44	7000	220.82	83.68	241.50
1995–1996	83,800	91.60	7900	249.21	94.27	272.06
1996–1997	72,520	79.27	7290	208.99	100.52	290.10
1997–1998	69,820	76.32	7150	225.55	102.40	295.53
1998–1999	72,135	78.85	7170	226.18	135.00	389.61
1999–2000	72,451	79.20	9290	293.06	173.00	499.28

Source: Data from various issues of "Spices Statistics", Spice Board, Cochin and Agricultural production statistics, Ministry of Agriculture, Govt. of India, Delhi.

Table 1.2 Statewise area, production and productivity of cardamom in India

Year	Variables	Kerala Actual	(%)	Karnataka Actual	(%)	Tamil Nadu Actual	(%)	India
1970–1971	Area (ha)	55,190	67.19	28,220	30.81	8070	8.81	91,480
	Production (tons)	2130		805	25.39	235	7.41	3170
	Productivity	38.59		28.53		29.12		34.65
1980–1981	Area (ha)	56,380	60.01	28,220	30.03	9350	9.95	93,950
	Production (tons)	3100	70.45	1000	22.73	300	6.82	4400
	Productivity	54.98		28.22		32.09		46.83
1990–1991	Area (ha)	43,826	53.74	31,605	38.75	6123	7.51	81,554
	Production (tons)	3450	72.63	800	16.84	500	10.53	4750
	Productivity	78.72		25.31		81.66		58.24
1997–1998	Area (ha)	43,050	61.66	21,410	30.66	5360	7.68	69,820
	Production (tons)	5430	75.94	1240	17.34	480	6.71	7150
	Productivity	126.13		57.92		89.55		102.41
1999–2000*	Area (ha)	41,522	57.31	25,882	35.72	5047	6.97	72,451
	Production (tons)	6550	70.51	1950	20.99	790	8.50	9290
	Productivity	213		103		205		172
		126.13						

Source: Data from various issues of "Spices Statistics", Govt. of India, Spices Board, Cochin.

Notes
Yield is estimated by dividing total production with area.
* Midterm estimate.

rock bottom during the same period. In 1985–86 cardamom export was 3272 tons while in 1989–90 the quantity exported was just 173 tons, in 1994–95 the export was only 251 tons, current export is to the tune of 260 tons/year. The export earnings have come down from Rs. 53.46 crores in 1985–86 to Rs. 7.6 crores during 1994–95.

Table 1.3 Production and export of cardamom from Guatemala

Year	Cultivated area (ha)	Production (mt)	Productivity (kg/ha)	Export (mt)
1985	32,336	7348.32	90.89	6173.50
1986	38,333	8845.20	92.33	7978.82
1987	41,418	10,591.56	102.29	11,489.69
1988	42,656	10,432.80	97.83	11,303.71
1989	43,000	11,340.00	105.49	11,076.91
1990	43,000	11,340.00	105.49	11,113.20
1991	43,000	12,201.84	113.51	13,163.47
1992	43,000	12,474.00	116.04	13,240.58
1993	47,472	12,927.60	114.57	14,442.62
1994	45,133	14,969.80	126.13	13,213.37
1995	47,472	15,603.84	131.48	13,920.98
1996	47,472	16,329.60	137.59	21,255.70
1997	119,540	16,692.48	139.64	14,020.78
Total		11,576.70		12,491.79

Source: Various publications of Banco De Guatemala and Spices Board, India.

Cardamom cultivation has picked up in certain localities in Papua New Guinea. Here cardamom has been grown under virgin forests and the cultivation is exclusively with private estate owners. These cardamom growing areas are also having an evenly distributed rainfall throughout the year, productivity is very high; even upto 2000–2500 kg/ha dried cardamom produce have been reported from certain estates (Krishna, 1997). The total production was about 313 tons in 1985, but declined later to about 54 mt in 1993. The present figure is about 68–70 mt.

In Tanzania cardamom was introduced at the beginning of the twentieth century by some German immigrants, and is being grown in certain localities like Amani and East Usambaras (Lawrence, 1978). The production was as high as 760 mt in 1973–74, but declined later to about 127 mt in 1984–85. This level continues even now. Sri Lanka is another small producer, contributing about 75 mt/year.

India has been the world's largest cardamom producer until 1979–80, when Guatemala came to the scene as the major rival and world leader in cardamom production. Now about 90 per cent of global trade in cardamom is the contribution from Guatemala. A comparison of the production of cardamom during the current decade is given in Tables 1.3 and 1.4.

Among the many causes that affected the cardamom production in India, the following are significant:

(i) The recurring drought year after year lasting for almost 6 months, combined with extensive deforestation and the resultant changes in the ecology of the cardamom habitats;
(ii) Crop loss due to diseases and pests;
(iii) Lack of intensive management of cardamom gardens.

In spite of the above limiting factors cardamom production has registered increase both in production and productivity mainly due to: (a) improved, high-yielding

Table 1.4 Production in major cardamom producing countries

Period	Per cent share in total by			World production (mt)
	India	Guatemala	Others*	
1970–71 to 1974–75	65.4	21.5	13.1	4678
1975–76 to 1979–80	53.7	34.5	11.8	6628
1980–81	42.9	48.8	8.3	10,250
1984–85	31.9	60.3	7.8	12,220
1985–86 to 1989–90	26.5	67.5	6.0	14,392
1990–91 to 1994–95	28.4	65.6	6.0	19,470
1995–96 to 1997–98	29.8	64.2	6.0	24,953

Sources: Cardamom Statistics, 1984–85, Govt. of India, Cardamom Board, Cochin.
Spices Statistics, 1991, Govt. of India, Spices Board, Cochin.
Spices Statistics, 1997, Govt. of India, Spices Board, Cochin.
All India Final Estimate of Cardamom – 1997–98, Govt. of India, Ministry of Agriculture.

Note
* Estimated figures (actual figures are not available).

Table 1.5 Cardamom exports from India (1970–1998)*

Year	Quantity exported (mt)	Export value ('000 Rupees)	Export as per cent total production
1980–81	2345	347,539	53.30
1981–82	2325	301,969	56.71
1982–83	1032	163,690	35.59
1983–84	258	54,423	16.13
1984–85	2383	648,653	61.10
1985–86	3272	534,599	69.62
1986–87	1447	184,953	38.08
1987–88	270	34,003	8.44
1988–89	787	103,736	18.52
1989–90	180	30,668	5.81
1990–91	379	102,224	7.98
1991–92	544	155,741	10.88
1992–93	190	75,057	4.47
1993–94	387	145,483	5.86
1994–95	257	76,261	3.67
1995–96	375	131,000	4.75
1996–97	226	86,967	3.10
1997–98	297	106,371	4.15
1998–99	475	252,121	6.62
1999–2000	550	276,035	5.92

Source: Various issues of Spices Statistics, Spices Board, Govt. of India.

Note
* Midterm estimate by Spices Board.

clones, (b) improved agrotechnology developed by the research organizations, and (c) better awareness of phytosanitation and control of diseases and insect pests.

Cardamom export from India was oscillating between 787 mt (in 1988–89) and 180 mt (1989–90). The export crossed the 500 mt mark in 1991–92 (544 mt) and 1999–2000 (550 mt) (Table 1.5). During this period there was an increase in unit price from Rs. 125.94/kg (1987–88) to Rs. 395.04/kg (1992–93). In 1996–97 the unit price was Rs. 383.88/kg while the unit price now (1999) is Rs. 450/kg. The cost of production in Guatemala is reported to be only half of that in India, and hence it is not possible for India to compete with Guatemala in the international market.

India has a very large domestic market for cardamom, consuming around 7000 mt/year. Some recent survey has put the domestic market in India to be around 7300 mt, valuing around Rs. 220 crores. Apart from individual and household consumption, cardamom is consumed by several industries, including biscuits, cardamom flavoured tea, and milk, cardamom as a component in a variety of herbal medicine, pan masala, food mixes etc. The industrial consumption of cardamom in India is estimated to be around 2050–2100 mt/year by 2000 AD. The demand in the hotel, bakery, and fast food sector is expected to be around 1250 mt by the 2000 AD. The total domestic demand for cardamom is estimated to be around 9500 mt by 2000 AD (George and John, 1998).

3   R & D EFFORTS IN CARDAMOM

The research and development efforts in cardamom were initiated after independence. The first centre for cardamom research was established at Pampadumpara in Idukki district of Kerala in 1956, and in 1957 a centre was started at Mudigere in Chikmagalur district of Karnataka. Both these centres are now with the respective state Agricultural Universities. Though the work in cardamom was initiated in 1956–57, the impetus came only with the establishment of All India Coordinated Cashew and Spices Improvement Project in 1971, which from 1985 became the All India Coordinated Research Project on Spices. The main objectives of this project are to improve spice crops by conducting research on various aspects of crop improvement, management and protection besides developing technology for multiplication of high yielding planting material for distribution to growers. The All India Coordinated Research Project on spices has initiated research programmes on germplasm collection, evaluation, crop improvement and development of agrotechnology and plant protection of cardamom through a network approach.

A full-fledged research institute – Indian Cardamom Research Institute (ICRI) was established in 1978 by Spices Board, functioning under the Ministry of Commerce. This institute is located at Myladumpara in the Idukki district, the major cardamom growing area of Kerala. In 1974 Cardamom Research Centre under Central Plantation Crops Research Institute, Kasargode has started functioning at Appangala in Coorg (Kodagu) district of Karnataka, mainly for conservation of germplasm and developing suitable production technology for the Coorg area. This centre later (1986) got amalgamated with the National Research Centre for Spices (currently the Indian Institute of Spices Research), and is now working mainly on virus disease management, breeding for high yield and resistance and developing high production technology for the area. During the past two decades eleven high yielding improved clones have been developed for cultivation.

The present R&D efforts are directed towards: (i) developing high yielding, location specific cultivars for the major cardamom growing areas, (ii) evolving high production technology for the major growing tracts, (iii) evolving lines tolerant or resistant to clump rot and virus diseases, (iv) selection/breeding for high quality and (v) evolving drought- and heat-resistant lines. Another area that has attracted much attention is post-harvest technology, industrial utilization and product development. Research organizations such as Central Food Technological Research Institute, Mysore and Regional Research Laboratory, Thiruvananthapuram (Trivandrum) and certain private industrial houses are working in this area. Cardamom is finding use as an important flavour in a variety of food products, beverages, and oral formulations of medicines. In addition, cardamom is popular as an after food mouth flavourant in India. Indeed many people keep a stock of cardamom for this purpose even during their travel.

The Western Ghats, the centre of diversity for cardamom, has undergone much climatological changes during the past century due to the devastation of forest habitat. As a result, prolonged drought became an annual feature in the cardamom growing areas, leading to sharp decline in cardamom production and productivity. In view of this situation, the need of the hour is the development of lines tolerant or resistant to heat and moisture stress. The major advantage of other cardamom growing countries, where the yield levels are much higher than that in India, is the well distributed rainfall throughout the year. In India also, in certain areas in Coorg district where cardamom plantations are irrigated, high yield levels have been recorded, indicating moisture availability as a major limiting factor in cardamom production.

If India wants to recapture the lost glory of the Queen of Spices, there should be concerted attempts to achieve a quantum jump in productivity not only through high-yielding varieties, but also through the process of constraint alleviation, mainly in respect of drought, pest and diseases. As the global requirements are growing, the future seems to be bright both for Guatemala and India.

REFERENCES

George, C.K. and John, K. (1998) Future of cardamom industry in India. *Spice India*, 11(4), 20–24.

Holttum, R.E. (1950) The Zingiberaceae of the Malay Peninsula. *Garden's Bull.*, 13, 236–239.

Krishna, K.V.S. (1997) Cardamom plantations in Papua New Guinea. *Spice India*, 10(7), 23–24.

Lawrence, B.M. (1978) Major tropical spices – cardamom (*Elettaria cardamomum*). *Essential oils*, 105–155.

Linschoten, J.H. Van (1596) *Voyage of John Huygen Van Linschoten in India*. Vol. II, 86–88 (Quoted by Watt 1872).

Ludlow, E. (1868) *Memoirs of Cardamom Cultivation in Coorg*. (Quoted by Watt 1872.)

Mabberley, D.J. (1987) *The Plant Book*, Cambridge University Press, Cambridge.

Mahindru, S.N. (1982) *Spices in Indian Life*, Sultanchand & Sons, New Delhi.

Mollison, J.W. (1900) Cardamom cultivation in the Bombay Presidency. *Agric. Ledger*, 11 (Quoted by Ridley 1912.)

Nair, C.K., Natarajan, P., Jayakumar, M. and Naidu, R. (1991) Rainfall analysis of the Cardamom tract. *J. Plantation Crops* (suppl.), 18, 184–189.

Parry, J.W. (1969) *Spices*, Vol. I, Chemical Pub. Co., New York.

Ridley, H.N. (1912) *Spices*, Macmillan & Co. Ltd., London.

Sakai, S. and Nagamasu, H. (2000) Systematic studies of Bornean Zingiberaceae. 11. *Elettaria* of Sarawak. *Edinb. J. Bot.*, 57, 227–243.

Watt, G. (1872) *Dictionary of Economic Products of India*, Vol. 3, pp. 227–236.

2 Botany, crop improvement and biotechnology of cardamom

K.J. Madhusoodanan, K. Pradip Kumar
and P.N. Ravindran

1 TAXONOMY

Cardamom (*Elettaria cardamomum* Maton) belongs to the monocotyledonous family Zingiberaceae (ginger family) of the natural order Scitaminae. The genus *Elettaria* consists of about six or seven species distributed in India, Sri Lanka, Malaysia and Indonesia; only *E. cardamomum* is economically important (Holttum, 1950; Willis, 1967; Mabberley, 1987).

ELETTARIA Maton: Maton, Trans. Linn. Soc. London 10: 250, 1811; Bentham in Bentham & Hooker f., Gen. Pl. 3: 646, 1883; Baker in Hooker f., Fl. Brit. India 6: 251, 1892; Trimen, Handb. Fl. Ceylon 4: 260, 1898; Schumann in Engler, Pflanzenr. 4(46): 267, 1904; Fischer, Bull. Bot. Surv. India 9: 179, 1921, in Gamble, Fl. Pres. Madras 8: 1491, 1928; Holttum, Gard. Bull. Singapore 13: 236, 1950; Ramamoorthy in Saldanha & Nicolson, Fl. Hassan Dist. 767, 1976; Burtt & Smith in Dassanayake, Rev. Handb. Fl. Ceylon 4: 528, 1983; Matthew & Britto in Matthew, Fl. Tamilnadu Carnatic 2: 1616, 1983; Manilal, Fl. Silent Valley 312, 1988.

Type species: *Elettaria cardamomum* (Linn.), Maton

Etymology: The generic epithet *Elettaria* is derived from Rheed's *Elettari*. *Elathari* (modern transcription of Rheed's name) is still used for the seeds of *E. cardamomum* (*thari* means granules).

Holttum (1950) has provided the following description for *Elettaria*:

> Rhizome stout or fairly stout, the intervals between leaf-shoots often short. Leaf-shoots tall, with many blade-bearing leaves; petioles short. Inflorescences from rhizome close to the base of a leaf-shoot, long and slender, prostrate, either just at the surface of the ground or just below it (not bearing roots), protected by alternate fairly large scale-leaves, in the axils of which cincinni arise, their attachment being sometimes supra-axillary. Cincinni short, bearing a close succession of tubular bracts, each of which encloses entirely the next flower and also the next bract; the flowers in two close rows on one side of the composite axis of the shoot, all pointing in the same direction, and curved, opening in succession. Calyx tubular, split about 1/4 of its length down one side, shortly 3-toothed; in some species joined at the base to the corolla-tube about as long as calyx; lobes not very broad, subequal, the upper with a concave apex. Labellum as in *Amomum*, with yellow median band and red stripes, sometimes so curved that it stands as a hood over the top of the flower. Staminodes none, or short and narrow. Filament of anther very short, broad.

Anther longer than filament, the connective produced at the apex into a small crest. Stigma small, in close contact with the distal end of the pollen sacs. Fruit globose or ellipsoid, thin-walled, smooth or with longitudinal ridges when ripe.

Burtt and Smith (1983) provided the following taxonomic description for *E. cardamomum*.

Elettaria cardamomum (L.) Maton in Trans. Linn. Soc. 10: 254, 65(1811); K. Schum. Pflanzenr., 268 (1904).

Leafy shoot up to 4 m high; petioles up to 2.5 cm long; lamina up to ca. 1 m × 15 cm, lanceolate, accuminate, lightly pubescent or glabrous below; ligule up to 1 cm long, entire. Inflorescence usually borne separately on a prostrate, (or erect or semierect) stalk up to 40 cm (or more in certain cases). Bracts 2–3 × 0.8–1 cm, lanceolate, acute glabrous, rather persistent but becoming fimbriate with age. Cincinni many flowered. Bracteoles up to 2.5 cm long, tubular, mucronate, glabrous. Calyx up to 2 cm long, 2- or obscurely 3-lobed, lobes mucronate. Corolla tube about as long as calyx; lobes 1–1.5 cm long, rounded at the apex, the dorsal tube widest. Labellum white, streaked violet, 1.5–2 × 1 cm at widest part, obovate, obscurely 3-lobed, narrowed at the base. Lateral staminodes inconspicuous, subulate. Anther sessile, thecae ca. 1 cm long, parallel, connective prolonged into a short, entire crest. Ovary 2–3 mm long, glabrous. Fruit is a capsule, oblong or more or less globose (Fig. 2.1)

The genus contains only a few species, the important ones are *E. cardamomum* and *E. major* (*E. ensal*) from South India and Sri Lanka respectively.

1.1 Nomenclature

Elettaria cardamomum (Linn.) Maton, Trans. Linn. Soc. London 10: 254, t. 5. 1811; Baker in Hooker f., Fl. Brit. India 6: 251, 1892; Trimen, Handb. Fl. Ceylon 4: 261, 1898; Schumann in Engler, Pflanzenr. 4(46): 268, 1904; Fischer, Bull. Bot. Surv. India 9: 679,

Figure 2.1 A cardamom plant bearing flowers and fruits.

1921, in Gamble, Fl. Pres. Madras 8: 1491, 1928; Burtt and Smith, Notes Roy. Bot. Gard. Edinburgh 31: 182, 1972, in Dassanayake, Rev. Handb. Fl. Ceylon 4: 529, 1983; Ramamoorthy in Saldanha and Nicolson, Fl. Hassan Dist. 767, 1976; Burrt in Manilal, Bot. Hist. Hort. Malab. 147, 1980; Matthew and Britto in Matthew, Fl. Tamilnadu Carnatic 2: 1616, 1983; Manilal, Fl. Silent Valley 312, 1988; Nicolson *et al.*, Interpret. Rheede Hort. Malab. 317, 1988; Ramachandran and V.J. Nair, Fl. Cannanore 471, 1988.

Type: *Elettaria* Rheede, Hort. Malab. 11: 9, t. 4–5.

Amomum cardamomum Linn. sp. Pl. 1: 1, 1753.

Alpinia cardamomum (Linn.) Roxburgh, Asiat. Res. 11: 355, 1810, Pl. Coast Coromandel 3: 19, t. 226, 1819, Fl. Indica 1: 37, 1820; Graham, Cat. Pl. Bombay 206, 1839; Dalzell and Gibson, Bombay Fl. Suppl. 86, 1861.
Elettaria Rheede, Hort. Malab. 11: 9, t. 4–5. 1692.
Syn: *Elettaria cardamomum* var. *minus* Watt, Commercial Products of India, 512 (1908); *Elettaria cardamomum* var. *miniscula* Burkill in Kew Bull. 1930, 35, *nomen nutum*. Var. *major* Baker in Hook. F. Pl. Brit. Ind. 6: 251 (1892). Trimen, Hand. Pl. Ceyl. 4: 261 (1898).

The essential feature of the inflorescence of *Elettaria* is a prostrate axis bearing 2-ranked sheathes with a cincinnus in the axil of each. The inflorescence is often referred as panicle (being a branched one) and the main axis (peduncle) is racemose in nature (having indefinite growth). Hence the inflorescence is a panicle of cincinni. Floral characters are similar to *Amomum*, but strikingly different in the nature of inflorescence. The genus *Elettariopsis* is very closely related to *Elettaria*. According to Holttum (1950) some of the species included under *Elettariopsis* and *Cyphostigma* resemble so closely to *Elettaria* and that they also should come under the genus *Elettaria*. Holttum (1950) is of opinion that *E. cardamomum* and Malaysian *Elettaria* represent parallel developments from different points of origin in the *Alpinia* stock.

The Malaysian species *E. longituba* Holttum (Syn. *Elettariopsis longituba*) is one of the largest species of the genus. Its flowers appear singly at longer intervals and each cincinnus contains only a very few flowers. It seems that the cincinnus stops flowering as soon as a fruit is formed. The fruits are large, but having no value (Holttum, 1950). Sakai and Nagamasu (2000) in their studies on Bornean Zingiberaceae described six species of *Elettaria* (*E. stolonifera*, *E. kapitensis*, *E. surculosa*, *E. linearicrista*, *E. longipilosa* and *E. brachycalyx*. See Chapter 16).

1.2 Varieties

Based on the nature of panicles, three varieties of cardamom are recognized (Sastri, 1952). The var. *Malabar* is characterized by prostrate panicle and var. *Mysore* possesses erect panicle. The third type var. *Vazhukka* is considered a natural hybrid between the two and its panicle is semi-erect or flexuous (Fig. 2.2).

1.2.1 Var. Malabar

Plants are medium in size and attain 2–3 m height on maturity. The dorsal side of leaves may be pubescent or glabrous. Panicles are prostrate and the fruits are globose – oblong shaped. This variety is better suited to areas of 600–1200 m elevation. *Malabar* type is relatively less susceptible to thrips infestation. It can thrive under low rainfall conditions.

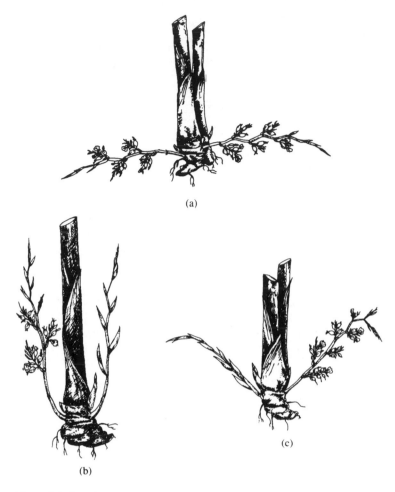

(a)

(b)

(c)

Figure 2.2 Panicle types in cardamom: (a) var. *Malabar* having prostrate panicle; (b) var. *Mysore* showing erect panicle; (c) var. *Vazhuka* having flexuous (semi-erect) panicle.

1.2.2   Var. Mysore

Plants are robust and attain 3–4 m in height. Leaves are lanceolate or oblong-lanceolate, glabrous on both sides. Panicles are erect and the capsules are ovoid, bold and dark green in colour. They are better adapted to altitudes ranging from 900–1200 m from mean sea level (msl) and thrive well under assured, well-distributed rainfall conditions.

1.2.3   Var. Vazhukka

This is considered to be a natural hybrid of vars. *Malabar* and *Mysore* and exhibits various characteristics intermediate to both. Plants are robust like *Mysore*. Its leaves are deep green, oblong-lanceolate or ovate, panicles are semi-erect (flexuous) in nature and capsules are bold, globose or ovoid in shape.

Table 2.1 gives the comparative features of the three varieties.

Table 2.1 Difference between the three varieties of cardamom

Parameters	Malabar	Mysore	Vazhukka
Adaptability	Lower elevation (600–1000 m msl)	Higher elevation (900–1200 m msl)	Higher elevation (900–1200 m msl)
Tolerance to drought	Withstands long dry spell (4–6 months)	Needs well distributed rain	Needs well distributed rain
Plant stature	Dwarf (2–3 m)	Tall (3–5 m)	Tall (3–5 m)
Leaf	Short petiole	Long petiole	Long petiole
Panicle	Prostrate	Erect	Semi-erect
Bearing nature	Early, short span of flowering	Late, long-flowering span	Late, long-flowering span
Capsule colour	Pale/golden yellow	Green	Green

Source: Sudharsan *et al.* (1991).

1.2.4 *Other varieties*

Two more varieties *mysorensis* and *laxiflora*, have been recognized based on some morphological characters.

Var. mysorensis – A robust, tall plant possessing either glabrous or pubescent leaves. This variety has flexuous panicles. The flowers are produced in short racemes. The capsules are bold and distinctly three angled.

Var. laxiflora – A less robust plant and are not tall as the former. Leaves are glabrous with short petioles. This variety has flexuous, lax decumbent panicles. The flowers are produced in 4–40 short lax racemes. The capsules are variable, oblong–oblong fusiform.

In India, a number of other cultivars of cardamom are also recognized. In general they can be considered as ecotypes of var. *Mysore*, *Malabar* or *Vazukka*. Most common among them are *Munjarabad*, *Bijapur*, *Kannielam*, *Makaraelam*, *Thara* and *Nadan*.

1.3 The Sri Lankan wild cardamom (*E. ensal* Abheywickrama)

Lot of confusion prevails over the botanical identity of Sri Lankan wild cardamom and the Indian cardamom varieties mentioned above. Cardamom varieties have been named differently by various authors.

E. cardamomum var. *minus*
E. cardamomum var. *miniscula*
E. cardamomum var. *major*
E. cardamomum var. *majus*
E. cardamomum var. *minor*

Ridley (1912) in one of the earliest descriptions writes:

> There are two forms of varieties of the plant, viz., var. *minus* with narrower and less firm leaves and globose fruits from 1/5 to 1/10 inches long, grayish yellow or buff in colour. This is confined to South India. Var. *majus* with shorter stems, broader leaves and oblong fruit, 1–2 inches long and rather narrower than the Malabar fruit, distinctly three sided, often arched and dark grayish brown when dry,

the seeds larger and more numerous and less aromatic. This is the Ceylon Cardamom and is peculiar to that country.

Owen (1901) in his notes on cardamom cultivation in Ceylon mentions three varieties, which he calls the indigenous *Ceylon*, the *Malabar* and *Mysore*. He says that the first two can be easily distinguished by the colour of the stem. The *Malabar* plant is green or whitish at the base of the leafy or aerial stem, while in the *Ceylon* plant the base has a pink tinge. He also mentions that the *Mysore* form is robust, its panicles are borne perpendicularly from the bulbs and the fruit grows in clusters of five or seven. This form is better suited for higher elevations.

   E. cardamomum var. *major* was described earlier as *E. major* Sm., (in Rees Cyclop., 39, 1819); but this name did not find favour with cardamom workers. Many subsequent authors indiscriminately used the terminology and even started mentioning var. *Mysore* as var. *major*. While doing the floristic study of Sri Lanka, Abheywickrama (1959) coined the name *E. ensal* for the Ceylon wild cardamom (from *Zingiber ensal* under which the plant was described by Gaertner (1791)). But Burtt (1980) is of opinion that the differences are not basic enough to segregate this into a new species. However, Bernhard *et al.* (1971) and Rajapakse *et al.* (1979) provided chemical evidence substantiating the distinct nature of Sri Lankan wild cardamom (see Chapter 16 also).

2   ANATOMY, EMBRYOLOGY, CYTOLOGY

2.1   Anatomy

Tomlinson (1969) and Mercy *et al.* (1977) carried out preliminary anatomical studies on aerial stem, rhizome, leaf sheath and root of cardamom plants. Aerial stems of leafy shoot has the typical monocot structure with numerous closed collateral endarch vascular bundles scattered in the ground parenchyma (Fig. 2.3). The stem is solid. There is a thin sub-epidermal layer of sclerenchymatous cells separating the inner vascular bundles from the outer ones. Each vascular bundle is surrounded by a prominent bundle sheath of sclerenchyma cells. There are two–three large metaxylem vessels and a few protoxylem vessels. A protoxylem lacuna is present in some bundles. In the parenchyma cells, rhomboidal crystals of calcium oxalate are found occasionally.

   Rhizome is sharply differentiated into an outer cortex and a central core by a plexus of irregular, congested vascular bundles (Fig. 2.4). Root traces are inserted in this plexus. In the peripheral cortex, there are only a few vascular bundles while numerous bundles are seen in the subcortical and central core regions, where they are irregularly distributed. They are closed, collateral and endarch as in the aerial stem. In each bundle, there are several protoxylem and metaxylem vessels. The bundles are supported with sclerenchymatous bundle caps. The outermost cell layers of the cortex have developed into a kind of periderm, which is a common feature in the underground rhizomes of Zingiberaceae.

   The sheathing base of the leaf has single layered upper and lower epidermis. Patches of hypodermal sclerenchyma occur associated with both upper and lower epidermal regions, but they are more prominent below the upper epidermis. Vascular bundles are arranged in a single row alternating with large schizogenous airspaces. Each bundle has a bundle sheath of sclerenchyma, a large patch of phloem and xylem below the phloem (Fig. 2.5).

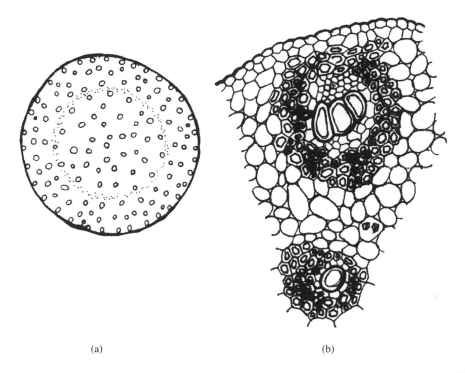

(a)   (b)

Figure 2.3 Anatomical features of cardamom stem: (a) TS of aerial stem-diagrammatic; (b) TS of aerial stem – a portion enlarged. See text for details (Source: Mercy *et al.*, 1977).

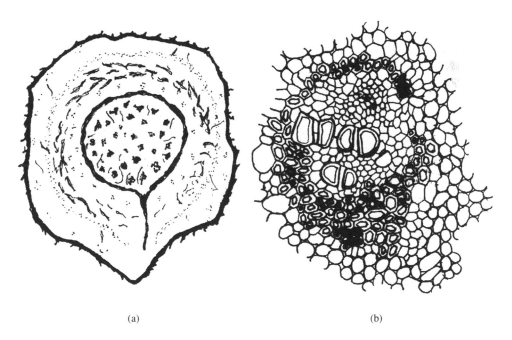

(a)   (b)

Figure 2.4 Anatomical features of cardamom rhizome: (a) TS of the rhizome diagrammatic; (b) TS of rhizome – a portion enlarged. See text for details (Source: Mercy *et al.*, 1977).

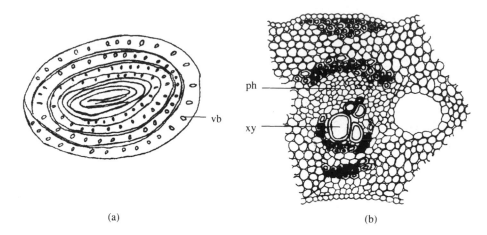

(a)   (b)

Figure 2.5 Anatomical features of cardamom leaf sheath: (a) TS of leaf sheath-diagrammatic view through the tip portion of the aerial shoot showing the cyclic layers of leaf sheath; (b) a portion enlarged. ph – phloem, xy – xylem, vb – vascular bundle (Source: Mercy *et al.*, 1977).

Cardamom plant has the characteristic fibrous root system of monocots. The root consists of epidermis that is single layered with a large number of epidermal hairs. Cortex and pith regions are large and parenchymatous. The endodermis shows well developed casparian strips (Fig. 2.6).

Cardamom leaf has the structure of a typical monocot leaf, bound by upper epidermis, hypodermis, single layered palisade tissue, spongy parenchyma and lower epidermis (Fig. 2.7). Stomata are distributed mostly on the lower epidermis only. Stomata are paracytic, having the guard cells flanked by some what elongated, conical shaped subsidiary cells (Fig. 2.7). The stomatal frequency (SF) showed variations (Krishnamurthy, 1989). The values for a few clones are given below (Value for 1 mm^2 and mean of 10 replications).

Clone	P1	P2	P3	P5	P6	P8	CL 258	CL 664	CL 668	CL 676	CL 757	V 179
SF		627	535	685	484	633	597	600	553	644	557	627	680

Pillai *et al.* (1961) studied the root apical organization in many Zingiberaceae members including cardamom. They found that the root apical organization of cardamom is similar to that in *Alpinia, Hedychium* etc. and consists of three sets of structural initials, one each for the root cap and plerome and a common zone for dermatogen and periblem (Fig. 2.9). The root cap has two distinct regions, the columella in the middle and a peripheral region around it, distinguishable by the different types of cell complexes. In the root body two histogens could be distinguished: (1) The plerome concerned with the formation of the stele and (2) the protoderm–periblem complex concerned with the formation of the outer shell to the stele including the periblem and the dermatogen. The protoderm–periblem complex is located outside the plerome and is composed of a single tier of cells. The cells of this zone located at the flanks exhibit T-divisions, which enables the tissue to widen out.

Dermatogen separates out as a distinct layer from the products of divisions of the protoderm–periblem complex at some distance on the flanks. This tier divides anticlinally, later forming the epidermis.

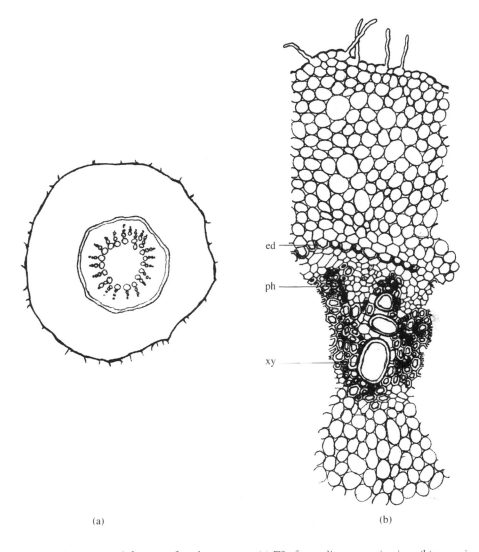

(a)   (b)

Figure 2.6 Anatomical features of cardamom root: (a) TS of root-diagrammatic view; (b) a portion
enlarged: xy – exarch xylem; ph – phloem; ed – endodermis (Source: Mercy *et al.*, 1977).

Hypodermis arises from the inner of the twin daughter cells of the T-divisions of the
protoderm–periblem initials. The cells composing this tissue vacuolated earlier than
the other cells of the cortex. Endodermis differentiates from the periblem cells. After
a few T-divisions, the innermost layer switches over to anticlinal divisions, ultimately
leading to the endodermis.

Plerome has at its tip a quiescent center, made of more or less isodiametric cells
(Fig. 2.8). On the sides of the plerome dome is the uniseriate pericycle. The
metaxylem vessel elements get differentiated near the plerome dome vacuolating
earlier. The isodiametric cells at the very center of the plerome divides like a rib-
meristem to give rise to the pith. Pillai *et al.* (1961) could distinguish at the root tip

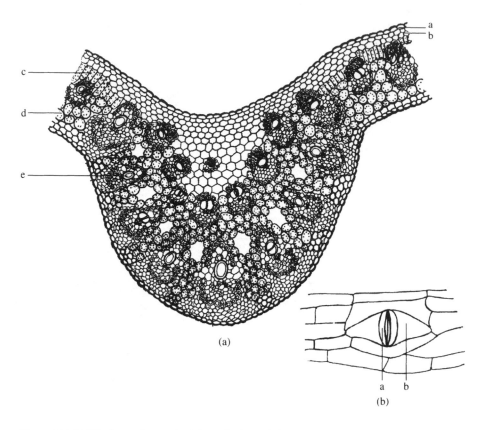

Figure 2.7 (a) TS of a cardamom leaf through the midrib: a – epidermis; b – hypodermis; c – single layered palisade; d – spongy parenchyma cells; e – vascular bundle; f – lower epidermis. (b) a portion of epidermis showing stoma: a – guard cell; b – subsidiary cell.

2 zones on cytophysiological grounds. The quiescent center found at the tip of the root body in characterized by:

(a)   cyloplasm lightly stained with pyronin-methyl green and haematoxylin;
(b)   smaller nuclei and nucleoli;
(c)   cell divisions less frequent;
(d)   vacuolation in most cells.

In median longisections this group of cells is in the shape of a cup with the brim forwards. This zone embraces and includes cells belonging to all the structural histogens of the root body and it is not structurally delimitable and gradually merges with the zone outside, the meristematic zone.

   The meristematic zone is shaped like an arch surrounding the quiescent center. This zone is composed of cells:

(a)   with cytoplasm more deeply stained with pyronin-methyl green and haematoxylin;
(b)   showing divisions more frequently;

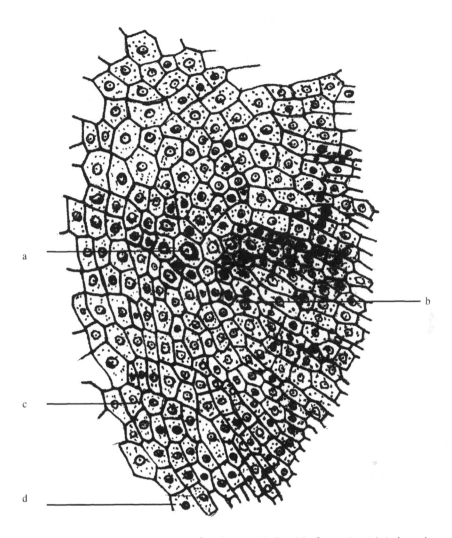

Figure 2.8 Root apical organization of cardamom. Median LS of root tip with independent plerome and root cap, and protoderm–periblem complex. a – the quiescent centre; b – meristematic zone; c – region where the *Kooper* and *Kappe* type divisions occur; d – dermatogen (Source: Pillai *et al.*, 1961).

(c) showing larger nucleoli and nuclei;

(d) where vacuolation is not prominent.

The root apical organization of cardamom falls under the category Type 2 of Easu (1953), which is the most prevalent type among the monocots.

2.2   Floral anatomy

Gregory (1936) and Thompson (1936) were the first to study the floral anatomy of cardamom. Pai (1965) carried out a reinvestigation of the cardamom floral anatomy, on which the following discussion is based. Two rings of bundles, each with six strands run in the floral axis beneath the ovary (Fig. 2.9a). The inner ring contains three large strands and three small radially flattened bundles alternating with each other. The latter travel initially outwards and thereafter upwards, opposite the loculi of the ovary. Simultaneously the bundles of the outer ring also travel out exhibiting extensive branching and traverse the ovary wall. The three large bundles of the inner ring, which are the placento-parietal bundles, extend laterally to form a temporary siphonostelic cylinder of vascular tissue (Fig. 2.9b). A little higher, the three bundles separate out, and divide into the outer parietal and inner placental strands (Fig. 2.9c). The former runs opposite the septa of the ovary in the ovary wall.

The placental bundles initially merge to form a triradiate axial strand (Fig. 2.9d). At the beginning of the ovuliferous zone, the composite placental cord divides into a few strands, chiefly located at the inner ends of the septa, which end in sending traces to the ovules (Fig. 2.9e). The small strands in the ovary wall end leaving a ring of nine strands in groups of three each, the median bundles of every set being the small median bundles of a sepal (Fig. 2.9f). The parietal bundles increase in size, divide into two or three strands, each one produces lateral branches, adjacent ones of which fuse to form an anastomizing vascular plexus on top of the ovary (Fig. 2.9g). From this plexus arise the vascular supplies to petals, androceium, glands and style. A transverse section above the level of the plexus shows: (1) an outer ring of nine bundles representing the vascular supply to the calyx, (2) and inner ring of about 15–17 strands representing the vascular supply to the petals and the androecium, (3) two prominent masses of vascular tissue in antereo-lateral positions for the two epigynous glands and (4) two traces flanking the arms of the V-shaped canal for the style. The canal is formed from the fusion of the axile cavities that in turn are derived from the narrow channels developed from the persistent loculi.

Glands and style are the first to separate from the outer column of tissue above the ovary (Fig. 2.9h). Calyx with the nine strands separates next from the inner floral tube. The two masses of vascular strands enter the two glands, and within them, extend in an anterio-posterior direction.

Nine of the 15–17 bundles at the base of the floral tube are arranged in groups of three each, two of which eventually enter the petal and one the fertile stamen. Higher up, the floral tube develops three ridges on its outer face, one posterior and the other two antereo-lateral (Fig. 2.9i). Some of the bundles in the floral tube divide and travel out into the ridges. The latter subsequently become the three petals (Fig. 2.9j).

The inner cylinder splits into a flat posterior and a thick anterior segment (Fig. 2.9k). The posterior segment supplies to the staminodes and the filament and eventually enter the connective and the anther. The anterior segment supplies to the labellum. This segment later divides into two, one each entering the two segments of the labellum. Based on the anatomical evidence Pai (1965) argued that the labellum in cardamom flower is a double structure.

2.3   Embryology

2.3.1   *Anther and male gametophyte*

Mature anther wall consists of 6–11 layers epidermis, endothecium, four or five middle layers and a secretary tapetum. In many cases through division of the primary tapetal

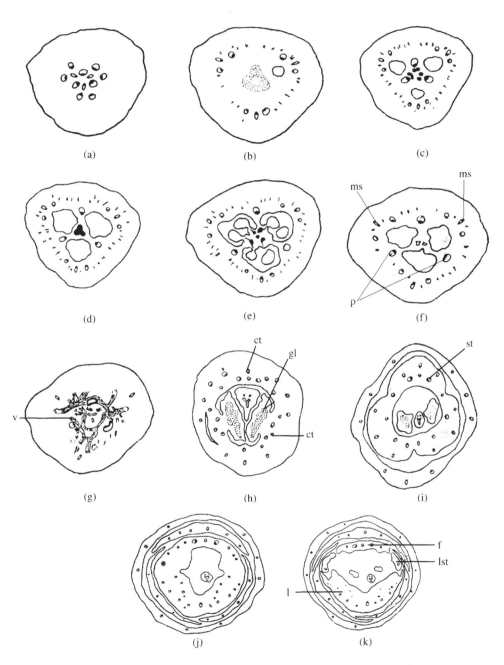

Figure 2.9 (a)–(k) Serial sections from the floral axis upwards through the cardamom flower: ct – median trace of petal; gl – gland; ms – median trace of a sepal; p – parietal bundle; st – staminal trace; v – vascular plexes; f – filament; l – labellum; lst – lateral staminode. See text for details (Source: Pai, 1965).

layer the tapetum becomes biseriate. The cells of the sporogenous tissue develop into microspore mother cells, they later undergo mitosis producing isobilateral tetrads. In microspores a generative cell is cut off near the wall, and later it separates from the wall and comes to lie in the cytoplasm of the vegetative cell. At maturity the microspore develops warty projections on the wall. The mature pollen grains are 2-celled at the time of shedding (Panchaksharappa, 1966).

2.3.2   *Ovule and female gametophyte*

Ovules are anatropous, bitegmic and crassinucellate, borne on an axile placenta. The inner integument forms the micropyle. In the ovular primordium, the hypodermal archesporial cell cuts off a primary parietal cell and a primary sporogenous cell. The former undergoes anticlinal, rarely transverse, divisions. The sporogenous cell enlarges into a megaspore mother cell, which undergoes meiosis forming megaspores.

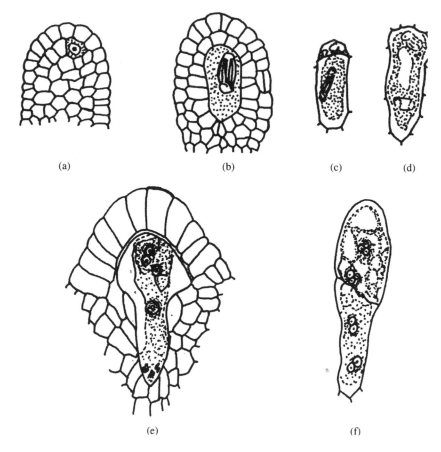

(a)   (b)   (c)   (d)

(e)   (f)

Figure 2.10 Embryosac development in cardamom. (a) LS of ovule primordium showing hypodermal archesporial cell; (b) LS of ovule showing megaspore mother cell at anaphase; (c) Functional megaspore with nucleus in divisions, degenerated megaspores at one end; (d) 2-nucleate embryosac with nuclei in division; (e) LS of ovule at mature embryosac stage; (f) Twin embryosacs (Source: Panchaksharappa, 1966).

The chalazal spore enlarges and becomes the embryosac. Its nucleus undergoes three successive divisions resulting in the 8-nuclete embryosac (Fig. 2.10). Rarely twin embryosacs were also observed (Panchakasharappa, 1966). The embryosac development is the polygonum type. In a mature embryosac the antipodals are ephemeral, and the secondary nucleus is situated in the narrow chalazal caecum (Panchaksharappa, 1966).

2.3.3   Cotyledon and epicotyl

Philip (1968) investigated the cotyledon and epicotyl development in cardamom, on which the following description is based. The zygote divides by a transverse wall to form a basal cell and a terminal cell, after which longitudinal walls are laid down in both cells (Fig. 2.11a). The mass of cells formed by subsequent divisions undergoes differentiation into basal and terminal tiers (Fig. 2.11b). One half of the terminal tier (c) grows at a faster rate than the other sector (e). Formation of a periclinal wall in (e) results in a deep staining subprotodermal cell (Fig. 2.11c,d). Through out the development sector (e) showed retarded growth compared to sector (c), thereby causing an elevation of this sector (Fig. 2.11e). Subsequently sector (c) outgrows (e) giving rise to a rather massive cotyledon (Fig. 2.11f–h). The boundary between the derivatives of the basal and terminal cells on the one hand and between those of the cotyledonary (c) and epicotylary (e) sectors on the other remain fairly distinct up to the stage in Fig. 2.11e. By the time the cotyledon attains full growth, the epicotylary sector undergoes just one or two divisions only and the whole sector appear stretched along the direction of growth. When increase in girth is initiated around the basal part of the cotyledon, the median longitudinal sections of the embryo appear to possess a notch on one side (Fig. 2.11h). At this stage it is possible to demarcate the lower boundary of the shoot apex. Subsequent development is similar to those outlined for other monocots.

2.3.4   Fruit and seed

The fruit is a capsule developed from an inferior ovary. It is more or less three-sided with rounded edges. The shape and size vary. In var. *Malabar* the fruits are short and broadly ovoid and dried fruits some what longitudinally wrinkled. In var. *Mysore* the fruits are ovoid to narrowly ellipsoid or elongate, the surface is more or less smooth (Fig. 2.12). The Sri Lankan wild cardamom (*E. ensal*) is much larger, elongate, angular and distinctly three sided. The dry pericarp is about 0.5–1 mm thick having rough woody texture. The capsule has three locules, septa is membranous and placentation axile. The seeds are five to eight in each locule and they adhere together to form a mass. Transverse section of a pericarp shows outer and inner epidermis consisting of small polygonal cells and a mesocarp of thin-walled, closely packed, parenchymatous cells. Vascular bundles traverse the mesocarp; each bundle consists of a few xylem vessels, phloem and a sclerenchymatous sheath partially surrounding the vascular elements. Many resin canal cells (oil cells) are found distributed in the mesocarp. The xylem vessels have spiral thickening. Some of the cells contain prismatic needle shaped calcium oxalate crystals (Fig. 2.13).

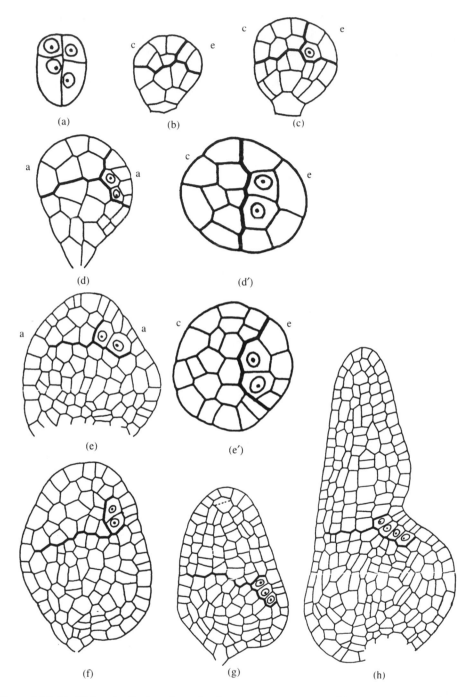

(a)   (b)   (c)

(d)   (d′)

(e)   (e′)

(f)   (g)   (h)

Figure 2.11 (a)–(h) Longitudinal sections of embryos at successive stage of development, the plane of section passing through the median part of the cotyledon and epicotylary meristem. Vertically oriented heavier lines indicate the boundary of the cotyledonary and epicotylary loci. Epicotylary cells are nucleated and cotyledonary cells non-nucleated. Horizontally oriented heavier lines delimit the derivatives of the short apical tier and the subjacent tiers (d) (d′), (e) (e′) – Tranverse sections at levels a–a as shown in Figs (d) and (e) respectively. Heavier line demarcates the boundary between the cotyledonary and epicotylary sectors (Source: Philip, 1968).

Figure 2.12 Fruit (capsules) of cardamom: ICRI-1 *Malabar* type; ICRI-2 *Mysore* type; PV-1 *Mysore* type; Fruits of *Njallani* green gold, a grower's selection, *Vazhukka* type; Fruits of *Vazhukka* type.

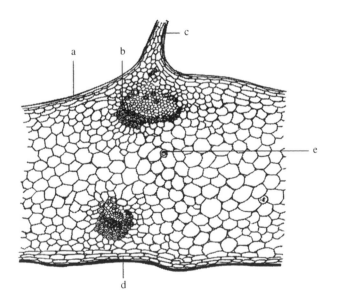

Figure 2.13 TS of pericarp: a – inner epidermis; b – vasculature; c – part of dissepiment; d – outer epidermis; e – oil cell (Source: Wallis, 1967).

Cardamom seed has externally an aril composed of a few layers of thin walled, elongated cells. In fully mature seeds these cells contain small oil globules. The testa consists of an epidermis composed of elongated fusiform cells, about 250–1000 μ long, in sectional view they are nearly square, about 18 μ wide and 25 μ high

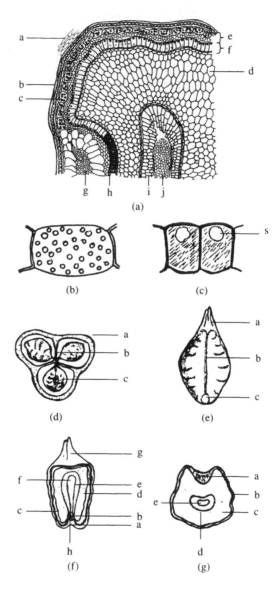

Figure 2.14 Cardamom seed TS: **(a)** a portion of TS of seed showing structural details: a – arillus; b – epidermis; c – oil cells; d – perisperm; e – outer integument; f – inner integument; g – raphae; h – selerenechyma; i – endosperm; j – embryo **(b)** an oil cell enlarged; **(c)** highly thickened cell of inner integument showing silica; **(d)** TS of cardamom fruit showing: a – pericarp; b – dissepiment; c – seed; **(e)** seed covered by arillus: a – arillus; b – raphae; c – operculum; **(f)** seed LS: a – hilum; b – operculam; c – oil cell layer; d – perisperm; e – endosperm; f – haustoria; g – arillus; h – embryo; **(g)** seed, TS: a – raphae; b – oil cell layer; c – perisperm; d – endosperm; e – embryo (Source: Wallis, 1967).

(Wallis, 1967). A layer of small, flattened parenchyma cells are found below the epidermis. Below this there is a layer of large rectangular cells, about 80–120 μ long and 20–45 μ wide and high and these cells are filled with globules of volatile oil. Interior to this layer of large cells, there are two or three layers of small parenchymatous cells. The above layers of cells together form the outer seed coat. These layers get widened around the raphe where the vascular strand is surrounded by large oil cells (Fig. 2.14).

The inner seed coat consists of two layers, the inner one consists of heavily thickened polygonal cells about 15–25 μ in length and breadth and 30 μ high (Wallis, 1960). These cells are so thickened that only a small lumen is found at the upper end, in which there is a globule of silica nearly filling the cavity. The inner layer of the inner seed coat consists of a narrow band of thin-walled cells (Wallis, 1967).

The kernel consists mostly of perisperm and a small endosperm. The perisperm consists of thin walled parenchymatous cells; measuring about 40–100 μ, each filled with starch grains. One or two prismatic, calcium oxalate crystals occur in each cell. The endosperm consists of thin walled, closely packed, parenchymatous cells, 20–40 μ in length, and containing pale yellow coloured deposits. On iodine staining the contents became deep blue, showing the presence of starch; and stains red with Millons reagent indicating the occurrence of proteins. The endosperm surrounds small, almost cylindrical embryo, which is made up of thin-walled cells. Parry (1969) as well as Trease and Evans (1983) also gives brief descriptions of the histology of cardamom seeds.

2.3.5   Nature of cardamom powder

Cardamom when powdered gives greyish brown powder with darker brown specks – it is gritty in texture having pleasant smell and flavour. The diagnostic characters of cardamom powder as given by Jackson and Snowdon (1990) are: (i) abundant starch grains, filling the cells of the periplasm, the individual starch grains are very small and angular, a hilum is not visible; (ii) the sclerenchymatous layer of the testa, composed of a single layer of thick-walled cells which in a mature seed are dark reddish brown in colour; each cell contains a module of silica; (iii) abundant fragments of the epidermis of the testa, composed of layers of yellowish brown, prosenchymatous cells, with moderately thickened pitted walls; (iv) the oil cells of the testa consisting of a single layer of large polygonal rectangular cells with slightly thickened walls and containing globules of volatile oil. This layer is found associated with the epidermis and hypodermis; (v) the parenchyma of the testa composed of several layers of small cells polygonal in surface view, with dark brown contents and slightly thickened heavily pitted walls; (vi) the abundant parenchyma of the perisperm and endosperm composed of closely packed, thin-walled cells; (vii) the fragment of the arillus is composed of very thin-walled cells; elongated and irregularly fusiform in surface view; (viii) calcium oxalate crystals, prismatic in shape, are found scattered in the cells of the perisperm and other cells and (ix) occasionally groups of xylem vessels are visible, having spiral thickening, associated with thin-walled parenchyma (Fig. 2.15).

The type of cardamom can be determined by counting the number of heavily thickened sclerenchymatous cells per square mm of a layer and using the standard figure for each type: Mysore-3310, Alleppey green-3790, Malabar-4600 (Wallis,

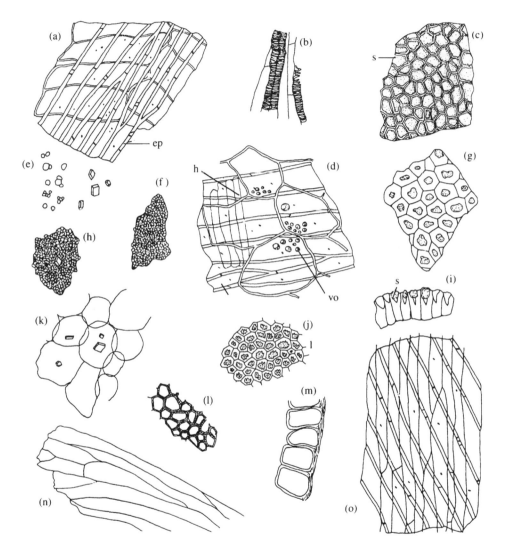

Figure 2.15 Microscopic details of cardamom seed powder: (a) epidermis of the testa (ep) in surface view with underlying oil cells; (b) a group of spirally thickened vessels and associated parenchyma; (c) sclerenchymatous layer of the testa in surface view, seen from above, showing silica nodules (s); (d) oil cells of the testa in surface view containing globules of volatile oil (vo), with underlying hypodermis (h) and epidermis (ep); (e) starch granules; (f) prisms of calcium oxalate; (g) sclerenchymatous layer of the testa in surface view, seen from below; (h) perisperm cells containing starch granules and prisms of calcium oxalate; (i) part of the sclerenchymatous layer of the testa in sectional view showing silica nodules (s); (j) sclerenchymatous layer of the testa from an immature seed, in surface view; (k) parenchyma of the perisperm from which the starch has been removed, showing prisms of calcium oxalate in some of the cells; (l) parenchyma of the testa in surface view; (m) epidermis of the testa in sectional view; (n) arillus in surface view; (o) arillus with underlying epidermis of the testa in surface view (Source: Jackson and Snowdon, 1990).

1967). The Ceylon wild cardamom contains 3020 sclerenchymatous cells per mm of layer.

2.4 Cytology

Gregory (1936) was the first to study the cytology of cardamom. According to him the basic chromosome number is $x = 12$ and $2n = 48$, indicating a balanced tetraploid nature of the plant. Reports of Ramachandran (1969) and Sudharshan (1989) also confirmed the findings (Fig. 2.16a,b). However, Chandrasekhar and Sampath Kumar (1986) observed variation in number as well as in morphology of chromosome of var. *Mysore* and var. *Malabar* and concluded that aneuploidy as well as structural alterations in chromosome has contributed to the varietal differentiation. According to them, from a karyological standpoint, var. *Mysore* stands at a higher rank in the evolutionary ladder. Meiosis is quite normal and pollen fertility high. Even the induced tetraploids

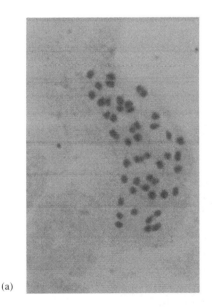

(a)

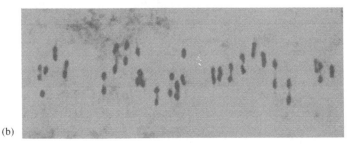

(b)

Figure 2.16 (a) Mitotic metaphase from a root tip squash showing $2n = 48$; (b) a late meiotic metaphase showing 24 bivalents.

(2n = 4x = 96) showed very limited chromosome abnormalities and had good pollen
fertility and seed set (Sudharshan, 1989).

3   GROWTH, FLOWERING AND FRUIT SETTING

3.1   Panicle production, panicle nature and flowering

Tillers emerge from axils of underground stem and from their bases vegetative buds
emerge almost throughout the year. However, majority of the vegetative buds
are produced during January–March. The linear growth of tiller increases with the
onset of Southwest monsoon and growth rate slows down with the cessation of rains.
Linear growth pattern of tiller is similar in all cultivars. It takes almost 10 months
for a vegetative bud to develop and about a year for the panicle to emerge from
the newly formed tillers (Sudharshan *et al.*, 1988). Kuruvilla *et al.* (1992) carried
out a round the year study on the phenology of tiller and panicle in three varieties
of cardamom.

Panicles emerge from the swollen bases of tillers (Fig. 2.17). Rarely terminal panicles borne on the aerial shoots are also observed. Generally, 2–4 panicles emerge from
the base of a tiller. Pattanshetty and Prasad (1976) and Parameswar (1973) have made
detailed observations on panicle production, its growth, duration of flowering etc.
Vegetative shoots require a period of 10–12 months to attain maturity for

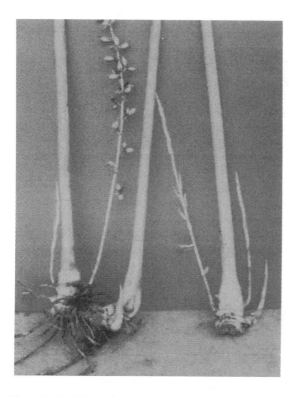

Figure 2.17 Origin of panicle from the base of tiller.

producing the reproductive buds and the newly emerging panicles take a period of 7–8 months for complete growth. Flowering in cardamom commences with the onset of monsoon. Pattern of flowering varies with regions (agro-climatic zones) and cultivars. Flowers appear on the panicles after 4 months and flowering continues for a period of 6 months. These panicles grow either erect (var. *Mysore*), prostrate and parallel to the ground (var. *Malabar*) or in a semierect (flexuous) manner (var. *Vazhukka*). Each inflorescence (panicle) possesses a long cane-like peduncle having nodes and internodes. Each node has a scale leaf in the axil of which flowers are borne on a modified helicoid cyme (cincinnus) (Fig. 2.18a,b), thus panicle is branched. Multiple branching of panicles occurs in certain cultivars. In such cases the central peduncle undergoes further branching (secondary and tertiary branches) producing multi-branched panicles. Such branching can either present at the lower portion of the main peduncle or top portion alone or throughout the peduncle (Fig. 2.19). Panicles bear leafy bracts on nodes and flowers are produced in cluster (cincinnus) in the axils of bracts. [The cluster of flowers was described as raceme by earlier cardamom workers. But it is not correct. Each cluster is a cincinnus (as described by Holttum, 1950). A cincinnus is a modified helicoid cyme]. It takes about 90–110 days for the first flower in a fresh panicle to open irrespective of the variety. The cincinni and capsules are formed during their 4th and 5th months respectively after

Figure 2.18 Portion of panicles showing flowers and fruits: (a) young panicles showing scale leaves and flowers emerging from the axils; (b) cincinni showing fruits; (c) one cincinnus showing fruits and a flower.

Figure 2.19 Multibranched panicles of cardamom.

the panicle initiation (Kuruvilla *et al.*, 1992). Capsule formation increased till August and thereafter declined slowly (Table 2.2). The flowers have the typical morphology of zingiberaceous flower. Flower opening commences from 3.30 a.m. and continues till 7.30 a.m. Anther dehiscence is between 7.30–8.30 a.m. in the morning. Flowers wither within a day. Normally, flowering is seen all the year round on panicles produced during the same year as well as on panicles produced in the previous year. Flowering is spread over a period of 6 months from May–October in India when majority of the cardamom areas receive the southwest monsoon. Almost 75 per cent of flowers are produced during June–August. The time required to reach full bloom stage from flower bud initiation ranges from 25–35 days and capsules mature in about 120 days from the full bloom stage (Krishnamurthy *et al.*, 1989).

3.2   Flower morphology

Flowers are bisexual, zygomorphic, and open in succession from base towards apex. It is white in colour, 30–35 mm long and with the central lip streaked blue or violet and trimerous in condition (Fig. 2.20). The labellum is oval and indistinctly 3-lobed. The corolla is unequally 3-lobed with large one at the posterier side. Fertile stamen is one and is without connective appendages but prolonged into a short crest. Anthers are adnate to the filament, 0.6–0.7 cm, 2-lobed dehiscing vertically. The filament is distinct and slender and deeply grooved. The ovary is inferior, trilocular with axial placentation. The style is undivided, stigma is small and is held above the anther. Structure of cardamom flower is pre-disposed for insect pollination as indicated by the prominent labellum, stigma positioning above anthers, and the presence of nectar glands. Fruits are ellipsoidal or almost spherical, non-dehiscent, fleshy and green or pale green in colour and leathery when dry. The flower morphology does not vary with biotypes or varieties.

Table 2.2 Panicle development and its growth pattern*

Month	Malabar			Vazhukka			Mysore		
	Length of panicle (cm)	No. of cincinni	No. of capsules	Length of panicle (cm)	No. of cincinni	No. of capsules	Length of panicle (cm)	No. of cincinni	No. of capsules
January	3.76	–	–	6.77	–	–	4.83	–	–
February	6.34	–	–	9.20	–	–	7.23	–	–
March	7.91	–	–	12.50	–	–	11.00	–	–
April	11.85	–	–	13.02	–	–	16.60	–	–
May	17.52	13.63	–	22.76	11.06	–	23.50	14.93	–
June	18.80	14.93	18.56	23.73	14.70	4.20	23.66	17.13	14.06
July	18.40	15.40	34.30	22.13	15.73	8.16	21.30	16.13	21.86
August	17.63	15.93	39.23	18.86	11.36	8.13	23.66	16.76	29.70
September	18.40	13.76	31.23	16.46	9.83	3.96	24.36	16.63	19.93
October	18.52	14.71	3.81	25.07	14.40	2.60	24.96	16.15	11.29
November	18.41	14.43	1.52	25.20	14.40	0.67	25.11	16.07	3.74
December	18.90	14.44	0.33	25.87	14.20	0	25.44	16.07	0.48

Note
* Kuruvilla *et al.* (1992).

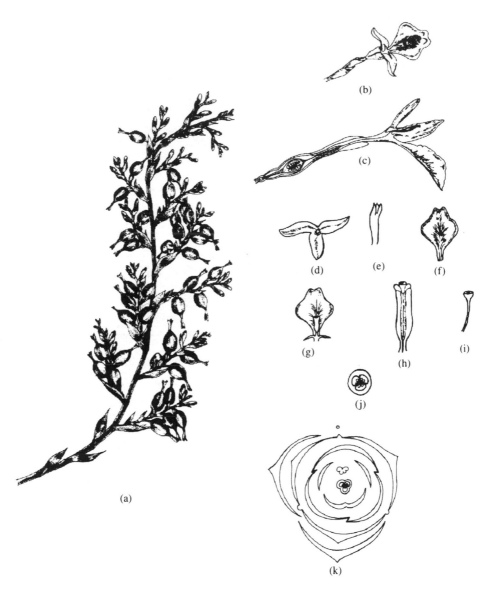

Figure 2.20 Cardamom flower structure: (a) a portion of spike; (b) a flower; (c) LS of flower; (d) corolla; (e) calyx tube; (f) labellum; (g) labellum along with fertile stamen and staminode; (h) stamen showing stigma located above the anther; (i) stigma; (j) TS of overy; (k) floral diagram.

Pai (1965) has concluded from anatomical evidence that the labellum of cardamom flower is a double structure. He showed that the mid anterior strand in the labellum is a fusion product of the marginal traces of the two component members belonging to the inner androecial whorl. Parameswar (1973) and Parameswar and Venugopal (1974) are of the view that labellum is made up of three modified anthers. Biometrical observations did not reveal any difference among the three varieties in the mean length of flower and the length of different floral parts such as flower stalk, corolla tube, anther and style.

3.3 Palynology and pollination biology

Pollen grains are rich in starch and 2-celled at the time of shedding. Moniliform refractive bodies are observed in some pollen grains. Exine develops warty projections that are spinescent (Panchaksharappa, 1966). Pollen fertility is maximum at full bloom stage and low at the beginning and the end of flowering periods (Venugopal and Parameswar, 1974). Size of pollen grains varies from 75 to 120 microns in different varieties of cardamom. Pollen grains loose their viability quickly, only 6.5 per cent remained viable upto 2 h and none after 6 h of storage (Krishnamurthy *et al.*, 1989).

Pollen fertility is comparatively lower during the early and late phases of flowering. Pollen germinates in 10 per cent sucrose solution, addition of 200 ppm boric acid improves germination and tube growth (Parameswar and Venugopal, 1974). Kuruvilla *et al.* (1989) found that 15 per cent sucrose and 150 ppm boric acid favoured pollen germination and tube growth, and that the ideal temperature is 15–20 °C. Low concentration (5–10 ppm) of coconut water, GA, cycocel and ethrel enhanced pollen germination and tube growth significantly.

Cardamom plants have bisexual flowers. The position of the stigma and the anthers are such that pollination does not take place by itself without the help of some external pollinating agents. Stigma is funnel shaped with cilia around a small cavity which is present on its top and is held sufficiently above the position of anther so as to eliminate the chances of pollen mass reaching the stigmatic surface in the natural course. Honeybees (*Apis cerana, A. indica* and *A. dorsata*) visit cardamom flowers during flowering season for collecting nectar and pollen and they do help in attaining over 90 per cent of pollination. The stigma remains receptive right from 4 a.m. on the day of flower opening depending upon the environmental conditions. It is reported that receptivity of stigma is highest between 8 a.m. and 12 noon (Krishnamurthy *et al.*, 1989, Kuruvilla and Madhusoodanan, 1988). Parvathi *et al.* (1993) and Belavadi *et al.* (1998) noticed the peak receptivity around 12 noon and coincides with peak pollinator activity. Stone and Willmer (1989) carried out a detailed study on pollination of cardamom in Papua New Guinea (PNG). There the commonest foragers are *Apis mellifera*, and to some extent *A. sapiens. A. mellifera* showed a change in the foraging behaviour over time, initially collecting pollen and switching over to nectar later in the day. The foraging starts around 7 h and reached a peak around 10 h. The pollen collecting activity declined to a very low level by 12.30 h. By this time the vast majority of stigmas got pollinated.

Belavadi *et al.* (1997) has carried out some interesting observations on cardamom flower structure and honeybee pollination. In cardamom flowers nectar is present in the corolla tube which is 23 ± 2.08 mm long (range 21.48–30.4 mm), through which the style passes. The honeybees (*Apis cerana* and *A. dorsata*) despite their short tongue lengths (14.5 mm and 5.5 mm respectively) drew nectar up to 11.45 ± 2.65 and 11.65 ± 1.85 mm respectively. Controlled experiments using capillary tubes of similar dimensions showed that the depth of feeding by the two bee species corresponded to their tongue length when there was no style. When a style is introduced the depth of feeding increased with increase in style thickness. The presence of a style inside the corolla tube helped bees to draw more nectar from the cardamom flowers.

The mean number of flowers open per day per bush is 34.5. Mean proportion of flowers per bush visited by each *A. mellifera* is 25 per cent, independent both of the numbers of flowers present on a plant and the time of day, hence the mean number of flowers visited is only 8.6. Pollen production per flower is reported to be 1.3 ± 0.2 mg, and this quantity gets reduced to 0.6 ± 0.2 after the visit of a bee, indicating that during the first foraging, about 50 per cent of the pollen is removed. Cardamom nectar is having pH of 7, and contains 55–100 mmol/l of glucose. The concentration of amino acids (at 8 a.m) is 3 mM. The nectar volume varied greatly over time. The initial volume, per flower at dawn was around 1.6 µl, and by 11.00 the volume increases to 209 µl. This increase is due to active secretion by the nectaries at the base of the corolla tube. After foraging by a bee, the nectar volume falls rapidly (Stone and Willmer, 1989). In one location in PNG the number of *A. mellifera* visiting each flower was about 31 times in a day, while in another location the average visit was only around 10.3 in a day. The fruit set in the first area was much higher than that in the second area. Bee-pollinated fruits were found to contain average of 11 seeds (Chandran *et al.*, 1983) in South India while in PNG the average number of seeds/capsule was 13.8.

Belavadi *et al.* (1993, 1997, 1998), Parvathi *et al.* (1993), Belavadi and Parvathi (1998) have carried out detailed studies on pollination ecology and biology of cardamom in a cardamom-cropping system situation of Karnataka, India. Here the pollinator activity starts by around 7.30 in the morning and continues till 6.30 in the evening, the peak is between 11.00–13.00 h. Bees start appearing on cardamom clumps when the temperature is around 21 °C. Individual foragers of *A. cerana* made 4–7 trips to a single patch of flowers in a day and the number of flowers visited on each successive trip progressively increased. Individual forager visited 157–514 flowers in a day. On a clear sunny day a flower is visited as many as 120 times, on a cloudy day 57 times and a rainy day 20 times on an average. Mean number of flowers visited by a bee at a given clump is 12.32, when the mean number of flowers/clump is 30.

The above workers (Belavadi *et al.*, 1998; Belavadi and Parvarthi, 1998) also calculated the number of honeybee colonies required for effective pollination of cardamom. Based on a cardamom population of 3000 plants/ha planted at 1.8 m apart, there will be approximately 60,000 flowers available for pollination daily. Based on the pollinator activity a minimum of three colonies/ha is needed, assuming that a colony will have about 5000 foragers. They also suggested based on pollinator activity an isolation distance of 15 m for seed production gardens (Belavadi *et al.*, 1993).

3.4   Fruit set

The fruit is globose or ellipsoid, thin walled, smooth or with longitudinal ridges when ripe. Varietal variations have been noticed in fruit shape. The fruit colour is green and turn golden yellow on ripening. Seeds are white when unripe and turn to brown on ageing and become black on maturity and their numbers per capsule vary between 10–20 in different genotypes. A thin mucilaginous membrane (aril) covers the seeds. The extent of fruit set is highest when the atmospheric humidity is very high in the cardamom tract and the setting is scanty in summer months even under irrigated condition. In general, fruit set percentage is high in young plants and when the plants surpass the economic life span, the fruit setting declines to 50 per cent or even less.

4 PHYSIOLOGY

4.1 Photosynthesis and related aspects

Total leaf area is very closely associated with total photosynthesis and dry matter accumulation. Hence accurate estimation of leaf area and canopy density are important. Korikanthimath and Rao (1983) reported a reliable method for leaf area estimation based on linear measurements of intact leaves followed by appropriate regression analysis. The leaf area factor differed among the varieties.

The light fraction that the leaves absorb has a direct bearing on crop growth and canopy development. Laboratory studies on photosynthetic efficiency in cardamom (cvs. PV-19 and PR 107) indicated that efficiency was greater at low light intensities than at higher ones. Photochemical process was favoured by low light compensation point. The translocation pattern showed that rhizome was the major sink followed by panicle and roots. Unlabelled leaves did not receive much of the labelled photosynthates from labelled leaves (Vasanthakumar *et al.*, 1989).

Kulandaivelu and Ravindran (1992) studied the photosynthetic activity of three cardamom genotypes, measured as the rate of O_2 liberated by isolated chloroplasts. Results showed drastic reduction in photosynthetic rates in plants exposed to warm climate. As much as 60–80 per cent decrease in the level of total chlorophyll was noticed in all the three varieties tested.

Light requirements for cardamom nursery is about 55 per cent of the normal (Ranjithakumari *et al.*, 1993) and at this light intensity growth and tiller production are best.

4.2 Growth regulators and fruit setting

A high percentage of fruits (capsules) shed before they reach maturity. Parameshwar and Venugopal (1974) observed about 80 per cent fruit drop. Various reasons are attributed to this low fruit set and high fruit drop, such as climatic factors (temperature, wind, humidity), nutritional deficiencies, injuries, competition for resources, soil nutrient status, pests, diseases etc. (Kuttappa, 1969) and physiological factors (Parameshwar and Venugopal, 1974).

Table 2.3 Effect of growth regulators on fruit set

Treatment		Mean fruit set (%)	Fruit weight (g)
Control		43.20	0.80
NAA	25 ppm	60.75	0.85
	50 ppm	61.62	0.83
	75 ppm	67.04	0.71
GA	25 ppm	37.72	0.78
	50 ppm	48.73	0.83
	75 ppm	54.11	0.90
2,4-D	2.5 ppm	69.80	0.80
	5.0 ppm	65.12	0.79
	7.5 ppm	47.56	0.82
CD at 5%		10.69	NS
CD at 1%		14.18	–

Growth regulators are useful in increasing fruit set. Krishnamurthy *et al.* (1989) investigated the effect of NAA, GA and 2,4-D on fruit set and fruit weight (Table 2.3). Significant differences could be noticed in the case of fruit set. Tissue concentration of auxins was highest 36 h after pollination (315 mg/g) and then declined to 80 mg/g 30 days after pollination. The fall in auxin activity resulted in the formation of an abcission zone resulting in the shedding of immature capsules. Application of 40 ppm NAA or 4 ppm 2,4-D decreased the capsule drop and also led to an increase in yield (Vasanthakumar and Mohanakumaran, 1988). Gibberellic acid (GA) at 25, 50, 100, 150, 200, 250, 300 ppm and 2,4-D at 2–5 and 10 ppm were sprayed on cardamom plants and studied (Pillai and Santhakumari, 1965). The treatments increased panicle length, particularly GA at 50 ppm, and the highest fruit set was in 200 ppm. IBA and IAA failed to increase fruit set (Nair and Vijayan, 1973). Siddagangaih *et al.* (1993) studied the effect of chloromequat, daminocide, ethepon and malic hydrazide (250 ppm). Daminozide (1500 ppm), chloromequat (250 ppm) and ethepon (100 ppm) significantly enhanced tiller production and other vegetative characters when applied on 7-month-old seedlings, but had little effect on other morphological characters.

4.3   Moisture stress

Cardamom tracts of Kerala and Karnataka experience a drought period of 4–5 months, and large-scale losses occur as a result of drought, especially in Idukki, Palakkad and Wynad districts of Kerala. Identification of lines having resistance to moisture stress is an essential pre-requisite in improving cardamom cultivation in India. At the Regional Research Station in Mudigere, Cardamom genotypes were screened with the aim of identifying drought-tolerant clones. Clones differ in their drought susceptibility. Krishnamurthy *et al.* (1989) reported chlorophyll stability index of the three varieties (vars. *Malabar, Mysore* and *Vazhukka*) as well as the related taxa (*Hedychium flavescence* and *Amomum subulatum*). Among the varieties, *Malabar* had higher CSI per cent (43.14), the *Mysore* had the lowest (24.5) and the *Vazhukka* had medium (31.14) CSI. Electrolyte leakage was 66.65 per cent in *Mysore* 69.42 in *Vazhukka*, and 43.90 per cent in *Malabar*. The same authors investigated the electolyte leakage in some clones and got the following result:

Clone	P1	P2	P3	P5	P6	P8	CL 258	CL 664	CL 668	CL 670	CL 757	P5 V 179
EL %	20.9	41.2	17.8	23.9	29.8	29.2	31.7	33.6	17.8	26.0	21.8	27.2

Krishnamurthy *et al.* (1989) reported variations in leaf area index (LAI) recorded during the stress period. Measurements recorded for 3 years indicated significant variations in LAI (Table 2.4). Clones P-6 (mean 8.2 for the 4 months March–June) and CL-757 (7.09) had higher LAI than the other clones tested.

4.4   Dry matter accumulation and harvest index

Information on dry matter partitioning and harvest index are useful to breeders to set their methodology for developing desired ideotype. Krishnamurthy *et al.* (1989) studied the dry matter accumulation under drought spells over a period of 3 years. Significant variations were recorded among the clones studied. The DMA during the drought season of March–June varied from 161 g/plant to 279.5 g/plant (P-6). At the end of the drought

Table 2.4 LAI and DMA in some cardamom clones

	LAI mean	DMA mean
P-1	6.4	231.8
P-2	4.6	170.7
P-3	5.6	060.8
P-5	6.7	247.5
P-6	8.2	279.5
P-8	5.7	204.9
CL-258	5.5	190.2
CL-757	7.1	260.7
V-179	7.0	216.0
CL-664	5.6	220.5
CL-668	5.7	204.5
CL-670	5.9	233.0
Mean	6.2	221.7

spell (June) the DMA ranged from 195 g/plant (P-2) to 391 g/plant (P-6). The variations were highly significant, the same clones (P-6 and CL-757) exhibited higher DMA as well as LAI (Table 2.4). Korikanthimath and Mulge (1998) carried out studies with 12 clones planted under the trench system of planting adopting a spacing of 1.8 × 0.6 m. Dry matter partitioning and harvest index were studied at the end of the 20th month. Dry matter content of roots, rhizome, panicles, capsules, tillers and leaves of different

Table 2.5 Harvest index and percentage distribution of dry matter in different components of plants in selected clones

Clones	Roots	Rhizome	Panicle	Capsule	Tiller (aerial shoot)	Leaves	Harvest index*
Sel.1	5.92	16.62	0.23	2.76	42.13	32.3 9	0.028
Sel.2	10.42	13.03	0.47	3.78	55.69	16. 59	0.038
Sel.3	6.32	11.79	0.32	0.28	56.61	24.6 6	0.003
Sel.4	5.45	10.58	0.44	5.97	59.99	23.5 7	0.062
Sel.5	5.68	9.70	0.28	1.77	62.49	20.09	0.018
Sel.6	5.13	12.66	0.34	3.41	53.36	25.0 9	0.034
Sel.7	9.73	13.03	0.57	6.23	34.07	36.3 6	0.050
Sel.8	5.89	17.02	0.10	2.36	47.53	27.1 9	0.023
Sel.9	5.81	12.21	0.97	9.08	51.18	20.7 4	0.091
Sel.10	5.00	8.23	0.59	4.43	57.05	24.7 0	0.044
Sel.11	5.26	7.89	0.53	3.21	61.80	21.3 0	0.032
Sel.12	6.92	14.57	0.55	5.85	45.16	26. 95	0.059
Local Check	2.68	5.36	0.21	0.68	71.81	19.27	0.006
SE m ±	0.419	0.497	0.015	0.189	1.205	0. 8199	0.00173
F test	**	**	**	**	**	**	**
CD at 5%	1.203	1.429	0.044	0.550	3.460	2. 360	0.0054
CV at 5%	6.80	4.20	3.50	4.90	2.3	3.3	4.6

Source: Korikanthimath and Mulge, 1998.

Notes
* HI = Dry weight of capsules/Total dry weight.
** F test significant at $p = 0.01$.

clones varied significantly. Total dry matter varied from 2579 g to 4853 g. The per cent distribution of dry matter and harvest index is given in Table 2.5. The dry matter yield of capsules ranged from 13.2 g (Sel.3) to 234.3 g in Sel.9. The percentage distribution of dry matter towards capsules was highest in Sel.9 (9.08), in contrast the local check gave a value of 0.68 per cent. Harvest index differed significantly among the entries tested, the highest was in Sel.9 (0.091), followed by Sel.4 (0.062), again the lowest value was in the local check (0.06). Partitioning of photosynthates to various plant parts is controlled by the genetic make up of the genotype. Partitioning efficiency reflected in the per cent dry matter distribution towards capsules in clones such as Sel.9, Sel.7 etc. and their higher harvest index have clearly indicated their superior yielding ability, and their usefulness in crop improvement programmes.

5   CROP IMPROVEMENT

Use of genetically superior planting materials and cultivation adopting improved agro-techniques are the two important means to enhance crop productivity. The lack of superior genotypes and frequent vagaries of drought and unprecedented upsurge of diseases and pests are some of the major factors that contribute to low productivity in cardamom. Varieties with high yield potential, superior capsule quality and wider adaptability are essential to increase productivity. Selection of clones having resistance/tolerance to major diseases and pests as well as drought also should get due importance in planned breeding programme of this crop.

5.1   Germplasm

Cardamom, being a cross-pollinated crop and propagated mostly through seeds, natural variability is fairly high. An assembly of diverse genetic stocks of any crop is the raw material from which a new variety can be moulded to suit the requirements of farmers and end users. Hence, collection, conservation, evaluation and exploitation of germplasm deserve utmost importance in breeding strategies. In 1950s, two surveys were conducted in cardamom growing areas in India to record the genetic resources and wild populations (Mayne, 1951) and to understand the geographical distribution and environmental impact on cardamom (Abraham and Thulasidas, 1958). Thereafter, explorations for germplasm collection are made by as many as six research organizations in India and the total number of accessions presently available with different centres are 1350 (Madhusoodanan et al., 1998, 1999). Earlier documentation was based on an old descriptor (Dandin et al., 1981) and a key for identification of various types has been formulated (Sudharshan et al., 1991). During 1994, a detailed descriptor for cardamom was brought out by the International Plant Genetic Resources Institute (IPGRI), Rome, Italy. Among the collections, genotypes having marker characters include terminal panicle, narrow leaves, pink coloured tillers, compound panicles, elongated pedicel etc. Asexuality, cleistogamy and female sterility are a few of the variations observed among the collections.

Conservation of cardamom genetic resources under *in situ* situation does not exist, though natural population occurs in protected forest areas, especially in the Silent Valley Biosphere Reserve, where a sizable population of cardamom plants in its

Figure 2.21 Panicle variation in cardamom: (a) panicle variation in var. *Malabar*; (b) panicle types from var. *Mysore*; (c) panicle types from var. *Vazhukka*.

natural setting exists. *Ex situ* conservation of cardamom germplasm is being undertaken mainly by four organizations. Table 2.6 gives the present holdings in these centres.

Ex situ conservation in cardamom is being maintained as field gene banks, and they are used for preliminary evaluation, maintenance as well as for characterisation. Characterisation involves morphological, agronomical as well as chemical characters. Many variations in morphological and chemical characters and in yield have been recorded in these collections (Zachariah and Lukose, 1992; Zachariah *et al.*, 1998).

Ex situ conservation is always at risk due to a variety of reasons, mainly biotic and abiotic stress factors. The prevalence of virus diseases is a serious threat to *ex situ* conservation of germplasm. An alternative is *in vitro* conservation and establishment of

Table 2.6 Germplasm holdings of cardamom in India

Centre	Cultivated germplasm	Wild and related taxa
IISR Regional Station, Appangala, Coorg (Karnataka)	314	13
Indian Cardamom Research Institute, Myladumpara, Idukki Dt., Kerala	600	12
Cardamom Research Centre, Pampadumpara, Idukki Dt., Kerala (Kerala Agricultural University)	72	15
Regional Research Station, Mudigere, Chickmagalur, Karnataka	236	7

Table 2.7 Growth characteristics of different cardamom types

Sl. No	Panicles		Leaves	Pseudostem			Capsules		
	Type	Nature	(Pub. or Nonpub.)	Height	Colour	Size	Shape	Cross section
1	E	Sim.	G	T	Green	Bold	Oval	Angular
2	"	"	"	"	"	Small	Globular	Circular
3	"	"	"	D	"	Medium	Oval	Angular
4	"	"	"	"	"	Small	Globular	Circular
5	SE	"	"	T	"	"	Oval	Angular
6	"	"	"	"	"	"	Globular	Circular
7	"	"	"	D	"	"	Oval	Angular
8	"	"	"	"	"	"	Globular	Circular
9	"	"	Pub.	T	"	"	Oval	Angular
10	"	"	"	"	"	"	Globular	Circular
11	"	"	"	D	"	"	Oval	Angular
12	"	"	"	"	"	"	Globular	Circular
13	P	"	"	T	"	Bold	"	Angular
14	"	"	"	"	"	Medium	"	Circular
15	"	"	"	D	"	"	"	"
16	"	"	"	"	"	Small	"	"
17	"	"	G	T	"	"	"	"
18	"	"	"	"	"	Medium	Oval	Angular
19	"	"	"	"	"	"	Oblong	"
20	"	"	"	"	"	Bold	Globular	"
21	"	"	"	D	"	Small	"	Circular
22	"	"	"	"	"	Medium	Oval	Angular
23	"	"	"	"	"	"	Oblong	"
24	"	"	"	"	"	Bold	Globular	"
25	"	"	"	"	Pink	Medium	Oval	"
26	"	MB	"	"	Green	"	"	"

Source: Krishnamurthy *et al.*, 1989.

Notes
E – Erect panicle; SE – Semierect; P – Prostrate; Sim. – Simple; MB – Multibranched; T – Tall; D – Dwarf;
G – Glabrous (Nonpubescent) leaf; Pub. – Pubescent leaf.

an *in vitro* gene bank. Protocols have been developed for *in vitro* conservation and germplasm by slow growth (see section 6.4).

5.1.1 Genetic variability

Krishnamurthy *et al.* (1989) classified the germplasm accessions of cardamom available at the Regional Research Station, Mudigere. They recognized 26 distinct types based on characters such as leaf pubescence, height and colour of aerial stem, panicle type etc. (Table 2.7).

At the Cardamom Research Centre of the Indian Institute of Spices Research in Kodugu district of Karnataka state, a study was carried out to assess the variability among 210 germplasm collections assembled from all major cardamom growing areas (Padmini *et al.*, 1999). These observations indicated that in general the var. *Vazhukka* and var. *Mysore* are more robust than var. *Malabar*. The total number of tillers as well as bearing tillers per plant, leafy stem diameter and number of leaves are more in *Vazhukka* and *Mysore* than in *Malabar*. The mean number of panicles/plant is higher in *Malabar* than the other two. Characters such as panicle number, nodes/panicles, internode length and capsule length exhibited high co-efficient of variation (Padmini *et al.*, 1999). Spike variability is given in Fig. 2.21a,b,c.

Among the *Malabar* accessions, cv per cent was the highest for number of panicles/plant (78.8), it was highest in Acc.75 (139), capsule breadth was maximum in Acc.40 (3.2 cm). Acc.58 recorded the highest number of bearing tillers per plant (27). Table 2.8 gives the data on variability in the var. *Malabar*. In the var. *Mysore* the characters having highest variability are panicles/plant (74.8 per cent) and internode length of panicles (69 per cent). Acc.98 recorded the highest number of panicles (135), internodes were shortest in Acc.109 (1.3 cm). Moderate variability existed for bearing tillers/plant, capsule breadth, capsule length, total tillers/plant etc. (Padmini *et al.*, 1999, Table 2.9). In var. *Vazhukka* the highest co-efficient of variation was recorded for number of panicles/plant (74.5 per cent), followed by number of bearing tillers/plant

Table 2.8 Variability in cardamom var. *Malabar*

Characters	Mean	Range	CV (%)
Plant height (cm)	1.67	0.58–2.19	2.96
Total tillers/plant	28.92	0.66–61.33	39.49
Productive tillers/plant*	8.65	0.00–27.00	56.30
Leafy stem diam. (cm)	1.08	0.32–1.99	23.15
No. of leaves/plant	208.93	10.00–411.80	43.64
Leaf length (cm)	46.96	19.00–64.00	15.74
Leaf breadth (cm)	7.64	3.00–11.33	17.80
No. of panicles/plant	42.65	0.00–139.33	78.76
Panicle length (cm)	35.18	0.00–80.83	43.69
No. of nodes/panicles	16.71	0.00–29.00	36.39
Internode length (cm)	3.05	0.00–5.13	32.79
Capsule length (cm)	0.69	0.00–1.50	42.03
Capsule breadth (cm)	0.43	0.00–3.20	69.77

Source: Padmini *et al.* (1999).

Note
* Tillers associated with panicle production.

Table 2.9 Variability in cardamom var. *Mysore*

Characters	Mean	Range	CV (%)
Plant height (cm)	1.72	0.91–2.28	18.60
Total tillers/plant	31.78	13.66–67.90	30.28
Productive tillers/plant	9.58	5.15–11.00	36.01
Leafy stem diam. (cm)	1.26	0.083–1.81	19.84
No. of leaves/plant	250.08	45.08–421.70	32.83
Leaf length (cm)	49.16	35.00–61.00	14.87
Leaf breadth (cm)	8.44	5.15–11.00	17.06
No. of panicles/plant	40.18	1.00–135.60	74.81
Panicle length (cm)	41.73	21.30–74.80	28.80
No. of nodes/panicle	20.18	7.33–30.67	25.07
Internode length (cm)	3.78	1.30–4.63	69.31
Capsule length (cm)	0.83	0.30–1.43	36.14
Capsule breadth (cm)	0.47	0.20–1.05	40.43

Source: Padmini *et al.* (1999).

Table 2.10 Variability in cardamom var. *Vazhukka*

Characters	Mean	Range	CV (%)
Plant height (cm)	1.81	1.38–2.20	13.25
Total tillers/plant	31.65	16.33–42.33	23.19
Productive tillers/plant	9.90	1.33–33.00	62.32
Leafy stem diam. (cm)	1.30	1.05–1.56	12.31
No. of leaves/plant	246.49	115.00–409.00	28.461
Leaf length (cm)	49.26	42.00–57.67	9.28
Leaf breadth (cm)	8.06	6.33–11.33	14.76
No. of panicles/plant	37.25	2.00–138.70	74.50
Panicle length (cm)	44.71	28.63–76.70	25.27
No. of nodes/panicle	20.74	11.50–27.33	24.30
Internode length (cm)	3.46	1.90–4.67	18.21
Capsule length (cm)	0.86	0.46–1.37	30.23
Capsule breadth (cm)	0.47	0.30–0.73	23.40

Source: Padmini *et al.* (1999).

(62 per cent). Acc.138 registered the highest panicle number. Moderate to low variability existed for capsule length, number of nodes/panicles, capsule breadth etc. (Padmini *et al.*, 1999, Table 2.10). Observations on natural variations in morphological and yield parameters under the cardamom growing situation of Idukki, Kerala were recorded by George *et al.* (1981). Among the characters highest variability was observed in panicle characters (Anonymous, 1958, 1986, 1987).

George *et al.* (1981) collected 180 accessions from the wild as well as from cardamom growing areas of Western Ghats. They could isolate distinct morphotypes and 12 eco-types showing heritable adaptations. The following were some of the observations made by the above workers. Vars. *Mysore* and *Vazhukka* were more vigorous than

Malabar, reaching even 6 m height. One clone (Acc.1: 6) had very narrow leaves (only 3 cm width). In two accessions tillers had characteristic pink and pale green colours. Though in general each tiller had two panicles, accessions having three and four panicles per tiller were present especially among the *Munzerabad* clones. Another known as 'Alfred clone', produced both basal and terminal panicles. Panicle length was highly variable among the accessions, ranging from 30–200 cm, the mean being 140 cm in vars. *Mysore* and *Vazhukka* and 80 cm in var. *Malabar*. Some accessions produced multiple branched panicles. Flower/fruit number varied from 4–36 per cincinnus. Variations were noticed in fruit shape as well. Acc.1: 175 had extra bold, triangular fruit; Acc.1: 89 (*Munzerabad*) had round fruits. Cv. PV_1, Acc.1: 14 and 1: 36 (Walayar type) had elongated fruit where the length was more than double the diameter. Green capsule weight varied from 132 g (Acc.1: 91) to 40 (Acc.1: 6), seed number per fruit varied from 5–31.

Plants having multibranched panicles or compound panicles occur in small proportions in the segregating populations of certain lines such as CL-37. Padmini *et al.* (2000) studied the variability among the compound panicle types, which are mostly having *Vazhukka* type of inflorescence. Among the compound panicles, proximal branching is more prevalent than the distal or entire branching types. The contribution of such branching towards yield (weight of fresh and dry capsules) varied from 12–41 per cent. Branching did not influence other yield or quality characters.

5.2 Genetic upgradation

Cardamom is amenable to both sexual and vegetative propagations; hence techniques such as selection, hybridization, mutation and polyploid breeding are used as means for genetic upgradation of the crop. Studies on certain aspects on crop improvement in cardamom have also been carried out in Sri Lanka (Melgode, 1938), Tanzania (Rijekbusch and Allen, 1971), Guatemala (Rubido, 1967) and PNG (Stone and Willmer, 1989). The paths for crop improvement in cardamom are given in Figs. 2.22 and 2.23.

5.2.1 Selection

The breeding strategies employed in the clonal selection pathway is given in Fig. 2.22. Gopal *et al.* (1990, 1992) carried out correlation and path analysis study in cardamom. They reported that dry weight of capsules per plant was positively correlated with the other polygenic characters like height of tillers (0.88), productive tillers per plant (0.78), panicles per plant (0.998), capsules per panicle (0.998), fresh weight of capsules per plant (0.99), length of panicles (0.87), nodes per panicle (0.96) and internodal length (0.63) of panicle. Number of panicles per plant, fresh weight of capsules per plant, nodes per panicle and internodal length in the panicle showed significant positive direct effects on yield. Panicles per plant showed maximum direct effect on yield followed by fresh weight of capsules per plant. The above workers concluded that panicles per plant, fresh weight of capsules per plant, nodes per panicle and internodal length of panicle were useful characters for improvement in yield of cardamom. Patel *et al.* (1997, 1998) also suggested use of traits like panicles per bearing tiller, panicles per clump, recovery ratio and capsules per panicle as criteria for selection for yield in cardamom. In a study using 12 genotypes these workers found that yield per clump had significant and positive correlation with capsules

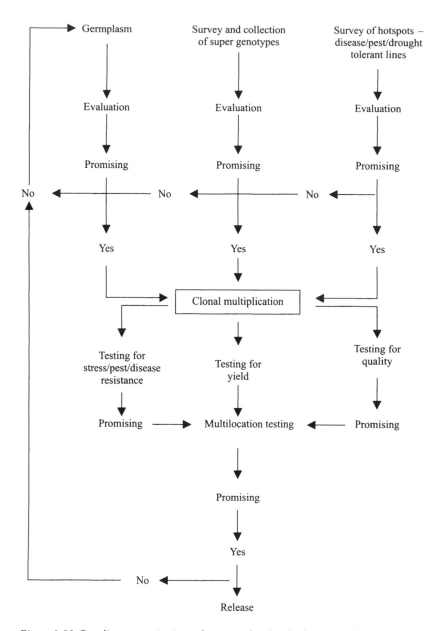

Figure 2.22 Breeding strategies in cardamom – the clonal selection pathway.

per panicle (0.967), cincinni per panicle (0.645), tillers per clump (0.639), panicle length (0.559), panicles per clump (0.537), bearing tillers per clump (0.340), vegetative buds per clump (0.309) and recovery ratio (0.224). Negative correlation was observed between fresh yield per clump and dry capsules per kg (−0.486). The above workers concluded that capsules and cincinni per panicle, bearing tillers and panicles per clump, panicle length and vegetative buds per clump are significant attributes primarily responsible for high yield of cardamom, and selection for improvement should be based on these attributes (Patel *et al.*, 1997, 1998).

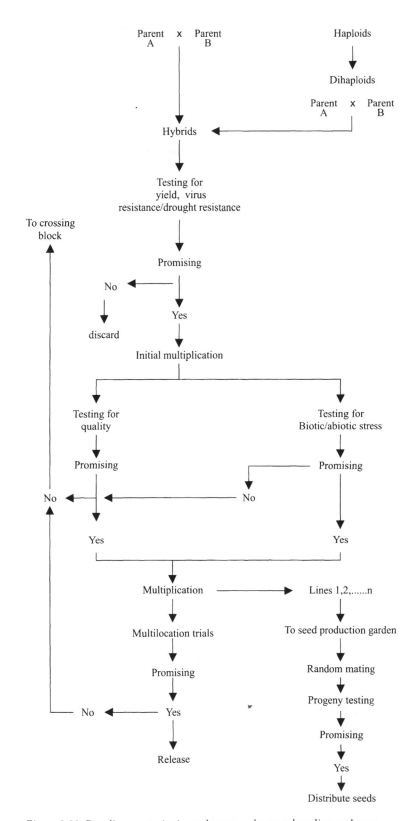

Figure 2.23 Breeding strategies in cardamom – the cross breeding pathway.

(a)

(b)

(c)

(d)

Figure 2.24 Some of the elite clonal selections released for cultivation: (a) ICRI-1 *Malabar* type; (b) ICRI-2 *Vazhukka* type; (c) *Njallani* green gold, *Vazhukka* type; (d) CCS-I *Malabar* type.

Systematic evaluation of germplasm accessions in India during the 1980s resulted in the identification and release of some elite clonal selections (Fig. 2.24). Initial collections for desirable traits were made from planters' fields as well as from wild habitats based on certain parameters. In order to isolate elite clones, germplasm collections were subjected to initial evaluation trials followed by comparative yield trials and multilocation testings under various agro-ecological conditions. Little studies have been taken up for selection of seedlings having precocity in bearing. The results available indicate that this phenomenon is not positively correlated with yield in cardamom (Madhusoodanan and Radhakrishnan, 1996). Selection for drought tolerance has also been attempted. The initial results indicated that the promising drought-tolerant genotypes are low yielders.

So far, twelve selections are available in India for commercial cultivation (Table 2.10). The performance of these selections are much superior to the conventionally used clones with regard to their yield and capsule characters. A sharp increase in national productivity of cardamom in India from 53 kg/ha (1989–90) to 173 kg/ha (1999–2000) has been recorded. It is of significance that during 1988–89 the area under the crop was 81,113 ha but declined to 72,450 ha by 1999–2000 and total production has increased to 9290 mt (1999–2000) from 3100 mt (1989–90). The major factors that contributed to this substantial increase in productivity and production are the use of improved clones during replanting the senile gardens and adoption of high production technology. Apart from the cardamom cultivars released from Research Institutes, some of the farmers have also carried out selection of superior clumps from populations. Cv. *Njallani Green Gold* is one such superior selection by a farmer, which became very popular in Idukki district. This is the best yielding line now available, having high yield potential. Growers have harvested up to 2475 kg/ha dry cardamom under good management (John 2000) (Table 2.11).

Table 2.11 Elite selections of cardamom

Sl. No.	Selections	Plant type	Average yield (kg/ha)	Oil (%)	Distinguishing characteristics
1	Mudigere-1	Malabar	275	8	Compact plant, suitable for high density planting. Tolerant to hairy caterpillars and white grubs. Short panicle, oval, bold pale green capsules. Pubescent leaves.
2	Mudigere-2	Malabar	476	8	Capsule round, high oil (8%), suitable for hill zones of Karnataka.
3	PV-1	Malabar	260	6.8	An early maturing variety with slightly ribbed light-green capsules. Short panicle, close cincinni, elongated capsules.
4	CCS-I (IISR Suvasini)	Malabar	409	8.7	An early maturing variety suitable for high density planting. Long panicle, oblong bold, parrot-green capsules.
5	ICRI-1	Malabar	656	8.3	An early maturing profusely flowering variety. Medium long panicle, globose, extra bold dark green capsules.
6	ICRI-2	Mysore	766	9,0	Performs well under irrigated conditions. Suitable for higher altitude. Medium long panicle, oblong bold and parrot-green capsules.

7	ICRI-3	Malabar	599	6.6	Early maturing type, non-pubescent leaves. Oblong, bold parrot-green capsules.
8	RR1 (IISR Avinash)	Malabar	960	6.7	Selection from hot spot areas. Resistant to rhizome rot.
9	TDK-4	Malabar	456	6.4	Suitable for low-rainfall area, highly adapted to the lower Pulney hills of Tamil Nadu and areas having similar agroecology.
10	PV-2	Malabar	982	10.45	Selection from OP seedling progeny of PV-1. Early maturing, bold capsules, high dry recovery, tolerant to thrips, capsule rot and clump rot.
11	IISR Vijetha (NKE 12)	Malabar	643	7.9	For highly shaded and mosaic disease areas, capsule oblong, virus resistant.
Farmer's selection					
12	Njallani Green Gold**	Vazhukka	–	–	Clonal selection by a cardamom grower in Idukki district. High yielding. Capsules bold, over 70% of the cured cardamom above 7 mm.

Notes
* Yield up to 1600 kg/ha in farmers' field.
** Yield up to 2475 kg/ha in farmers' fields.

5.2.1.1 Selection in seedling progenies

Korikanthimath *et al.* (1999) carried out a selection programme in the seedling progenies of the clone CL-37, using a population size of 5000 plants. The yield ranged from 325 g/clump to 7555 g/clump (wet weight). Out of this 17.8 per cent were high yielders with a mean yield of 6537 g of capsules/clump (wet weight). The above workers computed Bartlet index (BI) to find out earliness among the progenies and found that six lines were very early flowering types, but they were also poor yielders, except one line that gave an yield of 5156 g/clump (wet weight) and having a Bartlet index of 0.7085. In the study population early lines were 29.7 per cent (BI: 0.6–0.7), medium types were 51.5 per cent (BI: 0.5–0.6), late types were 16.7 per cent (BI: 0.4–0.5) and very late types were 0.3 per cent (BI: less than 0.4). The strategy further involves the multiplication of the top 10 per cent and distribution to farmers after multilocational testing.

5.2.2 Selection for biotic stress tolerance

At IISR Cardamom Research Centre at Appangala (Karnataka), efforts have been made to survey and collect disease escapes from hot spot areas of *Katte* disease. Collections of Natural Katte Escapes (NKE) lines from such surveys were then subjected to artificial inoculation through the use of insect vectors. The plants that have not taken up infection even after repeated screening were field evaluated again in a hot spot area. Some of these resistant lines are also high yielding giving good yield, comparable to the released selections, both in yield and quality (IISR, 1997; Tables 2.12 and 2.13).

Table 2.12 Mean yield of disease resistant cardamom collections (fresh yield/plant in kg)

Entry	Mean yield	Entry	Mean yield
RR-1	1.631 a	NKE-34	0.912 efg
CCS-1	1.379 b	NKE-8	0.900 efg
M-1	1.211 bc	NKE-26	0.887 fg
NKE-12	1.169 c	NKE-72	0.872 g
NKE-3	1.158 c	NKE-4	0.852 g
NKE-19	1.131 cd	NKE-78	0.850 g
MB-3	1.097 cde	NKE-71	0.838 g
NKE-27	1.091 cde	NKE-28	0.795 g
NKE9	1.073 cdef	MA (control)	0.765 g
NKE-32	0.938 defg	NKE-11	0.759 g
NKE-31	0.921 efg	NKE-4	0.440 h

Source: IISR Annual Report (1997).

Note
LSD = 199.8 Figures followed by the same letter are not significantly different.

Table 2.13 Quality characters of NKE lines and RR1

Entry	Recovery of dry cardamom (%)	No. of seeds fruit	Husk–seed ratio	Bold capsules (%)
NKE-3	20.8	19.0	1:2.9	81
NKE-4	21.6	16.1	1:3.1	58
NKE-5	21.8	17.7	1:3.0	73
NKE-8	21.6	16.8	1:3.0	68
NKE-9	20.8	20.6	1:2.9	69
NKE-11	21.4	15.5	1:2.9	66
NKE-12[a]	22.0	20.4	1:3.0	77
NKE-19	21.6	19.2	1:3.0	80
NKE-26	19.8	18.2	1:2.8	68
NKE-27	19.6	20.0	1:2.6	75
NKE-28	19.8	18.2	1:2.8	73
NKE-31	20.8	17.6	1:3.1	81
NKE-32	21.4	16.0	1:2.1	77
NKE-34	22.0	18.0	1:3.5	72
NKE-71	21.4	16.9	1:3.0	67
NKE-72	20.5	16.1	1:2.8	74
NKE-78	21.4	9.9	1:3.1	76
MB-3	21.2	16.8	1:2.7	66
RR-1[b]	20.8	13.3	1:3.1	51
M-1	21.4	15.5	1:2.9	58
CCS-1[c]	21.6	17.9	1:3.0	89
MA (control)	21.4	15.9	1:3.0	80

Source: IISR Annual Report (1997).

Notes
a – IISR Vijetha; b – IISR Avinash; c – IISR Suvasini.

One particular collection has been shown to be resistant to rhizome rot caused by *Phytophthora* sp. This rhizome rot resistant (RR1) line was tested in comparative yield trials along with the NKE lines. This line has given consistently good yield in all the years, the best among the lines tested giving 18 per cent more yield than the released high yielding variety (Table 2.12). Its negative point is the lower per cent of bold capsules (Table 2.13).

For further studies and improvement the NKE lines were used in a diallele-crossing programme (IISR, 1998).

5.2.3 Selection for drought tolerance

Selection of cardamom genotypes were carried out using parameters such as relative water content, membrane leakage, stomatal resistance and specific leaf weight, and significant variations have been noted among cultivars (IISR, 1997).

5.2.4 Hybridization

The breeding strategies employed for the crossbreeding pathway for cardamom improvement is given in Fig. 2.23. The popular cardamom variety namely *Vazhukka* possibly originated as a natural cross between var. *Malabar* and var. *Mysore*. Since cardamom is amenable to both sexual and vegetative propagation, hybridization is a very useful tool for crop improvement. As only one species occurs in India, crossing in cardamom is confined to infra-specific levels. Because of its perennial, cross-pollinated and heterozygous nature, the conventional methods for evolving homozygous lines in cardamom are time consuming.

Various workers carried out, both intergeneric and intervarietal hybridizations. The former one was tried with an intention of transferring the disease resistance character. Such attempts have not been encouraging except in a report of fruit set in a cross with *Alpinia neutans* (Parameswar, 1977). All other intergeneric crosses involving *Amomum. Alpinia. Hedychium* and *Aframomum* were sterile (Krishnamurthy *et al.*, 1989; Madhusoodanan *et al.*, 1990).

Intervarietal and inter-cultivar hybridizations have been carried out for producing high yielding heterotic recombinants. A diallele cross involving six selected types having characters of early bearing, bold capsule, high yield, long panicle, leaf rot resistance and multiple branching was carried out and 30 cross combinations were made. All the hybrids were more vigorous compared to the parental lines (Krishnamurthy *et al.*, 1989). In another study, intervarietal hybridization has been carried out using different varieties of cardamom. This has resulted in cross combinations of 56 F1 hybrids. Evaluation of these hybrids led to the isolation of a few high yielding heterotic recombinants having an average yield of 470 to 610 kg/ha under moderate management (Madhusoodanan *et al.*, 1998, 1999).

5.2.5 Selection in polycross progenies

Chandrappa *et al.* (1998), carried out studies on the impact of selection in a polycross progeny population. Promising clonal selection of *Malabar* type cardamom (including the ruling variety Mudigere-1), were grown in isolation, and open pollinated varieties of these selections were evaluated. In the case of 34 per cent of the progenies the average yield was found to be significantly higher than the average of the control variety (Mudigere-1).

This yield increase varied from 1–149 per cent, and certain clones (691, 692, D_{11}, D_{19}) were more promising than others. The above workers found that improvements of yield in cardamom could be more effectively achieved through a polycross breeding programme. Compared to the maximum of 1663 g expected in the progenies of Mudigere-1, one selection from polycross seeds of Mudigere-1 yielded 2360 g (44 per cent higher), and another progeny from the line D_{237} gave 2670 g yield, which represents 60 per cent higher yield compared to the highest yield observed in the progeny of Mudigere-1. The frequency and per cent of progenies yielding above 500 g per plant is given in Table 2.14.

Chandrappa *et al.* (1998) found that based on the polycross progeny test, 38 per cent clones were poor in combining ability and hence can be rejected. They further suggested that lines with better combining ability (such as Mudigere-1, Cl-691, Cl-692, D-11, D-19, D-535, D-186, Dl-730, D-18) and having 46–149 per cent higher yield compared to the means of the polycross progeny and the checks, may be used for restricted polycross nursery for isolating higher yielding selections. They also found that a population of more than 3000 plants is essential for carrying out effective selection.

5.2.6   Polyploidy

Polyploids ($2n = 4x = 96$) were successfully induced in cardamom by treating the sprouting seeds with 0.5 per cent aqueous solution of colchicine (Sudharshan, 1987, 1989). The polyploid lines exhibited gigantism. Increased layer of epidermal cells, thicker cuticle and higher wax coating on the leaves found in the induced polyploid lines are characters generally associated with drought tolerance in nature. The meiotic behaviour of induced polyploids were almost normal and they had reasonably good fertility. In all yield characters the tetraploids were reported to be inferior to the diploids (Anonymous, 1986).

5.2.7   Mutation breeding

Attempts for induction of desirable mutants using physical mutagens viz., x-rays and gamma rays (^{60}Co source) and chemical mutagens (ethyl methane sulphonate and maleic hydrazide) have been made. Out of a large number of selfed and open pollinated progenies of M1 plants, which did not take infection after repeated cycles of inoculations with *Katte* virus vector, 12 plants were selected as tolerant to the disease (Bavappa, 1986). There are reports on sterility (Pillai and Santha Kumari, 1965) and

Table 2.14 Frequency and percentage of high-yielding progeny in the poly-cross progeny testing in cardamom

Green capsule yield (g/clump)	Frequency of progenies	Percentage
500–600	150	6.30
601–700	102	4.30
701–800	86	3.60
801–900	55	2.30
901–1000	38	1.60
1001–1500	61	2.60
1501–2000	20	0.84
2001–2500	3	0.12
2501–2700	1	0.04

Source: Chandrappa *et al.* (1998).

absence of macro-mutation in M1 generation and its progeny (Anonymous, 1987). No desirable mutant could so far been developed in cardamom.

6 BIOTECHNOLOGY

6.1 Micropropagation

Micropropagation offers tremendous scope for rapid vegetative propagation of elite clones or varieties, eliminating systemic pathogens such as viruses. Replanting of senile, seedling-raised plantations with selected high yielding clones multiplied through micropropagation can give 5–6-fold increase in the current average productivity of cardamom (Anonymous, 1996). Micropropagation technique can be used for the following applications in cardamom (Bajaj *et al.*, 1988).

  (i) increase in the propagation rate of plants;
 (ii) availability of plants through out the year (in all seasons);
(iii) protection of plants against pests and pathogens under controlled conditions;
 (iv) uniform clones;
  (v) uniform productivity of secondary metabolities.

In cardamom, various tissue culture approaches have been made use of, such as (i) through callusing, (ii) through adventitious bud formation and (iii) through enhanced axillary branching (Table 2.15).

Table 2.15 Tissue culture work in cardamom

Explant	Media composition	In vitro response	Reference
Vegetative bud	MS + 1 mgl^{-1} BA, 0.5 mgl^{-1} NAA	Multiple shoots, *in vitro* rooting	Nirmal Babu *et al.* (1997)
	MS + 0.5 mgl^{-1}BA, 0.5 mgl^{-1} kin, 2 mgl^{-1} biotin 0.2 mgl^{-1} calcium pantothenate, 5% coconut milk	Multiple shots, *in vitro* rooting	Nadgauda *et al.* (1983)
Rhizome of TC plants	MS + 1 mgl^{-1} 2,4-D, 0.1 mgl^{-1} NAA, 7 mgl^{-1} BA, 0.5 mgl^{-1} kin	Callus	Lukose *et al.* (1993)
Immature panicles	MS + 0.5 mgl^{-1} NAA, 0.5 mgl^{-1} kin, 1 mgl^{-1} BA, 0.1 mgl^{-1} calcium pantho thenate, 0.1 mgl^{-1} folic acid, 10% coconut milk	Conversion of floral primordium into vegetative buds	Kumar *et al.* (1985)
Callus derived from vegetative buds	MS + 10% coconut milk, 2–5 mgl^{-1} BA	Regeneration of plantlets	Rao *et al.* (1982)
	MS + 1 mgl^{-1} 2,4-D, 0.1 mgl^{-1} NAA, 1 mgl^{-1} BA, 0.5 mgl^{-1} kin	Organogenesis and regeneration of plantlets	Lukose *et al.* (1993)

The first report of cardamom tissue culture was by Rao *et al.* (1982), who achieved regeneration of plants from callus cultures. Nadgauda *et al.* (1983) achieved a multiplication ratio of 1:3 when sprouted buds were cultured on MS medium supplemented with BAP (0.5 mg/l), Kn (0.5 mg/l), IAA (2 mg/l), calcium pantothenate (0.1 mg/l), biotin (0.1 mg/l) and coconut water (5 per cent). The plantlets were successfully rooted and grown in field. Kumar *et al.* (1985) reported direct shoot formation from inflorescence primordium when cultured using MS medium containing NAA, kinetin and BAP; and they also could get plantlets rooted. Reghunath and Bajaj (1992) has given a detailed treatment of micropropagation methods in cardamom. Lukose *et al.* (1993) used MS medium containing 20 per cent coconut water, 0.5 mg/l NAA, 0.2 mg/l IBA, 1.0 mg/l of 6-benzyladenine and 0.2 mg/l kinetin. The plantlets are rooted in White's basal medium containing NAA 0.5 mg/l and hardened in soil-vermiculite mixture. Other reports include those of Priyadarshan and Zachariah (1986), Vatsya *et al.* (1987), Priyadarshan *et al.* (1988), Reghunath (1989), Reghunath and Gopalakrishnan (1991), Nirmal Babu *et al.* (1997) and Pradip Kumar *et al.* (1997). Commercial multiplication of cardamom is being carried out by biotech companies, such as A.V. Thomas & Co. (Cochin) and Indo-American hybrid seeds (Bangalore) etc.

6.1.1 Propagation through callus culture

Reghunath and Bajaj (1992) gives the methodology for cardamom micropropagation in detail. They have tried various explants such as shoot primordium, inflorescence primordium, immature inflorescence segments and immature capsules, and tested serial treatments with 95 per cent alcohol, 2–4 per cent sodium hypochlorite and 0.05–0.2 per cent mercuric chloride for decontamination of explants. Both MS and SH (Schenk–Hildebrandt) media at half and full strengths were tested, along with auxins such as NAA, IAA and 2,4-D alone and in combination. The cultures on 0.6 per cent agar were incubated at a light intensity of 1000 lux and 16 h photoperiod. The maximum callus production was reported in MS medium supplemented with 4 mg/l NAA and 1 mg/l BAP. This callus on sub-culturing on an auxin-free medium having 3 mg/l BAP and 0.5 mg/l kinetin started caulogenesis; each culture producing six–nine meristemoids, and they on culturing on the same medium produced shoots within 28 days. Coconut water (15 per cent) enhanced caulogenesis. The tissue culture cycle is given in Fig. 2.25.

6.1.2 Propagation through enhanced axillary branching method

Priyadarshan *et al.* (1992) tested 17 media formulations and using shoot primordia as explants. They got best results with MS medium fortified with IAA, BAP, kinetin, calcium pantothenate, biotin and coconut water. Reghunath and Bajaj (1992) has outlined the culture method, using shoot and inflorescence primordium explants, and the media tested were MS and SH. The SH medium was found to be better than full or half MS medium as it has given 31 per cent greater shoot dry weight. Liquid medium culture under agitation using a gyratory shaker produced 111.5 per cent more axillary branches than those cultured on semi-solid medium. The culture conditions include temperature of 23 ± 2 °C, light intensity of 3000 lux and 16 h photoperiod. The number of axillary branches was highest in medium containing 4 mg/l BAP and 0.5 mg/l NAA. Axillary branch production was enhanced by coconut water. Var. *Mysore* and var. *Vazhukka* produced more axillary branches than var. *Malabar* (Fig. 2.23).

Figure 2.25 Tissue culture of cardamom: (a) multiple shoot formation from rhizome explant; (b) somatic embryogenesis from suspension culture; (c) somaclonal variation appearing in culture, showing variation in leaf and plant morphology.

Three separate media have also been tested (IISR 1997) for multiplication, subculturing and rooting as given below.

Multiplication	Subculture	Rooting
MS basal medium +	MS basal +	White's basal (liquid) +
0.5 mg/l NAA	0.5 mg/l NAA	0.5 mg/l NAA
1.0 mg/l BAP	0.5 mg/l BAP	
0.2 mg/l IBA	0.2 mg/l kinetin	
200 mg/l coconut water	200 mg/l CW	30 g/l sucrose
30 gl/l sucrose	30 g/l sucrose	

Nirmal Babu *et al.* (1997) (Fig. 2.23) has given the following culture condition for micropropagation of cardamom,

Explants   : Rhizome bits with vegetative buds.
Surface sterilization   : washing in running water, detergent solution and then in
                             0.1 per cent $HgCl_2$
Incubation   : 22 ± 2 °C, 14 h photoperiod, 3000 lux
Medium   : (a) MS + 1 mg/l NAA for rooting
                             (b) MS + 1 mg/l BAP, 0.5 mg/l NAA for multiplication
                             and rooting, in one step

6.1.3   In vitro rooting and hardening

The excised axillary shoots can be rooted on a semi-solid medium of half strength MS salt and 0.5 per cent activated charcoal for 1 week, followed by subculturing in 1/2 MS medium containing 1.5 mg/l IBA under a light intensity of 3500 lux (Reghunath and Bajaj, 1992). Rooted shoots were transferred to MS 1/2 liquid medium containing only mineral salts and were then shifted to green house for hardening. For planting, vermiculite-fine sand (1:1) was found the best giving 92 per cent establishment (Reghunath and Bajaj, 1992).

6.2   Callus culturing and somaclonal variations

Callus regeneration protocols are important for generating somaclonal variations for future crop improvement use. An efficient system for callus regeneration is essential to produce large number of somaclones and such a system has been reported earlier by Rao *et al.* (1982) and was also standardized at IISR (Fig. 2.26, Ravindran *et al.*, 1997). High amount of variability was noticed among the somaclones for morphological characters in the culture vessel itself (Ravindran *et al.*, 1997). The most striking morphological variant is a needle leaf variant (Fig. 2.26) having small needle-shaped leaves, that multiply and root profusely in the same medium (Modified MS), but its rate of establishment in the nursery and field is reported to be low (Nirmal Babu, personal communication). Nirmal Babu and his colleagues have standardized cell culture system for large-scale production of callus through somatic embryogenesis for enhancing genetic variability. The somaclones are being subjected to evaluation for virus resistance and other characters (Nirmal Babu, unpublished).

6.3   Field testing of TC plants

The first report of the field performance of TC plants was that of Lukose *et al.* (1993), though earlier Nadgauda *et al.* (1983) mentioned about the field establishment of TC plants of cardamom. Lukose *et al.* (1993) carried out two statistically laid out trials to evaluate TC plants together with suckers and seedlings. Trial one was conducted with clone CL-37 and the trial two was with cv. *Mudigere* 1. Variations observed among TC plants, suckers and seedlings were non-significant for most of the vegetative characters as shown by analyzing pooled data of 4 years. Yielding tillers, panicles/plant, green capsule yield and cumulative yield were significantly more in TC plants, in both the trials. The earlier differences observed in growth characters disappeared in later years.

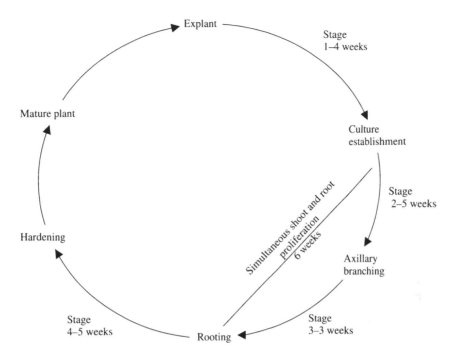

Figure 2.26 Tissue culture cycle in cardamom.

Sudharshan *et al.* (1997) reported the results of a large-scale evaluation carried out by ICRI. In one instance the performance of eight high-yielding micropropagated clones and open pollinated progenies were evaluated at 56 locations in an area of 37.5 ha. (Unfortunately suckers were not included in the trial). Variability was observed in the clonal population for vegetative characters. The overall variability in TC plants was 4.5 per cent as against 3 per cent in open pollinated seedling progenies for a given set of characters. Complete sterility was also reported in certain clones. Microcapsules were significantly more in TC plants, accounting for a major share of variation in these plants (8.4 per cent). However inspite of the occurrence of variations, TC plants yielded 34 per cent more than the seedlings. The causes of variations were attributed to adventitious bud formation during micropropagation via axillary buds, genetic instability of adventitious meristem and tissue culture induced disorganization of meristems (Sudharshan *et al.*, 1997). Chandrappa *et al.* (1997) tested eight tissue cultured cardamom selections against their suckers and two local checks. Of the clones TC 5, TC 6, and TC 7 were found to be promising, and they differed among themselves for yield and yield attributes. TC 5 was the best, recording superior values for most observations.

6.4 *In vitro* conservation

In vitro conservation is an alternative method for medium-term conservation. *In vitro* gene bank will be a safe alternative in protecting the genetic resources from epidemic diseases. Geetha *et al.* (1995) and Nirmal Babu *et al.* (1994, 1999a,b) reported conservation of cardamom germplasm in *in vitro* gene bank by slow growth. The above workers carried out various trials to achieve an ideal culture condition under which the growth is slowed down to the minimum without affecting the physiology or genetical make up of the plant. The slow growth is achieved by the incorporation of agents for increasing the osmotic potential of the medium, such as mannitol. They found that half strength MS without growth regulators and with 10 mg/l each of sucrose and mannitol was the best for *in vitro* storage of cardamom under slow growth. By using the above medium in screw capped vials the subculture interval could be extended to one year or more, when incubated in 22 ± 2 °C at 2500 lux of light and at 10 h photoperiod. Low temperature storage at 5 °C and 10 °C was found to be lethal for cardamom, as the cultures did not last more than three weeks (Geetha *et al.* 1995).

6.5   Other biotechnological work

6.5.1   *Isolation and culturing of protoplast*

Protoplasts could be isolated from mesophyll tissues collected from *in vitro* grown plantlets, achieving an yield of 35×10^5/g of leaf tissue on incubation in an enzyme solution containing 0.5 per cent macerozyme R10, 2 per cent Onozuka cellulase R10 and 9 per cent mannitol for 18–20 h at 25 °C in dark. (IISR, 1996; Geetha *et al.*, 2000). The yield of protoplasts from cell suspension culture was 1.5×10^5/g tissue, when incubated in 1 per cent macerozyme R10 and 2 per cent cellulose onozuka R10 for 24 h at 25 °C with gentle shaking at 53 rpm in dark. The viability of the protoplast was 75 per cent (mesophyll) and 40 per cent (cell suspension) respectively. The protoplasts on culturing developed into microcalli (Geetha *et al.*, 2000).

6.5.2   *Cryo-conservation*

Cryo-conservation of cardamom seed was attemped by Choudhary and Chandel (1995). They tried to conserve seeds at ultra-low temperature by either (i) suspending seeds in cryovials in vapour phase of liquid nitrogen (-150 °C) by slow freezing or (ii) by direct immersion in liquid nitrogen (-196 °C) by fast freezing. The result showed that seeds possessing 7.7–14.3 per cent moisture content could be successfully cryo-preserved and showed more than 80 per cent germination when tested after one year storage in vapour phase of liquid nitrogen (at -150 °C).

6.5.3   *Synthetic seeds*

The first report of the production of synseeds by encapsulation of shoot tips was by Ganapathy *et al.* (1994), who encapsulated shoot tips of cardamom cv. *Malabar* isolated from multiple shoots and encapsulated in 3 per cent (w/w) sodium alginate, with different gel matrices, and were subsequently cultured on MS medium. Sajina *et al.* (1997) reported the standardization of synseed production in many spices including cardamom. Synthetic seeds have many advantages over the normal micropropagation methods. This is an ideal system for conservation and exchange.

7 CONCLUSION

Inspite of the fact that cardamom is a native of South India, and is being used for the last many centuries, there are many gaps existing in our understanding of this crop plant. No in-depth study has gone into the botany of the crop. Areas such as developmental morphology and physiology are almost totally neglected. The entire area of production physiology needs thorough investigation. No information is available on the origin and interrelationships of cardamom, as it is represented by a single species in India. How far it is related to the Sri Lankan and Malaysian species? This aspect requires further studies.

In the area of crop improvement emphasis need be given in developing lines resistant to drought, as it is a major limiting factor to productivity increase. Exploitation of heterosis has not been attempted, and for this genetically homozygous lines have to be evolved. Attempts are going on at IISR Calicut in developing a protocol for anther culture and production of haploids and dihaploids in cardamom.

A priority area will be the breeding for disease resistance, especially against the devastating virus diseases. Biotechnology can contribute much to this area. Yet another area of great importance is the molecular characterization of cardamom germplasm. Concerted efforts are required in these areas. Alleviation of production constraints through conventional breeding or through molecular approaches will go a long way in increasing productivity and sustaining the production of this important spice crop.

REFERENCES

Abheywickrama, B.A. (1959) A provisional checklist of the flowering plants of Ceylon. *Ceylon J. Sci. (Biol. Sec.)*, 2, 119–240.

Abraham, P. and Thulasidas, G. (1958) South Indian cardamom and their agricultural value. *Tech. Bull.*, 79, ICAR, New Delhi, pp. 1–27.

Anonymous (1958) South Indian cardamoms – their evolution and natural relationships. *Tech. Bull.*, 57, ICAR, New Delhi.

Anonymous (1986) *Annual Report*, ICRI, Spices Board, India.

Anonymous (1987) *Annual Report*, ICRI, Spices Board, India.

Anonymous (1996) *Tissue Cultured Cardamom: Product Plan*, Final Report, Spices Board, India.

Bajaj, Y.P.S., Furmanowa, M. and Olszowska, O. (1988) *Biotechnology of the Micro-propagation of Medicinal and Aromatic Plants*. I., Springer Verlag, Berlin, pp. 60–103.

Bavappa, K.V.A. (1986) *Research at CPCRI. Tech. Bull.*, CPCRI., Kasaragod, India, p. 34.

Belavadi, V.V., Chandrappa, H.M., Shadakshari, Y.G. and Parvathi, C. (1993) Isolation distance for seed gardens of cardamom (*Elettaria cardamomum* Maton). In G.K. Veeresh, R. Umashanker and K.N. Ganeshan (eds) *Pollination in Tropics*, IUSSI, Bangalore, pp. 241–243.

Belavadi, V.V. and Parvathi, C. (1998) Estimation of honey bee colonies required for effective pollination in cardamom. *Proc. National Sym. Diversity of Social Insects & other Arthropods and the Functioning Ecosystems*, III Congress of IUSSI, Mudigere, p. 30.

Belavadi, V.V., Parvathi, C. and Raju, B. (1998) Optimal foraging by honey bees on cardamom. *Proc. XIII Plantation Crops Sym.*, Dec. 16–18, Coimbatore (Ab.).

Belavadi, V.V., Venkateshalu and Vivek, H.R. (1997) Significance of style in cardamom corolla tubes for honeybee pollination. *Curr. Sci.*, 73, 287–290.

Bernhard, R.A. (1971) Terpenoids of cardamom oil and their comparative distribution among varieties. *Phytochemistry*, 10, 177–184.

Burtt, B.L. (1980) Cardamoms and other Zingiberaceae in Hortus Malabaricus. In K.S. Manilal (ed.) *Botany and History of Hortus Malabaricus*, Oxford and IBH, New Delhi, pp. 139–148.

Burtt, B.L. and Smith, R.M. (1983) Zingiberaceae. In M.D. Dassanayake (ed.) *A Revised Hand Book to the Flora of Ceylon*, Vol. IV. Amerind Pub., New Delhi.

Chandran, K., Raja, P., Joseph, D. and Suryanarayana, M.C. (1983) Studies on the role of honeybees in the pollination of cardamom. *Proc. 2nd Int. Cof. Api. Trop. Climate*, pp. 497–504.

Chandrappa, H.M., Shadakshari, Y.G., Sudharsan, M.R. and Raju, B. (1997) Preliminary yield trial of tissue cultured cardamom selections. In S. Edison, K.V. Remana, B. Sasikumar, K. Nirmal Babu and S.J. Eapen (eds), *Biotechnology of Spices, Medicinal and Aromatic Plants*, Indian Soc. Spices, Calicut, India, pp. 102–105.

Chandrappa, H.M., Shadakshari, Y.G., Dushyandhakumar, B.M., Edison, S. and Shivashankar, K.T. (1998) Breeding studies in cardamom (*Elettaria cardamomum* Maton). In N.M. Mathew and C.K. Jacob (eds) *Developments in Plantation Crop Research*, Allied Pub., New Delhi, pp. 20–27.

Chandrasekhar, R. and Sampath Kumar, R. (1986) Karyological highlights on two cultivars of cardamom (*E. cardamomum* Maton). *J. Cytology and Genetics*, 21, 90–96.

Choudhary, R. and Chandel, K.P.S. (1995) Studies on germination and cryopreservation of cardamom (*Elettaria cardamomum* Maton) seeds. *Seed Sci. Biotech.*, 23, 235–40.

Dandin, S.B., Madhusoodanan, K.J. and George, K.V. (1981) Cardamom descriptor. *Proc. IVth Symp. Plant. Crops* (Placrosym-IV), CPCRI, Kasargode, pp. 401–406.

Esau, K (1953) *Plant Anatomy*, Mcgraw Hill, New York.

Evans, D.A., Sharp, W.R., Amamirato, P.V. and Yamada, Y. (1983) *HandBook of Plant Cell Culture* 1: *Techniques for Propagation and Breeding*, MacMillan, New York, USA.

Gaertner (1791) Quoted from Abheywickrama (1959).

Ganapathy, T.R., Bapat, V.A. and Rao, P.S. (1994) *In vitro* development of encapsulated shoot tips of cardamom. *Biotechnology Techniques*, 8, 239–244.

Geetha, S.P., Manjula, C. and Sajina, A. (1995) *In vitro* conservation of genetic resources of spices. *Proc. Kerala Sci. Congress*, Palakkad, pp. 12–16.

Geetha, S.P., Nirmal Babu, K., Rema, J., Ravindran, P.N. and Peter, K.V. (2000) Isolation of protoplasts from cardamom (*Elettaria cardamomum* Maton) and ginger (*Zingiber officinale* Rosc.). *J. Spices and Aromatic Crops*, 9, 23–30.

George, E.F. and Sherrington, P.D. (1984) *Plant Progation by Tissue Culture*, Exegenetics Limited, England.

George, K.V., Dandin, S.B., Madhusoodanan, K.J. and Koshy John (1981) Natural variations in the yield parameters of cardamom (*Elettaria cardamomum*). In S. Visveshawara (ed.) *Proc. PLACROSYM 1981*, pp. 216–231, ISPS, Kasaragod, India.

Gopal, R., Chandraswamy, D.W. and Nayar, N.K. (1990) Correlation and path analysis in cardamom. *Indian J. Agric. Sci.*, 60, 240–242.

Gopal, R., Chandraswamy, D.W. and Nayar, N.K. (1992) Genetic basis of yield and yield components in cardamom. *J. Plantation Crops*, 20(suppl.), 230–232.

Gregory, P.J. (1936) The floral morphology and cytology of *Elettaria cardamomum* Maton. *J. Linn. Soc.*, 50, 363–391.

Holttum, R.E. (1950) The Zingiberaceae of the Malay Peninsula. *Gard. Bull.* Singapore, 13, 1–249.

IISR (1996) *Annual Report for 1995–96*, Indian Institute of Spices Research, Calicut, India.

IISR (1997) *Annual Report for 1996–97*, Indian Institute of Spices Research, Calicut, India.

IISR (1998) *Annual Report for 1997–98*, Indian Institute of Spices Research, Calicut, India.

Jackson, B.P. and Snowdon, D.W. (1990) *Atlas of Microscopy of Medicinal Plants. Culinary Herbs and Spices*, Belhaven Press, UK.

John, K. (2000) *Njallani green gold* at cardamom productivity helm with precision farming technique. In *Spices and Aromatic Plants – Challenges and opportunities in the new century*, ISS, Calicut, India, pp. 105–106.

Korikanthimath, V.S. and Mulge, R. (1998) Assessment of elite cardamom lines for dry matter distribution and harvest index. *J. Med. Aromatic Plants*, 20, 28–31.

Korikanthimath, V.S. and Rao, G.S. (1983) Leaf area determination in cardamom (*Elettaria cardamomum*). *J.Plantation Crops*, 11, 151–153.

Korikanthimath, V.S., Mathew, M.J., Mulge, R., Hegde, R. and Hosmani, M. (1999) Studies on selection of elite lines of CL-37 seedling progenies of cardamom. *J. Spices and Aromatic Crops*, 8, 57–61.

Krishnamurthy, K., Khan, M.M., Avadhani, K.K., Venkatesh, J., Siddaramaiah, A.L., Chakravarthy, A.K. and Gurumurthy, B.R. (1989) *Three Decades of Cardamom Research at R.R.S., Mudigere. Tech. Bull.*, University of Agricultural Sciences, Bangalore, India.

Kumar, K.B., Kumar, P.P., Balachandran, S.M. and Iyer, R.D. (1985) Development of clonal plantlets in cardamom (*Elettaria cardamomum*). *J. Plantation Crops*, 13, 31–34.

Kuttappa, K.M. (1969) Capsule shedding in cardamom. *Cardamom News*, 3(5), 2–3.

Kulandaivelu, G. and Ravindran, K.C. (1992) Physiological changes in cardamom genotypes exposed to warm climate. *J. Plantation Crops*, 20(suppl.), 294–296.

Kuruvilla, K.M. and Madhusoodanan, K.J. (1988) Effective pollination for better fruit set in cardamom. *Spice India*, 1(6), 19–21.

Kuruvilla, K.M., Sudharshan, M.R., Madhusoodanan, K.J., Priyadarshan, P.M., Radhakrishnan, V.V. and Naidu, R. (1992) Phenology of tiller and panicle in cardamom (*Elettaria cardamomum* Maton). *J. Plantation Crops*, 20(suppl.), 162–165.

Lukose, R., Saji, K.V., Venugopal, M.N. and Korikanthimath, V.S. (1993) Comparative field performance of micropropagated plants of cardamom (*Elettaria cardamomum*). *Indian J. Agric. Sci.*, 63, 417–418.

Mabberley, D.J. (1987) *The Plant Book*, Cambridge Uni. Press, Cambridge.

Madhusoodanan, K.J., Kuruvilla, K.M. and Priyadarshan, P.M. (1994) Cardamom improvement. In K.L. Chadha and P. Rethinam (eds) *Advances in Horticulture*, 9, *Plantation Crops*, Malhotra Publication House, New Delhi, India, pp. 121–129.

Madhusoodanan, K.J., Kuruvilla, K.M., Radhakrishnan, V.V. and Potty, S.N. (1998) Cardamom hybrids for higher yield and better quality capsule. *Spice India*, 11(3), 6–7.

Madhusoodanan, K.J. and Radhakrishnan, V.V. (1996) Cardamom breeding in Kerala. *Breeding of Crop Plants in Kerala*, University of Kerala, Trivandrum, India, pp. 73–81.

Madhusoodanan, K.J., Radakrishnan, V.V. and Kuruvilla, K.M. (1999) Genetic resources and diversity in cardamom. In B. Sasikumar, B. Krishnamoorthy, J. Rema, P.N. Ravindran and K.V. Peter (eds) *Biodiversity, Conservation and Utilization of Spices, Medicinal and Aromatic plants*. IISR, Calicut, pp. 68–72.

Madhusoodanan, K.J., Sudharshan, M.R. and Kuruvilla, K.M. (1988) Improved crop variety for higher yield and better capsule quality in cardamom. *Spice India*, 1(6), 10–13.

Madhusoodanan, K.J., Sudharshan, M.R., Priyadarshan, P.M. and Radhakrishnan, V.V. (1990) Small cardamom – Botany and Crop Improvement. In *Cardamom Production Technology*, ICRI, Spices Board, India, pp. 7–13.

Mayne, W.W. (1951) Report on cardamom cultivation in South India. ICAR. *Tech. Bull.*, p. 50.

Melgode, E. (1938) Cardamom in Ceylon – Part 1 *'Tropical Agriculture*, 91, 325–328.

Mercy, S.T., Babu, L.C. and George, M.K. (1977) Anatomical studies of the aerial stem, rhizome, leaf sheath and root of *Elettaria cardamomum* Maton. *Agric. Res. Jour. Kerala*, 15, 10–12.

Nadgauda, R., Mascarenhas, A.F. and Madhusoodanan, K.J. (1983) Clonal multiplication of cardamom (*E. cardamomum* Maton) by tissue culture. *J. Plantation Crops*, 11, 60–64.

Nair, K.C. and Vijayan, P.K. (1973) A study on the influences of plant hormones on the reproductive behaviour of cardamom. *Agric. Res. J. Kerala*, 11, 85.

Nair, M.K. (1993) Origin and introduction of plantation crops. *J. Plantation Crops*, 21, 1–7.

Nirmal Babu, K., Geetha, S.P., Manjula, C., Ravindran, P.N. and Peter, K.V. (1994) Medium term conservation of cardamom germplasm – An *in vitro* approach. *II Asia Pac. Conf. Agric. Biotech. (Abst.)*, Madras, India, p. 51.

Nirmal Babu, K., Geetha, S.P., Minoo, D., Rajalakshmi, K., Girija, P., Ravindran, P.N. and Peter, K.V. (1999a) *In vitro* approaches for conservation and exchange of spices germplasm. In

B. Sasikumar, B. Krishnamoorthy, J. Rema, P.N. Ravindran and K.V. Peter (eds) *Biodiversity, Conservation & Utilization of Spices, Medicinal and Aromatic Plants*, IISR, Calicut, India, pp. 45–53.

Nirmal Babu, K., Geetha, S.P., Minoo, D., Ravindran, P.N. and Peter, K.V. (1999b) *In vitro* conservation of cardamom (*Elettaria cardamomum* Maton) germplasm. *Plant Gen. Resources News Letter*, 119, 41–45.

Nirmal Babu, K., Ravindran, P.N. and Peter, K.V. (1997) *Protocols for micropropagation of Spices and Aromatic Crops*. Indian Institute of Spices Research, Calicut, India.

Owen, T.C. (1901) Notes on cardamom cultivation. Colombo.

Padmini, K., Venugopal, M.N. and Korikanthimath, V.S. (1999) Biodiversity and conservation of cardamom (*Elettaria cardamomum* Maton). In B. Sasikumar, B. Krishnamoorthy, J. Rema, P.N. Ravindran and K.V. Peter (eds) *Biodiversity, Conservation and Utilization of Spices. Medicinal and Aromatic Plants*, IISR, Calicut, pp. 73–78.

Padmini, K., Venugopal, M.N., Korikanthimath, V.S. and Anke Gowda, S.J. (2000) Studies on compound panicle type in cardamom (*Elettaria cardamomum* Maton). In N. Muralidharan and R. Rajkumar (eds) *Recent Advances in Plantation Crops Research*, Allied pub., New Delhi, pp. 97–99.

Pai, R.M. (1965) The floral anatomy of *Elettaria cardamomum* Maton, a re-investigation. *New Phytologist*, 64, 187–194.

Panchaksharappa, M.G. (1966) Embryological studies in some members of Zingiberaceae II. *Elettaria cardamomum, Hitchenia, Caulina* and *Zingiber macrostachyum*. *Phytomorphology*, 16, 412–417.

Parameswar, N.S. (1973) Floral biology of cardamom (*Elettaria cardamomum* Maton). *Mysore J. Agric Sci.*, 7, 205–213.

Parameswar, N.S. (1977) Intergeneric hybridisation between cardamom and related genera. *Curr. Sci.*, 6, 10.

Parameshwar, N.S. and Venugopal, R. (1974) Capsule setting studies in *Elettaria cardamomum* Maton. *Curr. Res.*, 3(5), 57–58.

Parry, J.W. (1969) *Spices*. Vol. II. Chemical Pub. Co., New York.

Patel, D.V., Kuruvilla, K.M., Madhusoodanan, K.J. and Potty, S.N. (1997) Regression analysis in small cardamom. *Proc. of the Symp. on Trop. Crop Res. and Dev. in India*, Trichur, India (in press).

Patel, D.V., Kuruvilla, K.M. and Madhusoodanan, K.J. (1998) Correlation studies in small cardamom (*Elettaria cardamomum* Maton). In N.M. Mathew and C.K. Jacob (eds) *Developments in Plantation Crops Research*, Allied Pub., New Delhi, pp. 16–19.

Pattanshetty, H.V. and Prasad, A.B.N. (1976) Blossom biology, pollination and fruit set in cardamom. In K.L. Chadha (ed.) *Proc. Intern. Sym. Subtropical and Tropical Horti.* (Vol. 1); Today and Tomorrow's publishers, New Delhi, pp. 262–268.

Parvathi, C., Shadakshari, Y.G., Belavadi, V.V. and Chandrappa, H.M. (1993) Foraging behaviour of honey bees on cardamom (*Elettaria cardamomum* Maton). In G.K. Veeresh, R. Umashanker and K.N. Ganesan (eds) *Pollination in Tropics*, IUSSI, Bangalore, pp. 99–103.

Philip, V.J. (1968) The origin of cotyledon and epicotyl in *Elettaria cardamomum* Maton. *Australian J. Botany*, 16, 427–431.

Pillai, P.K. and Santha Kumari, S. (1965) Studies on the effect of growth regulators on fruit setting in cardamom. *Agric. Res. J., Kerala*, 3, 5–15.

Pillai, S.K., Pillai, A. and Sachdeva, S. (1961) Root apical organization in monocotyledons – Zingiberaceae. *Proc. Indian Acad., Sci.*, (B), 53, 240–256.

Pradip Kumar, K., Mary Mathew, K., Rao, Y.S., Madhusoodanan, K.J. and Potty, S.N. (1997) Rapid propagation of cardamom through *in vitro* techniques. *Proc. 9th Kerala Sci. Cong.*, Trivandrum, India, p. 185.

Priyadarshan, P.M. and Zachariah, P.K. (1986) Studies on *in vitro* culture on cardamom (*Elettaria cardamomum* Maton – Zingiberaceae) progress and limitations. In *Proc. Inter. Congr. Plant Tissue and Cell Culture*, Minnesota, p. 107 (Ab).

Priyadarshan, P.M., Kuruvilla, K.M. and Madhusoodanan, K.J. (1988) Tissue culture technology: impacts and limitations. *Spice India*, 1(6), 31–35.

Priyadarshan, P.M., Kuruvilla, K.M., Madhusoodanan, K.J. and Naidu, R. (1992) Effect of various media protocols and genotypic specificity on micropropagation of cardamom (*Elettaria cardamomum* Maton). In N.S. Subba Rao, C. Rajagopalan and S.V. Ramakrishna (eds) *New Trends in Biotechnology*, Oxford IBH Publications, New Delhi, pp. 109–117.

Rajapakse, L.S. (1979) G.L.C. study of the essential oil of wild cardamom oil of Sri Lanka. *J. Sci. Food Agric.*, 30, 521–527.

Ramachandran, K. (1969) Chromosome numbers in Zingiberaceae. *Cytologia*, 34, 213–221.

Ranjithakumari, B.D., Kuriachan, P.M., Madhusoodhanan, K.J. and Naidu, R. (1993) Studies on light requirements of cardamom nursery. *J. Plantation Crops*, 21(suppl.), 360–362.

Rao, N.S.K., Narayana Swamy, S., Chacko, E.K. and Doreswamy, M.E. (1982) Regeneration of plantlets from callus of *Elettaria cardamomom* Maton. *Proc. Indian Acad. Sci.*, 91(B), 37–41.

Ravindran, P.N., Rema, J., Nirmal Babu, K. and Peter, K.V. (1997) Tissue culture and *in vitro* conservation of spices – an overview. In S. Edison, K.V. Ramana, B. Sasikumar, K. Nirmal Babu and S.J. Eapen (eds) *Biotechnology of Spices, Medicinal & Aromatic plants*, Indian Society for Spices, Calicut, pp. 1–12.

Reghunath, B.R. (1989) *In vitro studies on the propagation of cardamom (Elettaria cardamomum Maton)*. Ph.D Thesis, Kerala Agric. Univ. Trichur, India.

Reghunath, B.R. and Bajaj, Y.P.S. (1992) Micropropagation of cardamom. In Y.P.S. Bajaj (ed.) *Biotechnology in Agriculture and Forestry*, Vol. 19, *High-Tech and Mircopropagation* III, Springer-Verlag, Berlin, pp. 175–198.

Reghunath, B.R. and Gopalakrishnan, P.K. (1991) Successful exploitation of *in vitro* culture techniques for rapid clonal multiplication and crop improvement in cardamom. In *Proc. Kerala Sci. Cong.*, Kozhikode, pp. 70–71.

Ridley, H.N. (1912) *Spices*, McMillan & Co. Ltd., London.

Rijekbusch, P.A.H. and Allen, D.J. (1971) Cardamom in Tanzania. *Acta Hort.*, 21, pp. 144–150.

Rubido, J.F. (1967) Prospects of cardamom growing in Gautemala. *Revista Cafetera Gautemala*, 71, 10–13.

Sajina, A., Minoo, D., Geetha, S.P., Samsudheen, K., Rema, J., Nirmal Babu, K. and Ravindran, P.N. (1997) Production of synthetic seeds in a few spices crops. In S. Edison, K.V. Remana, B. Sasikumar, K. Nirmal Babu and S.J. Eapen (eds) *Biotechnology of Spices, Medicinal and Aromatic plants*, Indian Society of Spices, Calicut, pp. 65–69.

Sastri, B.N. (1952). *The Wealth of India – Raw Materials*, D–E, pp. 150–160.

Siddagangaih, Krishnakumar, V. and Naidu, R. (1993) Effect of chloremquat, daminozide, ethephon and maleic hydrazide on certain vegetative characters of cardamom (*Elettaria cardamomum* Maton.) Seedlings. *J. Spices and Aromatic Crops*, 2, 53–59.

Stone, G.N. and Willmer, P.G. (1989) Pollination of cardamom in Papua New Guinea. *J. Agric. Res.*, 28(4), 228.

Sudharshan, M.R. (1987) Colchicine – induced tetraploids in cardamom (*Elettaria cardamomum* Maton). *Curr. Sci.*, 56(1), 36–37.

Sudharshan, M.R. (1989) Induced polyploids in cardamom. *J. Plantation Crops*, 16, 365–369.

Sudharshan, M.R., Bhatt, S.S. and Narayanaswamy, M. (1997) Variability in the tissue cultured cardamom plants. In S. Edison, K.V. Ramana, B. Sasikumar, K. Nirmal Babu and S.J. Eapen (eds) *Biotechnology of Spices, Medicinal & Aromatic plants*, Indian Society for Spices, Calicut, pp. 98–101.

Sudharshan, M.R., Kuruvilla, K.M. and Madhusoodanan, K.J. (1988) Periodicity phenomenon in cardamom and its importance in plantation management. *Spice India*, 1(6), 14–20.

Sudharshan, M.R., Kuruvilla, K.M. and Madhusoodanan, K.J. (1991) A key to the identification of types in cardamom. *J. Plantation Crops*, 18, 52–55.

Thompson (1939) Cited from Pai (1965).

Tomlinson, P.B. (1969) *Anatomy of the Monocotyledons*. Vol. 3, Commelinales and Zingiberales. C.R. Metcalf (ed.) Oxford University Press, pp. 343–358.

Trease, G.E. and Evans, W.C. (1983) *Pharmacognosy* (12 edn). Baillere Tindall, London.

Venugopal, R. and Parameswar, N.S. (1974) Pollen studies in cardamom (*Elettaria cardamomum* Maton). *Mysore Jour. of Agric. Sci.*, 8, 203–205.

Vasanthakumar, K. and Mohanakumaran, N. (1998) Synthesis and translocation of photosynthates in cardamom. *J. Plantation Crops*, 17, 96–100.

Vasanthakumar, K., Mohanakumaran, N. and Wahid, P.A. (1989) Rate of photosynthesis and translocation of photosynthates in cardamom. *J. Plantation Crops*, 17(2), 96–100.

Vatsya, B., Duiesh, K., Kundapurkar, A.R. and Bhaskaran, S. (1987) Large scale plant formation of cardamom (*Elettaria cardamomum*) by shoot tip cultures. *Plant Phy. Biochem.*, 14, 14–19.

Wallis, T.E. (1967) *Text Book of Pharmacognosy* (4 edn), J & A Churchill, London.

Willis, J.C. (1967) *A Dictionary of the Flowering Plants and Ferns* – 7th edition (revised by H.K. Airey Shaw), Cambridge University Press.

Zachariah, T.J. and Lukose, R. (1992) Aroma profile of selected germplasm accessions in cardamom (*Elettaria cardamomum* Maton). *J. Plantation Crops*, 20(suppl.), 310–312.

Zachariah, T.J., Mulge, R. and Venugopal, M.N. (1998) Quality of cardamom from different accessions. In N.M. Mathew and C.K. Jacob (eds) *Developments in Plantation Crops Research*, Allied Pub., India, pp. 337–340.

3 Chemistry of cardamom

T. John Zachariah

1 INTRODUCTION

Cardamom is used as a flavouring material in three forms; whole, decorticated seeds and ground. The spice is distilled for essential oil and solvent extracted for oleoresin. In the international trade generally whole cardamom is the item of commerce, while trade in decorticated seed is small and that of ground spice negligible.

Essential oil of cardamom is the source for its aroma and flavour. Research so far carried out concerned mainly with the composition of the oil. As early as 1908 there were reports that cardamom oil contained terpinene, sabinene, limonene, 1,8-cineole, α-terpineol, α-terpinyl acetate, terpinen-4-yl formate and acetate and terpinen-4-ol (Guenther, 1975). The characteristic odour and flavour of cardamom is determined by the relative composition of the components of volatile oil.

Dried fruit of cardamom contains steam-volatile oil, fixed (fatty) oil, pigments, proteins, cellulose, pentosans, sugars, starch, silica, calcium oxalate and minerals. The major constituent of the seed is starch (up to 50 per cent) while in the fruit husk it is crude fibre (up to 31 per cent). The constituents of the spice differ among varieties, variations in environmental conditions of growth, harvesting, drying procedures and subsequent duration as well as conditions of storage. The main factor that determines the quality of cardamom is the content and composition of volatile oil, which governs the odour and flavour. The colour of the fruit does not generally affect the intrinsic organoleptic properties. However faded fruit colour generally indicate a product stored for a longer period and possibility of deterioration in the organoleptic properties through evaporation of the volatile oil (Purseglove *et al.*, 1981).

Cardamom oil is produced commercially by steam distillation of powdered fruits. The yield and organoleptic properties of the essential oil so obtained are dependent upon many factors. Fruits from recent harvest yield more oil than the one stored for a longer period. Many workers have shown that at least 4 h extraction is essential to get full recovery of oil.

Industrial production of cardamom oleoresin is relatively on a smaller scale. Solvent extraction yields about 10 per cent of oleoresin and the content is dependent upon the raw material and solvent used. Cardamom oleoresin contains about 52–58 per cent volatile oil (Purseglove *et al.*, 1981). Oleoresin is used for flavouring and is normally dispersed in salt, flour, rusk or dextrose, before use.

2   COMPOSITION

Main components of cardamom volatile oil are listed in Table 3.1 and the trace components are listed in Table 3.2. The volatile oil is extracted from the seeds and the husks hardly give 0.2 per cent oil. Even though the public perception about good quality cardamom is the greenish capsule, the appearance of the capsule has little to do with the recovery of volatile oil (Sarath Kumara *et al.*, 1985). Husk gives good protection and prevents the loss of oil from seeds, and loss of oil from dehusked seeds is found to be fast. Seeds start losing oil the moment husk is removed and this increases with the storage time.

Table 3.1 Main components of cardamom volatile oil

Components	Total oil (%)
α-pinene	1.5
β-pinene	0.2
Sabinene	2.8
Myrcene	1.6
α-phellandrene	0.2
Limonene	11.6
1,8-cineole	36.3
γ-terpinene	0.7
p-cymene	0.1
Terpinolene	0.5
Linalool	3.0
Linalyl acetate	2.5
Terpinen-4-ol	0.9
α-terpineol	2.6
α-terpinyl acetate	31.3
Citronellol	0.3
Nerol	0.5
Geraniol	0.5
Methyl eugenol	0.2
trans-nerolidol	2.7

Sources: Lawrence, 1978; Govindarajan, 1982.

Table 3.2 Trace components in cardamom volatile oil

Hydrocarbons	Acids	Carbonyls
α-thujene	Acetic	3-methyl butanal
Camphene	Propionic	2-methyl butanal
α-terpinene	Butyric	Pentanal
cis-ocimene	2-methyl butyric	Furfural
trans-ocimene	3-methyl butyric	8-acetoxy carvotanacetone
Toluene	Alcohols and Phenols	Cuminaldehyde
p-dimethylstyrene	3-methyl butanol	Carvone
Cyclosativene	*p*-methyl-3-en-l-ol	Pinole
α-copaene	Perillyl alcohol	Terpinene-4-yl-acetate
α-ylangene	Cuminyl alcohol	α-terpinyl propionate
γ-cadinene	*p*-cresol	Dihydro-α-terpinyl acetate
Δ-cadinene	Thymol	

Sources: Lawrence, 1978; Govindarajan, 1982.

Bleached cardamom tends to lose oil faster, as the husk becomes very brittle due to bleaching. Oil from freshly separated seeds or from whole capsules (seeds and husk) is almost identical (Govindarajan *et al.*, 1982).

Steam distillation is being adopted for extracting oil by most of the commercial units in India and elsewhere. Cryo-grinding using liquid nitrogen is ideal to prevent volatile oil loss during grinding. Supercritical extraction using liquid carbon dioxide is shown to extract more oil and the flavour is closer to natural cardamom.

In oil extraction, the early fractions are rich in low boiling terpenes and 1,8-cineole and the later fractions are rich in esters. Volatile oil content is highest 20–25 days before full maturity. Ratio of the two main components, 1,8-cineole and α-terpinyl acetate, determine the critical flavour of the oil. The volatile oil from var. *Malabar* represented by Coorg greens are "more camphory" in aroma, due to the relatively higher content of 1,8-cineole. This oil is reported to be ideal for soft drinks. It is known that the early fractions during distillation are dominant in low-boiling monoterpenes and 1,8-cineole. Techniques are available to remove these fractions by fractional distillation so that the remaining oil will have more α-terpinyl acetate, which contributes to the mildly herbaceous, sweet spicy flavour, that is predominant in the var. *Mysore* or the commercial grade commonly known as "Alleppey green" (Govindarajan *et al.*, 1982). Mathai (1985) evaluated 18 export grades (Agmark) of Indian cardamom for their chemical and physical qualities. The grades with heavier and bigger capsules Alleppey Green Extra Bold (AGEB) and Coorg Green Extra Bold (CGEB) were inferior in their flavour constituents to the medium grade (Alleppey Green Small, AGS). Chemical bleaching of the capsules reduced the amount of essential oil in the capsules. Vasanthakumar *et al.* (1989) reported that cardamom at the black seed stage or "karimkai" is ideal for consumption as well as for essential oil extraction. Gopalakrishnan *et al.* (1989) reported that thrips-affected cardamom capsules contained relatively higher, 1,8-cineole.

Nirmala Menon *et al.* (1999) extracted bound aroma compounds from fresh green cardamom and the free volatiles were isolated with ether:pentane (1:1) and the bound compounds with methanol. The major compounds in the aglycone fraction were identified as 3-methyl pentan-2-ol, linalool and *cis* and *trans* isomers of nerolidol and farnesol. Noleau and Toulemonde (1987) reported 122 compounds from cardamom oil cultivated in Costa Rica. They claim that among them 74 are reported for the first time (Table 3.3).

3   BIOSYNTHESIS OF FLAVOUR COMPOUNDS

3.1   Sites of synthesis

The accumulation or secretion of monoterpenes and sesquiterpenes is always associated with the presence of well-defined secretory structures such as oil cells, glandular trichomes, oil or resin ducts or glandular epidermis. A common feature of these secretory structures is an extracytoplasmic cavity in which the relatively toxic terpenoid cells and resins appear to be sequestered. This anatomic feature distinguishes the essential oil plants from others in which terpenes are produced as trace constituents that either volatilize inconspicuously or are rapidly metabolized. Several lines of evidence indicate that secretory structures are also primary sites of mono and sesquiterpene biosynthesis. (Francis and O'Connell, 1969).

Table 3.3 Compounds identified in *Elettaria cardamomum* from Costa Rica

Name of the compound	Occurrence (%)
Hydrocarbons	
4,8-dimethyl-1-1,3,7-nonatriene	tr
trans-4-*trans*-8,12-*trimethyl*-1,3,7,11-tridecatetraene	0.15
cis-β-ocimene	0.06
trans-β-ocimene	0.13
Myrcene	2.21
α-farnesene	tr
α-phellandrene	0.01
α-terpinene	0.17
γ-terpinene	0.55
Terpinolene	0.36
1,3,8-menthatriene	tr
1,4,8-menthatriene	tr
Limonene	0.20
β-elemene	0.03
Germacrene-D	0.03
Humulene	0.01
α-pinene	1.51
β-pinene	0.15
Camphene	0.02
α-fenchene	tr
α-thujene	tr
Sabinene	3.78
δ3-carene	tr
γ-cadinene	0.12
δ-cadinene	0.03
γ-murolene	0.07
α-selinene	0.08
β-selinene	0.23
Guaiene (tentative)	tr
Valencene	tr
β-caryophyllene	0.01
β-gurjunene	tr
Tricyclene	tr
p-cymene	tr
α,*p*-dimethylstyrene	tr
	9.91
Alcohols	
2-methylpropan-1-ol	tr
2-methylbutan-1-ol	0.01
2-methyl-3-buten-2-ol	tr
1-hexanol	tr
1-heptanol	tr
1-octanol	0.21
1-noanol	tr
1-decanol	tr
Dec-9-cen-1-ol	tr
Geraniol	2.66
Nerol	0.11
Linalol	5.91
Farnesol	0.01
trans-nerolidol	0.68
α-terpineol	2.97

δ-terpineol	0.07
Terpinen-4-ol	1.76
Isopiperitenol	tr
cis-carveol	0.05
trans-carveol	0.01
Perillyl alcohol	tr
4-thujanol	tr
1,8-methadien-4-ol	tr
p-mentha-1(7),8-dien-2-ol	tr
trans-*p*-mentha-2,8-dien-l-ol	0.01
Globulol	tr
	14.46
Aldehydes	tr
trans-2-butenal	0.01
2-methylbutanal	0.01
3-methylbutanal	0.01
Hexanal	tr
Octanal	0.02
trans-oct-2-enal	tr
Nonanal	tr
Decanal	tr
trans-dec-2-enal	tr
cis-dec-4-enal	tr
trans-dodec-5-enal	0.10
trans-2-*cis*-6-dodecadienal	tr
Citronellal	tr
Geranial	0.43
Neral	0.33
Farnesal isomer	0.02
Farnesal isomer	0.03
	0.96
Ketones	
6-methyl-5-hepten-2-one	tr
Geranyl acetone	tr
Farnesyl acetone	tr
Undecan-2-one	tr
Camphor	tr
Piperiterone	tr
α-ionone	tr
Acids	
Hexadecanoic acid	tr
Octadecanoic acid	tr
Eicosanoic acid	tr
Esters	
Octyl acetate	0.01
Decyl acetate	0.01
Decadienyl acetate (tentative)	tr
Dodecyl acetate	tr
Dode-5-cenyl acetate (tentative)	tr
Geranyl acetate	0.68
Neryl acetate	tr
Linalyl acetate	1.96
Hydroxy-methyl acetate isomer	tr

Table 3.3 (Continued)

Name of the compound	Occurrence (%)
Hydroxy-methyl acetate isomer	tr
Hydroxy-methyl acetate isomer	tr
α-terpinayl acetate	39.31
4-terpinyl acetate	0.01
Bornyl acetate	0.03
4-thujyl acetate	tr
6-hydroxyterpinayl acetate (tentative)	tr
6-oxoterpinyl acetate (tentative)	tr
α-terpinyl propionate	tr
Geranyl propionate	0.01
Neryl propionate	tr
Ethyl 2-hydroxyhexanoate	tr
Menthyl geraniate	0.03
Menthyl 9,12-octadienoate (tentative)	tr
Menthyl cinnamate	tr
	42.05
Phenols	
Thymol	tr
Carvacrol	tr
Oxides	
1,8-cineole	31.79
2,3-dehydro-1,8-cineole	0.02
trans-epoxyocimene	tr
1,2-limonene epoxide	tr
cis-linalol oxide	0.02
trans-linalol oxide	tr
Perillene	tr
	31.83
Total	99.21

Source: Noleau and Toulemonde (1987).

3.2   Biological function

The monoterpenes and sesquiterpenes traditionally have been regarded as functionless metabolic waste products. Yet certain studies have shown that these compounds can play varied and important roles in mediating the interactions of plants with their environments. The monoterpenes 1,8-cineole and camphor, have been shown to inhibit the germination and growth of competitors and thus act as allelopathic agents.

3.3   Early biosynthetic steps and acyclic precursors

All plants employ the general isoprenoid pathway in the synthesis of certain essential substances. The mono and sesquiterpenes are regarded as diverging at the C_{10} and C_{15} stages respectively in biosynthetic pathways. This, now well-known pathway, begins with the condensation of 3-acetyl-CoA in two steps to form hydroxymethyl-glutaryl-CoA which is reduced to mevalonic acid, the precursor of all isoprenoids. A series of phosphorylations

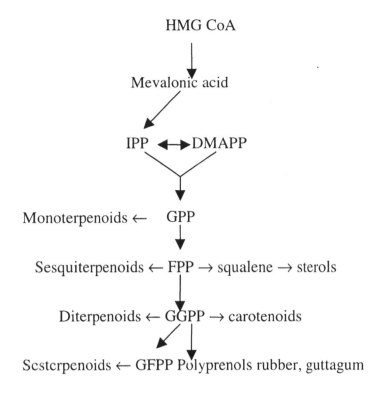

IPP   – Isopentenyl pyrophosphate HMG CoA
DMAPP   – Dimethyl allyl pyrophosphate
FPP   – Farnesyl pyrophosphate
GGPP   – Geranyl geranyl pyrophosphate
GFPP   – Geranyl farnesyl pyrophosphate
HMG-CoA – Hydroxymethyl-glutaryl-CoA

Figure 3.1 General pathway of isoprenoid biosynthesis.

and decarboxylation with the elimination of the C-3 oxygen function (as phosphate) yields isopentenyl pyrophosphate (IPP) (Mc Caskill and Croteau, 1995). This is isomerized to dimethylallyl pyrophosphate (DMAPP). This in turn leads to the synthesis of geranyl pyrophosphate (GPP) and farnesyl pyroposphate (FPP) (Fig. 3.1).

A number of monoterpene cyclases have been studied in detail especially the one responsible for the synthesis of α-terpinene, γ-terpinene and 1,8-cincole. Other cycliza-tions of interest are cyclization of geranyl pyrophosphate to limonene and cyclization of geranyl pyrophosphate to sabinene, the precursor of C3oxygenated thujene-type monoterpenes. The biosynthesis of thujene monoterpenes (such as 3-thujene) involves photooxidation of sabinene and also involves α-terpineol and terpinen-4-ol as inter-mediates (Croteau and Sood, 1985).

Although mechanisms of cyclization of geranyl phosphate and farnesyl pyrophos-phate to the corresponding monoterpenes and sesquiterpenes are not similar, the limited information available suggests that monoterpene and sesquiterpene cyclases are incapable of synthesizing larger and smaller analogs.

Pinene biosynthesis has been studied extensively. Three monoterpene synthases (cyclases) catalyze the conversion of GPP. Pinene cyclase I converts FPP into bicyclic (+)-α-pinene, (+)-β-pinene and to monocyclic and acyclic olefins (Bramley, 1997). The biosynthesis of monoterpenes, limonene and carvone, proceeds from geranyl diphosphate. Geranyl diphosphate is cyclized to (+)-limonene by monoterpene synthase. This intermediate is either stored in the essential oil ducts without further metabolism or is converted by limonene-6-hydroxylase to (+)-*trans* carveol. This is oxidized by a dehydrogenase to (+)-carvone (Bouwmeester *et al.*, 1998). Turner *et al.* (1999) demonstrated the localization of limonene synthase. Studies in peppermint (Gershenzon *et al.*, 2000) suggested that monoterpene biosynthesis is regulated by genes, enzymes and cell differentiation.

The biosynthesis of 1,8-cineole is suggested from linalyl pyrophosphate (Clark *et al.*, 2000). 1,8-cineole (or eucalyptol) is a biosynthetic dead end in many systems thus allowing accumulation of large quantities of this compound in many plants. Other than cardamom oil 1,8-cineole is also found in essential oils of artemisia, basil, betel leaves, black pepper, carrot leaf, cinnamon bark, leaf and of course in eucalyptus and in many other essential oil-yielding plants.

Most of the processes of the terpenoid biosynthesis are associated with cell organelles. Calcium and magnesium play important roles in sesquiterpnes biosynthesis (Preisig and Moreau, 1994). McCaskill and Croteau (1995) indicate that cytoplasmic mevalonic acid pathway is blocked at HMG-CoA reductase and that the IPP utilized for both mono-terpene and sesquiterpene biosynthesis and are synthesized exclusively in the plastids.

3.4   Industrial production

Cardamom oil is extracted industrially by steam distillation. The distillation unit consists of a material holding cage, condenser and receiver for steam distillation and adopt condi-tions for obtaining acceptable quality oil. Usually lower-grade capsules harvested after full maturity is used for distillation. Such capsules are first dehusked by shear in a disk mill with wide distances between disks and seeds separated by vibrating sieves. The dehusked seeds are further crushed to a coarse powder (Govindarajan *et al.*, 1982).

The essential oil-containing cells in cardamom seeds are located in a single layer below the epidermis and fine milling will result in loss of volatile oil. Cryo-grinding using liquid nitrogen will be ideal to prevent loss of volatile oil. Study on steam distillation revealed that nearly 100 per cent of the volatile oil was recovered in about 30, 50 and 60 min respectively. A study of the composition of the fractions indicated that in the first 15 min, most of the hydrocarbons and 1,8-cineole distilled over, while only 25–35 per cent of the important aroma contributing esters was recovered during this time. A further distillation for 2 h. was required to recover remaining esters. Hence a distillation time of 2–3 h. was essential for complete extraction of volatile oil. The Essential Oil Association (EOA) of America specification for typical cardamom oil is described in Table 3.4 (Anonymous, 1976).

The value of cardamom as a food and beverage additive depends very much on the aroma components which can be recovered as volatile oil. The volatile oil has a spicy odour similar to eucalyptus oil. The oil yield ranges from 3–8 per cent and it varies with varieties, maturity at harvest, commercial grade, freshness of the sample, green or bleached and distillation efficiency.

Table 3.4 Specification of cardamom volatile oil

Definition, source	Volatile oil distilled from the seeds of *E. cardamomum* (Linn.) Maton; family Zingiberaceae: cardamom grown in South India, Ceylon, Guatemala, Indonesia, Thailand, South China
Physical and chemical constants	Appearance: colourless to very pale-yellow liquid Odour and taste: aromatic, penetrating, some what camphoraceous odour of cardamom, persistently pungent, strongly aromatic taste Specific gravity: 0.917–0.947 at 25 °C Optical rotation: +22 °–+44 ° Refractive index: 1.463–1.466 at 20 °C
Descriptive characteristics	Solubility: 70% alcohol in five volumes, occasional opalescence Benzyl alcohol: in all proportions Diethyl phthalate: in all proportions Fixed oil: in all proportions Glycerin: insoluble Mineral oil: soluble with opalescence Propylene glycol: insoluble Stability: unstable in presence of strong alkali and strong acids Relatively stable to weak organic acids; affected by light
Containers and storage	Glass, aluminium or suitably lined containers, filled full; tightly closed and stored in cool place, protected from light

Source: EOA No. 289 (Anonymous, 1976).

3.5   History

3.5.1   *Composition*

The first detailed analysis of cardamom oil was reported by Nigam *et al.* (1965) and the constituents were identified with the help of gas chromatography (GC) and infrared (IR) spectroscopy, using authentic reference compounds and published data. Ikeda *et al.* (1962) reported 23.3 per cent of the oil as hydrocarbons with limonene as a major component. They have also reported the presence of methyl heptenone, linalool, linalyl acetate, β-terpineol, geraniol, nerol, neryl acetate and nerolidol. Richard *et al.* (1971) identified the compounds present in commercial samples and compared them with that of the wild Sri Lankan cardamom oil (Table 3.5). Govindarajan *et al.* (1982) has elaborated the range of concentration of major flavour constituents, their flavour description and their effect on flavour use (Table 3.6). TLC (thin-layer chromatography), column chromatography and subsequently GC were employed for the separation of oil constituents. Fractional distillation, IR spectroscopy, mass spectrum and nuclear magnetic resonance (NMR) were adopted to identify the specific compounds. The major constituents identified were α-pinene, α-thujene, β-pinene, myrcene, α-terpinene, γ-terpinene and *p*-cymene. These were identified in the monoterpene hydrocarbon fraction of cardamom oil. In 1966 and 1967 itself different commercial cardamom samples were compared for its chemical composition (Lawrence, 1978).

Sayed *et al.* (1979) evaluated the oil percentage in different varieties/types of cardamom. Mysore and Vazhukka contained the highest percentage of oil (8 per cent). Percentage by weight of cardamom seeds in the capsules ranged from 68 to 75. Percentage of cardamom

Table 3.5 Approximate per cent composition of cardamom oils

Component	Cardamom varieties				
	Oil (NF)	Var. Malabar (Ceylon)	Var. Malabar (Guatemala)	Var. Mysore	Sri Lankan (wild)
α-pinene	0.59	1.1	0.71	1.4	13.0
Camphene	0.01	0.02	0.03	0.04	0.13
Sabinene	1.2	2.5	3.4	3.1	4.9
β-pinene	0.19	0.20	0.34	0.26	4.9
Myrcene + Terpinene	0.37	1.8	1.5	1.1	2.5
α-phellandrene	<0.01	<0.01	<0.01	<0.01	0.42
D-limonene	0.28	0.02	0.12	0.14	2.1
1,8-cineole	51.3	31.0	23.4	44.0	3.3
γ-terpinene	0.03	0.12	0.34	0.10	22.2
Linalool	1.4	2.1	4.5	3.0	3.7
DL-camphor	0.03	<0.01	0.01	<0.01	0.35
α-terpineol citronellal	0.13	<0.01	0.04	0.06	0.13
4-terpinenol	0.83	0.14	0.28	0.87	15.3
α-terpineol	3.3	1.4	1.9	1.5	0.86
Citronellol	0.02	<0.01	0.04	<0.01	0.01
Nerol	0.07	0.02	0.04	0.06	0.78
Linalyl acetate	0.74	3.3	6.3	3.1	0.31
Geraniol	0.57	0.27	0.38	0.25	0.34
α-terpinyl acetate	34.6	52.5	50.7	37.0	0.14
Geranyl acetate Bisabolene	0.18	0.08	0.13	0.15	1.5
trans-nerolidol	0.05	0.09	0.83	0.07	0.44
cis-nerolidol	0.65	0.23	1.6	0.28	0.37

Source: Richard et al., 1971.

Table 3.6 Composition of volatile oils of cardamom from different sources*

Sample component	Coorg (var. Malabar)	Alleppey green (var. Mysore)	Imported	Ceylon (Oleoresin)
α-pinene	1.4	1.6	0.8	1.1
Camphene	–	–	0.1	–
Sabinene	3.2	4.9	2.5	1.9
Myrcene	0.2	1.9	2.6	0.7
1,4-cineole	–	–	0.2	–
D-limonene	2.4	2.5	2.7	1.7
1,8-cineole	41.0	34.2	30.4	22.2
p-cymene	0.5	0.5	0.6	–
Methyl heptenone	1.2	0.1	0.4	–
γ-terpenene	–	–	–	0.6
Linalool	0.4	6.4	4.1	2.3
Linalyl acetate	1.6	3.1	2.8	5.3
β-terpineol	0.8	1.7	0.2	–
α-terpineol	0.8	1.7	4.3	2.0
α-terpinyl acetate	30.0	34.5	39.7	53.2
Neryl acetate	1.1	1.2	0.6	–
Geranyl acetate		–	1.1	–
Geranoil	0.7	0.7	1.2	1.1
Nerol	1.4	0.6	0.3	–
Nerolidol	0.3	0.7	1.6	–

Source: Govindarajan et al., 1982.

Note
* Composition as area (%).

seeds is positively correlated with volatile oil (r = 0.4365) on dry-seed basis where as percentage of husk to volatile oil is negatively correlated (r = −0.4365).

Detailed studies on the volatile oil revealed large differences in the 1,8-cineole content, as high as 41 per cent in the oil of var. *Malabar* and as low as 26.5 per cent in the oil from var. Mysore. While the α-terpinyl acetate content was comparable, the linalool and linalyl acetate was markedly higher in var. *Mysore*. The combination of lower 1,8-cineole with its harsh camphory note and higher linalyl acetate with its sweet, fruity-floral odour result in the relatively pleasant mellow flavour in the var. *Mysore*, represented by the largest selling Indian cardamom grade, Alleppey green. Zachariah

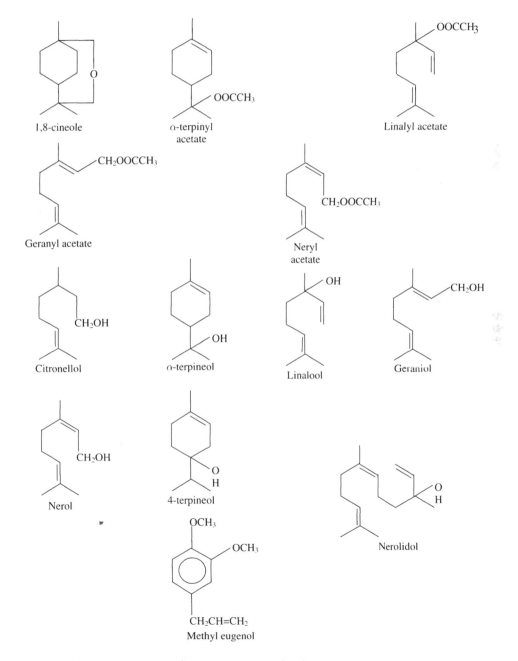

Figure 3.2 Structural formulae of major components of cardamom essential oil.

and Lukose (1992) and Zachariah *et al.* (1998) identified cardamom lines with relatively low cineole and high α-terpinyl acetate. An interesting observation is that lines APG 221 and 223 gave consistently higher oil yield (7.8 per cent) and high α-terpinyl acetate (55 per cent). The performance of APG 221 was consistent for about five seasons (Zachariah *et al.*, 1998).

Earlier gas chromatograms showed up to 31–33 peaks and up to 23 compounds were identified, while the improved procedure gave higher resolution with more than 150 peaks. All peaks have not been identified. All results confirm the earlier observations that 1,8-cineole and α-terpinyl acetate are the major components in cardamom oil. The chemical structures of major components are represented in Fig. 3.2.

Many workers used techniques like combination of fractional distillation, column and gas chromatography, mass spectrometry (MS), IR spectroscopy and NMR to identify the constituents of cardamom oil. Recently Nirmala Menon *et al.* (1999) studied the volatiles of freshly harvested cardamom by adsorption on AmberliteXAD-2, from which the free volatiles were removed by elution with pentane: ether and glycosidically-bound volatiles with methanol. Gaschromatographic–mass spectrometric analysis of the two fractions led to the identification of about 100 compounds. Among the free volatiles the important ones are 1,8-cineole, and α-terpinyl acetate. The less important ones are geraniol, α-terpineol, *p*-menth-8-en-2-ol, γ-terpinene, β-pinene, carvone oxide etc, while a large number of compounds were present in trace amounts. Among the aglycones the important ones are 3-methylpentan-2-ol, α-terpineol, isosafrole, β-nerolidol, *trans,trans*-farnesol, *trans,cis*-farnesol, *cis,trans*-farnesol, T-murrolol, cubenol, 10-epi-cubenol, *cis*-linalol oxide, tetrahydrolinalol etc (Table 3.7). Sixty-eight compounds were identified in the volatile fraction while sixty-one compounds were identified in the glycosidically bound fraction.

Table 3.7 Occurrence of major aglycones and free volatiles

Major compounds	Aglycones (%)	Free volatile (%)
3-methylpentan-2-ol	16.3	–
α-terpineol	6.2	3.4
β-nerolidol	6.7	0.2
trans,trans-farnesol	4.5	tr
trans,cis-farnesol	4.0	tr
cis,trans-farnesol	4.0	tr
Cubenol	2.6	tr
10-epi-cubenol	2.6	–
T-murrolol	2.4	–
Isosafrole	3.8	–
cis-linalool oxide	2.9	–
Tetrahydrolinalol	1.9	–
Cedrol	1.4	–
Geraniol	1.1	3.7
Linalol	1.1	
Oct-1-en-3-ol	1.0	
1,8-cincole	0.2	28.4
δ-terpinyl acetate	–	21.3
p-menth-8-en-2-ol	0.3	2.5
β-pinene	–	1.9
γ-terpinene	–	1.7
Nonan-5-one	–	1.6
Carvone oxide	–	1.8

Source: Nirmala Menon *et al.* (1999).

Table 3.8 Flavour characteristics of volatile components in cardamom

Components	Flavour description	Effect in flavour use	Range of concentration in cardamom oil (%)
Esters			
α-terpinyl acetate	Mildly herbaceous sweet spicy, piney variation in odour warm, mild spicy taste	To stretch cardamom herbal spice, imitation citrus and cherry peach flavours	34–52
Linalyl acetate	Sweet, floral, fruity odour and taste. Poor tenacity but stronger than terpinyl acetate	Fresh, sweet modifier in perfume and berry flavours	0.7–6.3
Geranyl acetate	Sweet, floral fruity with green; note stronger than geraniol	Sweetener, modifier in citrus fragrances and fruit, citrus and spice flavours	0.17–0.23
Neryl acetate	Very sweet; fruity floral	Effective in berry flavours, higher levels in perfumery	0.17–0.23
Ethers			
1,8-cineole	Fresh, camphoraeous cool and odour and taste very diffusive, poor tenacity	Refreshing effect extensively used in perfume and flavours	23–51
Alcohols			
Linalool	Floral, woody with citrusy note: creamy floral taste at low levels	Peculiar pleasant taste effect at low levels	1.4–4.5
α-terpineol	Delicately floral, sweet, lilac-like	Citrus and spice compositions	1.4–3.3
Citronellol	Fresh rosy odour and floral rosy bitter taste	Extensive use in perfumes and fruit flavours	<0.01–0.04
Geraniol	Floral, rosy with warm dry tones	Berry and sweet spicy flavours	0.29–0.64
Nerol	Sweet rosy, fruity	Berry and sweet spicy flavours	0.29–0.64
4-terpineol	Warm peppery wood with earthy, musty notes pleasantly green	Mainly used in citrus and spice compositons; warm, herbaceous effects	0.14–0.87
Nerolidol	Woody, floral slightly green	Excellent tenacity	0.28–1.6
Borneol	Dry camphoraceous woody, peppery	Background herbaceous camphory	–
Carbonyls			
Camphor	Warm minty odour, bitter warm and then cool mouth feel	Modifier diffusive	0.02–0.04
Citral	Powerful lemon fruity odour	Widely used at high dilutions in citrus and spice compositions	0.19–0.26
Citronellal	Powerful fresh green, citrusy, slightly woody	Citrus and spice composition	–

3.6   Evaluation of flavour quality

The flavour quality of food results from the interaction of its chemical constituents with the taste perception of the person using it. The most significant component of cardamom, as spice, is the volatile oil with its characteristic aroma described as sweet, aromatic spicy, camphory etc. Cardamom oil is richer in oxygenated compounds, all of which are potential aroma compounds (Table 3.8). Capillary column and gas chromatography has shown over 150 compounds in cardamom oil. While many of the identified compounds – alcohols, esters and aldehydes – are commonly found in many spice oils, the dominance of the ether 1,8-cineole and the esters α-terpinyl and linalyl acetates in the composition, make the cardamom volatile oil a unique one (Raghavan et al., 1991). The bitterness compound present in cardamom is α-terpinyl, present to the extent of about 0.8-2.7 per cent in the oil. Govindarajan et al. (1982) had described the ratio of esters to 1,8-cineole in different commercial types of cardamom oil (Table 3.9). In occasional samples, defective notes described as slightly "oxidized terpinic" were noted at high dilution levels, but were overshadowed by total cardamom aroma at higher level of concentrations. Samples markedly camphory (lacking sweet aromatic components) or high in defectives, oxidized terpinic, resinous, oily, earthy or bitter in flavour are rated poor and unacceptable. He suggested that quality grading of cardamom is possible by observing three major attributes of balance of profile, intensity/tenacity and absence of defects. The desirable and defective notes of cardamom oil are described in Table 3.10. General profile of Alleppey green cardamom is described in Table 3.11.

Sensory evaluation by trained panelists, for an overall quality ranking and for flavour profile, and correlating such data through different statistical tests will help to relate the data from instrumental analysis with the actual flavour profile. Sensory ratings and aroma profile by a trained panel for three true cardamom extracts and distilled oils were as follows: mature bleached cardamom extract was described as rich, fully representative of cardamom seed and rated better than distilled oil, which was described as of same intensity but less flowery, sweet and relatively harsh; the distilled oil was better than extracts from immature cardamom, which was described as weak, and more pungent.

Table 3.9 Ratios of esters and alcohols to 1,8-cineole in cardamom volatile oils

Source	Ratio of		
	α-terpinyl acetate	α-terpinyl + linalyl acetates	Esters + linalool + α-terpineol
Alleppey green (cv. Mysore)	1.30	1.59	1.77
Alleppey green (cv. Mysore)	1.10	1.19	1.43
Kerala, Ceylon (cv. Mysore)	0.83	0.91	1.03
Ceylon, commercial	1.21–1.77	–	–
Ceylon, from extract	1.67–2.40	–	–
Ceylon, green expressed (cv. Mysore)	1.69	1.80	1.91
Ceylon, green from extract	2.40	2.64	2.83
Ceylon, green expressed (cv. Mysore)	2.17	2.40	2.68
Coorg green (cv. Malabar)	0.73	0.77	0.80

Source: Govindarajan et al. (1982).

Table 3.10 Flavour profile of cardamom oil and extracts

Desirable notes	Defective notes
Fresh cooling	Unbalanced
CAMPHORACEOUS	Sharp/harsh
Green	Heavy
SWEET SPICY	Earthy
FLORAL	Oily (Vegetable) oxidized
WOODY/BALSAMIC	Resinous
Herbal	Oxidized terpinic
Citrus	
Minty	
Husky	
Astringent, weakly	Bitter

Source: Govindarajan *et al.* (1982).

Note:
The description in capitals are the dominant characteristics perceived, defectives arranged in order of increasing impact.

Table 3.11 Profile of oil of cardamom

Origin	Commercially distilled from Alleppey green cardamoms
Odour	Initial Impact
	Penetrating, slightly irritating
	Cineolic, cooling
	Camphoraceous, disinfectant like warm, spicy
	Sweet, very aromatic pleasing
	Fruity lemony, citrus-like
	Persistence
	The oil rapidly airs-off on a smelling – strip loosing its
	freshness and becoming herby, woody with a marked
	musty back-note
	Dry out
	No residual odour after 24 h

Source: Heath (1978).

Analysis of a Japanese cardamom oil sample indicated some new compounds like 1,4-cineole, *cis-p*-menth-2-en-1-ol and *trans-p*-menth-2-en-1-ol, all of them in low amounts of 0.1–0.2 per cent. Cardamom oil from Sri Lanka gave a high range of values for α-pinene plus sabinene, 4.5–8.7 per cent and linalool 3.6–6 per cent and a wider range for the principal components 1,8-cineole 27–36.1 per cent and α-terpinyl acetate 38.5–47.9 per cent (Govindarajan *et al.*, 1982).

Some compounds such as α-thujene, sabinene, *p*-cymene, 2-undecanone, 2-tridecanone, heptacosane or *cis* and *trans-p*-menth-2en-1-ols were rarely detected in cardamom samples. Components like camphor, borneol and citrals might modify the overall flavour quality of cardamom, mainly determined by a combination of terpinyl and linalyl acetate and cineole. Locations also do play a role in altering the concentration of linalool, limonene, α-terpineol etc.

Flavour quality depends on interaction of chemical constituents of food with human taste buds, and the perception of taste by individuals depends on different attributes. A meaningful judgement of quality has to establish a casual relationship of physical and chemical characteristics of food and their sensory perception and judgement by human assessors. According to many workers the ratio of 1,8-cineole to α-terpinyl acetate is a fairly good index of the purity and authenticity of cardamom volatile oil (Purseglove *et al.*, 1981). The ratio is around 0.7–1.4. Cardamom Research Centre, Appangala (Karnataka) under the Indian Institute of Spices Research (IISR), Calicut, Kerala, India, could collect many accessions from cardamom growing areas with the flavour ratio more than one. 1,8-cineole and α-terpinyl acetate together with terpene alcohols (linalool, terpinen-4-ol and α-terpineol) are important for the evaluation of aroma quality of cardamom. The oils from cv. *Malabar* exhibit the lowest flavour ratio while that from cv. *Mysore* has high flavour ratio.

Cardamom samples from Sri Lanka and Guatemala have higher ratios indicating their superiority in flavour, similar to that of the var. *Mysore*. The occurrence of components such as borneol and citral often modifies the flavour quality.

Pillai *et al.* (1984) made a comparative study of the 1,8-cineole and α-terpinyl acetate contents of cardamom oils derived from diverse sources (Table 3.12). His study indicated that Guatemalan cardamom oil is marginally superior to Indian cardamom oil due to the higher content of α-terpinyl acetate. The high concentration of 1,8-cineole makes the oil from Papua New Guinea poor. The above workers found fair degree of concordance in the IR spectra of oils irrespective of their origin. The IR spectra provide a fingerprint of the oil as it projects the functional groups and partial structures that are present. The spectrum also helps in tracking the ageing process of the oil (Fig. 3.3a–c).

Table 3.12 Concentration of 1,8-cineole and α-terpinyl acetate in cardamom oils from different origins

No.	Oil kind	Percentage	
		1.8-cineole	*α-terpinyl acetate*
1	Guatemala I	36.40	31.80
2	Guatemala II	38.00	38.40
3	Guatemalayan Malabar type	23.40	50.70
4	Guatemalayan I	39.08	40.26
5	Guatemalayan II	35.36	41.03
6	Synthite (Commercial grade)	46.91	36.79
7	Mysore-type (Ceylon)	44.00	37.00
8	Malabar-type (Ceylon)	31.00	52.50
9	Mysore I	49.50	30.60
10	Mysore II	41.70	45.90
11	Mysore	41.00	30.00
12	Malabar I	28.00	45.50
13	Malabar II	43.50	45.10
14	Ceylon type	36.00	30.00
15	Alleppey I	38.80	33.30
16	Alleppey green	26.50	34.50
17	Coorg green	41.00	30.00
18	Mangalore I	56.10	23.20
19	Mangalore II	51.20	35.60
20	Papua New Guinea	63.03	29.09
21	Cardamom oil (India origin)	36.30	31.30

Source: Pillai *et al.* (1984).

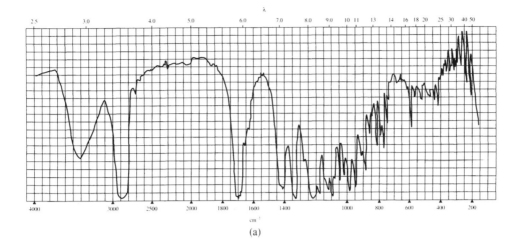

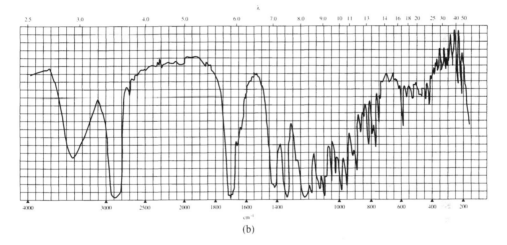

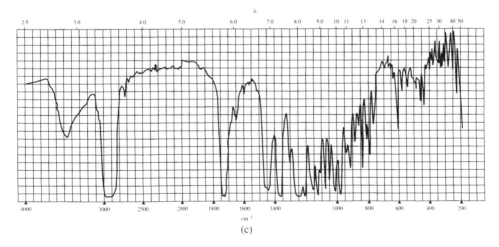

Figure 3.3 IR Spectrum of three samples of cardamom essential oil: (a) Guatemala; (b) Papua New Guinea; (c) Indian (Source: Pillai *et al.*, 1984).

The extraction methods like cryogenic grinding (Gopalakrishnan *et al.*, 1991) and supercritical extraction also influence the flavour profile. Such extraction techniques can extract the trace compounds that are otherwise lost in other methods of extraction.

3.7   Cardamom oleoresins and extract

The total solvent extract or oleoresin is known to reflect the flavour quality more closely than the distilled volatile oil. In the case of cardamom, oil more or less truly represents both flavour and taste. The stability of oleoresin depends on the changes that occur to the fat and terpenic components that are usually very susceptible to oxidative changes.

The existing studies led to the conclusion that there exists a clear-cut difference in the flavour profile of cardamom varieties, which in turn is influenced by agroclimatic conditions, post harvest processing and cultural practices.

Table 3.13 Chemical quality profile of some cardamom accessions from the germplasm assemblage at IISR

Acc. No.	Essential oil (%)	1,8-cineole (% of oil)	α-terpinyl acetate (% of oil)
APG 7	5.5	51	29
" 12	8.6	37	43
" 20	9.4	48	34
" 23	9.4	39	39
" 25	7.5	38	43
" 27	6.9	34	33
" 32	5.6	38	44
" 44	7.1	34	38
" 48	6.3	41	33
" 54	6.8	44	34
" 65	7.5	32	45
" 69	8.3	49	28
" 71	7.3	36	31
" 87	5.7	42	32
" 98	6.3	33	30
" 106	10.0	43	39
" 112	6.6	45	34
" 117	6.3	38	43
" 134	6.0	40	38
" 135	9.9	45	40
" 158	6.6	40	38
" 175	9.8	43	40
" 178	5.6	22	39
" 180	6.8	31	32
" 183	7.8	24	39
" 187	8.0	22	48
" 193	8.3	28	47
CCS-1	8.6	42	36
APG 215	6.0	23	55
" 217	6.0	25	51
" 218	7.8	24	52
" 221	7.8	37	40

3.8 Variability in composition

Analysis of germplasm collections conserved at the IISR Regional Station, Appangala, Coorg Dist. indicated variability in oil content and in the concentration of the two important components of the oil, 8-cineole and α-terpinyl acetate (Table 3.13). Selective breeding of the high-quality accessions having low 1,8-cineole and high α-terpinyl acetate, such as APG-221, will go a long way in enhancing the total flavour quality of the Indian cardamom.

3.9 Pharmaceutical properties of cardamom oil

Cardamom oil is shown to have antibacterial and antifungal action. Badei *et al.* (1991a,b) studied the chemical composition, physicochemical properties and anti-microbial activity of dried fruits of cardamom to assess the potential usefulness of cardamom oil as a food preservative. The antimicrobial effect of the oil was tested against 9 bacterial strains, 1 fungus and 1 yeast, the oil was 28.9 per cent as effective as phenol. The minimal inhibitory concentration of the oil was 0.7 mg/ml and it was concluded that cardamom oil could be used at a minimal inhibitory concentration range of 0.5–0.9 mg/ml without any adverse effect on any flavour.

Cardamom oil is effective as an antioxidant for cottonseed oil as assessed by stability, peroxide number, refractive index, specific gravity and rancid odour. The effect is increased by increasing the content of cardamom oil in cottonseed oil from 100 to 5000 ppm. Organoleptic evaluation showed that addition of up to 1000 ppm cardamom oil did not affect the characteristic odour of cottonseed oil.

3.10 Fixed oil of cardamom seeds

In addition to the volatile oil, cardamom seeds also contain fixed oil (fatty oil). The composition of the fatty oil has been investigated and found to contain mainly oleic and palmitic acids (Table 3.14). Gopalakrishnan *et al.* (1990) based on HNMR and MS studies reported that non-saponifiable lipid fraction of cardamom consisted mainly of

Table 3.14 The composition of the fixed (fatty) oil of cardamom seed

Fatty acid	Total fixed oil (%)
Oleic	42.5–44.2
Palmitic	28.4–38.0
Linoleic	2.2–15.3
Linolenic	5.8
Caproic	5.3
Stearic	3.2
Hexadecanoic	1.9
Caprylic	5.3
Capric	<0.1–3.8
Myristic	1.3–1.4
Arachidic	0.2–2.1
Hexadecanoic	1.9
Pentadecanoic	0.4
Lauric	0.2

Source: Verghese (1996).

waxes and sterols. Waxes identified were n-alkanes (C_{21}, C_{23}, C_{25}, C_{27}, C_{29}, C_{31} and C_{33}) and n-alkenes (C_{21}, C_{23}, C_{25}, C_{27}, C_{29}, C_{31} and C_{33}). In the sterol fraction, β-sitosterol and γ-sitosterol are reported. Phytol and traces of eugenyl acetate were also identified in cardamom.

4   CONCLUSION

The cardamom oil is a wonderful gift of nature, constructed delicately by kaleidoscopic permutations and combinations of terpenes, terpene alcohols, esters and other compounds, which even defy precise analytical techniques. Preparing a "synthetic cardamom oil" from its components having identical sensory qualities is well beyond the human capabilities even now. The sensory analysis, which is regarded by food scientists as the touchstone of quality, is sensitive to concentrations ranging from 10^{-8} to 10^4 ppm. The superiority of the Alleppey green is its superior sensory qualities, a better total perception of the flavour, which need not necessarily be dependent on the relative concentration of any one component. However, the natural quality is often lost during the extraction process, during storage and post-harvest handling. Cryo-grinding and super critical fluid extraction are alternatives for improving flavour quality.

Indexing of genetic resources for flavour quality and incorporation into breeding programmes the superior quality genotypes can go a long way in improving the overall flavour quality of cardamom. Chemical fingerprinting of the cardamom genotypes available in the germplasm conservatories using IR/GC-MS/or NMR spectral characters as well as by sensory evaluation is needed to pickup the really wonderful genotypes in terms of flavour quality.

REFERENCES

Anonymous (1976) Specification, Oil of Cardamom, EOA No. 289 (1976) Essential oil Association of United States of America, New York.

Badei, A.Z.M., El-Akel, A.T.M. and Morsi, H.H.H. (1991) Evaluation of chemical, physical and antimicrobial properties of cardamom essential oil. *Bull. Faculty of Agri., University of Cairo*, 42(1), 183–197.

Badei, A.Z.M., Morsi, H.H.H. and El-Akel, A.T.M. (1991) Chemical composition and antioxidant properties of cardamom essential oil. *Bull. Faculty of Agri., University of Cairo*, 42(1), 199–215.

Bouwmeester, H.J., Gershenzon, J., Konings, M.C.J.M. and Croteau, R. (1998) Biosynthesis of the monoterpenes limonene and carvone in the fruit of caraway. I. Demonstration of enzyme activities and their changes with development. *Plant Physiology*, 117, 901–912.

Bramley, P.M. (1997) Isoprenoid metabolism. In P.M. Dey and J.B. Harborne (eds) *Plant Biochemistry*, Academic Press, California, pp. 417–434.

Clark, G., Stuart, C. and Easton, M.D. (2000) Eucalyptol. *Perfumer & Flavourist*, 25, 6–16.

Croteau, R. and Sood, V.K. (1985) Metabolism of monoterpnes: Evidence for the function of monoterpene catabolism in pippermint (*Mentha piperita*) rhizomes. *Plant Physiol.*, 77, 801–806.

Francis, M.J.O. and O'Connell, M. (1969) The incorporation of mevalonic acid into rose petal monoterpenes. *Phytochemistry*, 8, 1705–1708.

Gershenzon, J., McConkey, M.E. and Croteau, R.B. (2000) Regulation of monoterpene accumulation in leaves of peppermint. *Plant Physiology*, 122, 205–213.

Gopalakrishnan, M., Narayanan, C.S., Kumaresan, D. and Bhaskaran, P. (1989) Physicochemical changes in cardamom infested with thrips. *Spice India*, 2(12), 18–19.

Gopalakrishnan, M., Narayanan, C.S. and Grenz, M. (1990) Non saponifiable lipid constituents of cardamom. *J. Agri. and Food Chemistry*, 38(12), 2133–2136.

Gopalakrishnan, M., Luxmi Varma, R., Padmakumari, K.P., Beena Symon, Howa Umma and Narayanan, C.S. (1991) Studies on cryogenic grinding of cardamom. *Indian Perfumer*, 35(1), 1–7.

Govindarajan, V.S., Shanti Narasimhan, Raghuveer, K.G. and Lewis, Y.S. (1982) Cardamom – Production, technology chemistry and quality. CRC *Critical Rev. Food Sci. Nutrition*, Vol. 16, p. 326, Florida.

Guenther, E. (1975) The Cardamom Oils. In *The Essential Oils*, Vol. V, Robort E. Krieger, Publishing Company, New York, pp. 85–106.

Heath, H.B. (1978) *Flavour Technology: Profiles. Products. Applications*, AVI Publishing Co., Inc., Connecticut, p. 113.

Ikeda, R.M., Stanley, W.L., Vannier, S.H. and Spitler, E.M. (1962) Monoterpene hydrocarbon composition of some essential oils. J. *Food Sci.*, 27, 455.

Lawrence, B.M. (1978) Major tropical spices – Cardamom (*Elettaria cardamomum*). In *Essential Oils*, pp. 105–155.

Mathai, C.K. (1985) Quality evaluation of the 'Agmark' grades of cardamom *Elettaria cardamomum. J. Sci. Food Agri.*, 36(6), 450–452.

McCaskill, D. and Croteau, R. (1995) Monoterpene and sesquiterpene biosynthesis in glandular trichomes of peppermint (*Mentha x piperita*) rely exclusively on plastid-derived isopentenyl diphosphate. *Planta*, 197, 49–56.

Nigam, M.C., Nigam, I.C., Handa, K.L. and Levi, L. (1965) Essential oils and their constituents XXVIII. Examination of oil of cardamom by gas chromatography. *J. Pharm. Sci.*, 54(5), 799.

Nirmala Menon, A., Chacko, S. and Narayanan, C.S. (1999) Free and glycosidically bound volatiles of cardamom (*Elettaria cardamomum* Maton var. *miniscula* Burkill). *Flavour Fragr. J.*, 14, 65–68.

Noleau, I. and Toulemonde, B. (1987) Volatile constituents of cardamom (*Elettaria cardamomum* Maton) cultivated in Costa Rica. *Flavour Fragr. J.*, 2, 123–127.

Pillai, O.G.N., Mathulla, T., George, K.M., Balakrishnan, K.V. and Verghese, J. (1984) Studies in cardamom 2. An appraisal of the excellence of Indian cardamom (*Elettaria cardamomum* Maton). *Indian Spices*, 21(2), 17–25.

Preisig, C.L. and Moreau, R.A. (1994) Effects of potential signal transduction antagonists on phytoalexin accumulation in tobacco. *Phytochemistry*, 36, 857–863.

Purseglove, J.W., Brown, E.G., Green, C.L. and Robbins, S.R.J. (1981) *Spices*, Vol. I, Longman, New York, pp. 174–228.

Raghavan, B., Abraham, K.O., Sankaracharya, N.B. and Shankaranarayana, M.L. (1991) Cardamom – studies on quality of volatile oil and product development. *Indian Spices*, 28(3), 20–24.

Richard, A.B., Wijesekera, R.O.B. and Chichester, C.O. (1971) Terpenoids of cardamom oil and their comparative distribution among varieties. *Phytochemistry*, 10, 177–184.

Sarath Kumara, S.J., Packiyajothy, E.V. and Jansz, E.R. (1985) Some studies on the effect of maturity and storage on the chlorophyll content and essential oils of the cardamom fruit (*Elettaria cardamomum*). *J. Sci. Food Agri.*, 36(6), 491–498.

Sayed, A.A.M., Korikanthimath, V.S. and Mathew, A.G. (1979) Evaluation of oil percentage in different varieties / types of cardamom. *Cardamom*, II(1), 33–34.

Turner, G., Gershenzon, J., Nielson, E.E., Froehlich, J.E. and Croteau, R. (1999) Limonene synthase, the enzyme responsible for monoterpene biosynthesis in peppermint is localized in leucoplasts of oil gland secretory cells. *Plant Physiology*, 120, 879–886.

Verghese, J. (1996) The world of spices and herbs. *Spice India*, 12–20.

Vasanthakumar, K., Mohanakumaran, N. and Narayanan, C.S. (1989) Quality evaluation of three selected cardamom genotypes at different seed maturity stages. *Spice India*, 2, 25.

Zachariah, T.J. and Lukose, R. (1992) Aroma profile of selected germplasm accessions in cardamom (*Elettaria cardamomum* Maton). *J. Plantation Crops*, 20(suppl.), 310–312.

Zachariah, T.J., Mulge, R. and Venugopal, M.N. (1998) Quality of cardamom from different accessions. In N.M. Mathew and C. Kuruvilla Jacob (eds) *Developments in Plantation Crops Research*, Allied Publishers Ltd., Mumbai, pp. 337–340.

4 Agronomy and management of cardamom

V.S. Korikanthimath

1 DISTRIBUTION

Cultivation of cardamom is mostly concentrated in the evergreen forests of Western Ghats in South India. Besides India, cardamom is grown as a commercial crop in Guatemala and on a small scale in Tanzania, Sri Lanka, El Salvador, Vietnam, Laos, Cambodia and Papua New Guinea (PNG). Earlier, India accounted for 70 per cent of the world production and now it is 41 per cent only, while Guatemala contributes around 48 per cent of the present world production. The total area under cardamom in India was around 1,05,000 ha till the 80s but has subsequently reduced to the present level of around 75,000 ha. It is mainly cultivated in the three southern states of India – Kerala, Karnataka and Tamil Nadu; contributing approximately 60, 31 and 9 per cent of total area respectively.

Cardamom is cultivated mostly under natural forest canopy, except in certain areas in Karnataka (North Kanara, Chickmagalur, Hassan) and Wynad district in Kerala where it is often grown as a subsidiary crop in arecanut or coffee gardens. The important areas of cultivation in India are Uttara Kannada, Shimoga, Hassan, Chickmagalur, and Kodagu (Coorg) districts in Karnataka state; northern and southern foot hills of Nilgiris, hill regions of Madurai, Salem and Tirunelveli, Annamalai and Coimbatore districts of Tamil Nadu state; Wynad and Idukki districts as well as in the Nelliampathy hills of Palghat district of Kerala.

2 CLIMATE

2.1 Altitude

The optimum altitudinal range for growing cardamom is from 600 to 1500 m above MSL (Anonymous, 1976, 1982). In South India, all cardamom plantations lie between 700 m and 1300 m above msl, rarely even up to 1500 m, but there the growth and performance are poor. Cardamom cultivation in India is restricted to the Western Ghat regions, which constitute an extensive chain of hills that lie parallel to the West Coast of peninsular India. The var. *Malabar*, the traditional cardamom of Karnataka, possess the capacity to be productive at lower elevations of 500–700 m above sea level. The other cardamom varieties, *Mysore* and *Vazhukka* are not productive below 700 m elevation (Abraham and Tulasidas, 1958). At lower elevation, vegetative growth is satisfactory, fruit production is poor. In Guatemala, cardamom is grown at varying altitudes, ranging from 90 to 1200 m above MSL. Most of the plantations in the south are at high altitudes, while in north the crop is grown both at low as well as at high altitudes (George, 1990). Cardamom is highly sensitive to elevation and the wrong choice of cultivar, or wrong location, can severely affect the growth and productivity. Cardamom is also highly prone to wind and drought damages and therefore, areas liable to be affected by such conditions are unsuitable (Mohanchandran, 1984).

2.2 Temperature

In Guatemala, conditions are ideal for growth and productivity of cardamom. Average annual temperature varies from 17–25 °C in the southern parts of cardamom growing areas, and from 18–23.5 °C in the north (George, 1990). In India, optimum growth and development is observed in the warm and humid conditions at a temperature range of 10–35 °C (Anonymous, 1976). The upper temperature limit will normally be around 31–35 °C. In the eastern side of the Western Ghats a combination of desiccating winds passing from the hinter lands of east and low humidity leads to desiccation and drying of plants. In such areas protective irrigation would be essential for retention of humid conditions for adequate growth, panicle initiation and setting of capsules (Korikanthimath, 1991a). It is noticed that rate of spread of *katte* disease will be more during summer than in rainy season. Cold conditions would result in almost poor or no setting of capsules. Thus extremes of temperature and wide diurnal variations are not conducive for healthy growth of cardamom plants.

2.3 Rainfall

Cardamom is grown in South India under rainfall conditions ranging from 1500 to 5750 mm. Climate of the area is determined by the rainfall and the year can be divided into winter, summer and monsoon seasons. November–February is characterized by relatively dry weather and cool temperature. The hot weather prevails from March to June and is marked by moderately high temperature and occasional showers. The south-west monsoon sets in during June and continues until early September. In the more westerly portions of the hills, the rains at this period are very heavy and continuous, but they decrease considerably in the eastern slopes, which experience strong west winds, much cloud and frequent light showers. After a short gap, the northeast rains commence and occasional rains continue up to December. This is a dry period in the more northerly and westerly areas, but is marked by heavy rains and overcast skies in the south and east (Mayne, 1951). In general, cardamom growing areas of Karnataka and many of the Idukki and Wynad areas of Kerala, experience a dry period extending from November–December to May–June. Such a long dry period of 6–7 months is in fact the major constraint facing cardamom production.

The national average yield of Indian cardamom is only 149 kg/ha as against 300 kg/ha in Guatemala and New Guinea. One of the main reasons for the increased yield in Guatemala is the well-distributed rainfall (Mohanchandran, 1984). In India, at present 75–80 per cent of the total area under cardamom is rainfed (Charles, 1986). With denudation of forests in many areas of Western Ghats, the normal congenial habitat for cardamom has been affected, destabilizing the ideal cool humid microclimate and the productivity of the crop.

Studies carried out on the influence of rainfall on yield of cardamom revealed that the contribution of variation in monthly rainfall to variation in yield of cardamom is higher than that of total rainfall and number of rainy days. In other words, yield is influenced more by the distribution of monthly rainfall than total rainfall and number of rainy days. Data collected from various cardamom estates have shown that in 10 out of 13 estates, highest yield was recorded when the annual rainfall was less than 2000 mm. In another survey data from 57 locations of the Coorg district have shown that in 42 cases more than 100 kg/ha dry capsule yield was obtained when the annual rainfall was less than 2000 mm. This indicates that total annual rainfall is not the criteria and further it has been suggested that an annual rainfall of 2000 mm evenly distributed may be optimum for cardamom cultivation (Subbarao and Korikanthimath, 1983; Ratnam and Korikanthimath, 1985). Most of the rainfall received during June–August would result in runoff leading to severe soil erosion and hence proper soil conservation measures are required to minimize soil and land degradation. Storing of runoff water during rainy periods in suitable farm ponds, tanks or embankments and recycling it during summer as protective irrigation coinciding with critical crop physiological stages offer great scope for evading total failure of crop stand and stabilizing yield of cardamom (Korikanthimath, 1987a; Cherian, 1977).

In Guatemala, rainfall conditions are much more favourable than in India. Here rainfall varies from 2000 to 5000 mm per annum in cardamom-growing belts, distributed evenly through out the year, though two peaks of high rainfall occur. Because of the absence of heat and drought stress, cardamom yields are much better in Guatemala, registering a national average of 300 kg/ha. Similar situation occurs in PNG also, leading to the harvest of very high yields (Krishna, 1982, 1997).

In India, prolonged drought in the first 6 months during 1983, when rain was most needed for cardamom plantations, brought about devastating effect leading to significant crop loss especially in exposed and partially shaded areas of Idukki district of Kerala. India's cardamom production came down to the lowest level of 1600 mt, there by giving a grim pointer to the need for combating recurring drought by proper soil and moisture conservation, mulching, adequate shade management along with provision for life-saving irrigation for sustaining yield.

3   SYSTEMS OF PLANTING

The mode of cardamom cultivation can be classified into four types (Mayne, 1951).

1   Kodagu (Coorg) Malay system;
2   North Kanara system;
3   Southern system;
4   Mysore (Karnataka) system.

3.1   Kodagu Malay system

It is restricted to Coorg district of Karnataka state. Small patches of forest land, a quarter-one sixth hectare in area, are cleared and planted with cardamom. Care is taken in selecting plots that face north or northeast for ensuring adequate lateral shade from surrounding forest trees. Seedlings from natural regeneration are thinned out and spaces filled in, or seedlings are raised in nurseries and transplanted in shallow pits, 1.5–2.5 m, apart. The areas are weeded periodically. Var. *Malabar* is the type commonly grown under this system. After a lapse of about 15 years, the area is left to natural forest growth, while the cultivation is shifted to another patch. A somewhat similar system was followed by the Madras (Tamil Nadu) Forest Department earlier and cardamom was collected as a minor forest product, the areas partially cleared by selection felling were used for such cultivation.

3.2   North Kanara system

It is followed in North Kanara, Shimoga and parts of Chickmagalur districts in Karnataka, where cardamom is grown as a secondary crop in arecanut gardens. Seedlings are raised in nurseries and planted in rows, 1.5–1.8 m apart; about 1200 seedlings are planted per hectare. The type grown is var. *Malabar*. Here cardamom seedlings are planted between arecanut palms on the margins of drains.

3.3   Southern system

This is the most important system for commercial production and accounts for about 90 per cent of the total cardamom crop in India. It consists of clearing selected areas of jungle land of all undergrowth, thinning out the overhead shade, planting cardamom seedlings at regular distance and cultivating according to a regular schedule. This system is adopted in Kerala, Tamil Nadu and parts of Karnataka. The size of the holdings vary widely, but the greater part of the production comes from holdings of 2–20 ha. In majority of areas of the south of Nilgiris, Kerala and parts of Tamil Nadu, the

Figure 4.1 General view of a cardamom plantation in the Western Ghat region. Cardamom is grown as an undercrop in forests utilizing the shade of the forest trees.

principal types of cardamom grown are var. *Mysore* and *Vazhukka*. The type grown in Karnataka is exclusively var. *Malabar* (Fig. 4.1).

3.4   Mysore system

In parts of coffee growing areas of Karnataka, cardamom is grown in isolated pockets, in ravines or in low-lying areas in coffee plantations. In these situations cardamom is found either as sole crop in narrow strips along the ravines or as scattered clumps interspersed with coffee plants.

4   PLANTING IN THE MAIN FIELD (ESTATE)

4.1   Preparation of land

Where cardamom is cultivated on a plantation scale in virgin forest areas, the initial work consists of clearing all undergrowth and thinning out overhead canopy in order to obtain an even density of shade. If land is sloppy, it is advisable to start clearing from top and work downwards. The shade will have to be regulated in such a way as to allow sunlight to filter through tree canopy almost uniformly. The bushes/shrubs and undergrowth are cleared and debris heaped in rows or piles to decay. Where slopes are steep, it is advisable to utilize such debris in such a way as to assist in checking any soil movement, which the exposure of the soil surface may facilitate. Contour terraces may be formed in cases where land is too steep. In arecanut gardens deep trenches and pits are dug among palms and filled up with fresh soil brought from neighbouring forest. The ground so prepared being utilized for planting cardamom. In marshy areas, adequate provisions should be made to drain off excess water by providing main and lateral drains depending on natural gradient of land.

4.2   Spacing

The spacing should be decided based on variety and duration of the crop in the field. Where it is intended to grow on a limited cycle, with regular replanting, it is obviously desirable to plant as closely as possible without unduly restricting the plants, so that early crops may be as large as possible. If a crop is only meant to last for 10 years, a commonly suggested crop cycle, only eight harvests are likely to be taken and the first two at least will be dependent on the number of plants per hectare. If, on the other hand, plantings are expected to remain in the field for longer periods, then too close planting will lead to the early overcrowding and crop reduction. This is of some importance, as the cardamom clumps tend to spread outwards as they get older and there will be gradual decline in new shoot production in the centre. This will lead to gradual reduction in yield as years advance.

In an experiment that was laid out in 1982 to study spacing, age of seedlings and their performance in relation to different levels of fertilizers, number of tillers, leaves/plant and height of plants were significantly influenced by treatments. Highest number of tillers (10.9/plant) and leaves (102.1/plant) were noticed with 18 months old seedlings planted at 2×1 m spacing and receiving 75:75:150 kg NPK and 100 kg neem cake per hectare (Korikanthimath, 1983b). In another spacing cum fertilizer study under rainfed conditions, the treatment differences were highly significant for number of tillers and number of leaves/plant. Highest number of tillers and leaves were noticed under 2.0×1.5 m spacing with fertilizer dose of 75:75:150 kg NPK/ha (Korikanthimath, 1982). The spacing commonly adopted for var. *Mysore* type is 3×3 m and for the less vigorous var. *Malabar* type 2×2 m between plants and rows (Anonymous, 1976). In the high production technology demonstration plots, cardamom seedlings planted at 2×1 m in hill slopes along the contour and 2.1×1.2 m in flat lands, gentle slopes and valley bottoms, yielded over 500 kg dry capsules/ha in just 2 years of planting (Korikanthimath and Venugopal, 1989). In a spacing trial carried out at Yercaud, Tamil Nadu, it was found that closely spaced (1×1 and 1.5×1.5 m) plants recorded better yield per unit area than plants at wider spacing of 2.5×2.5 and 2×2 m.

In sloppy lands, contour terraces may be made sufficiently in advance and pits taken along the contour. Depending on the slope, a distance of 4–6 m may be given along the slope between the contour lines. Close planting may be adopted along the contour.

4.3   Methods of planting

The systems of planting vary according to the land, soil fertility and the probable period over which the plantations are expected to last. In some places seedlings are planted in holes, just scooped out at the time of planting. In other areas, considerable care is taken in the preparation of planting pits. The spots where pits are to be dug are marked with stakes, soil dug out from pits, and they are filled with surface soil mixed with leaf mould, compost or cattle manure (Subbaiah, 1940).

A pit size of $60 \times 60 \times 45$ cm^3 is commonly used. Trials also were carried out in some estates with pits 90 cm^3 or even $120 \times 120 \times 30$ cm^3. In South and North Karnataka, pits are smaller, usually 45 cm^3 (Mayne, 1951).

In Kerala, for planting var. *Mysore* and *Vazhukka*, usually pits of $60 \times 45 \times 45$ cm are used. Normally pits are opened during April–May after the pre-monsoon showers. Pits are filled with a mixture of topsoil and compost or well rotten farmyard manure and 100 g of rock phosphate. In sloppy land, contour terraces are to be made sufficiently in advance and pits taken along the contour (Anonymous, 1985, 1986).

Most cardamom growing tracts are situated on hill slopes of Western Ghats. The undulating terrain and the heavy rainfall in the region increase the problem of soil erosion and run off losses of rainwater and nutrients. So it is imperative to conserve adequate soil moisture and make provisions for safe disposal of water. In view of this, background studies were conducted on systems of planting-cum-fertilizer levels in cardamom under rainfed condition at IISR (Cardamom Research Centre, Appangala) from 1985 onwards. In such trials the maiden yields obtained just 2 years after planting were the highest. The impact of systems of planting coupled with graded doses of fertilizers on the growth and yield of cardamom is given in Table 4.1. Korikanthimath (1989) reported greater moisture retention under the trench system than under the pit system of planting, and concluded that the trench system is better than the pit system.

Trenches may be taken at a width of 45 cm and depth of 30 cm to any running length across the slope or along the contour. Top 15 cm fertile soil may be taken out and kept separately on top (higher) side and the lower 15 cm depth soil excavated from trenches placed below the trench. The top 15 cm depth soil may be filled back in the trenches along with cattle manure. While closing the pits about 5 cm deep space may

Table 4.1 Yield of dry cardamom (kg/ha) as influenced by systems of planting and fertilisers levels

Systems of planting (A)	NPK fertilizers (kg/ha) (B)					Mean
	0:0:0	40:80:160	80:80:160	120:120:240	160:160:320	Mean
Pit	123.9	277.5	388.9	437.0	455.6	336
Trench	134.6	369.3	416.7	465.4	496.5	376
Mean	129.2	323.4	402.8	451.2	476.0	

Notes
SE/Plot – 88.0; Gen. Mean – 356.5; CV (%) – 24.7; CD for A 57.1; CD for B 90.3; CD for A × B 127.7.

be left at the top to facilitate application of fertilizers and mulches. Though trench system of planting costs about 35–40 per cent more of labour than pits, it is worth attempting because of the benefits of soil moisture conservation and its ultimate beneficial effect on plant growth and yield (Fig. 4.2). However, the pit method only is practicable in the low-lying areas where there is a possibility of continuous water stagnation.

4.4 Planting season

Planting season is decided based on the topography and rainfall pattern. Usually planting is done in June July. In areas experiencing heavy south-west monsoon, planting is either completed before July or is taken up in August–September, so as to avoid very heavy rains of July. Early planting gives better establishment and growth than late plantings (Mayne, 1951). In low-lying areas (such as valleys) planting should be done only after the heavy rains of July (Korikanthimath, 1980). In Mudigere area of Karnataka, Pattanshetty and Prasad (1972) reported better establishment and growth when seedlings are planted in August.

Figure 4.2 An young plantation established using the trench method of planting. The trenches are taken across the slope.

Cardamom suckers are planted from June to August on the surface or 15–20 cm deep below the surface. Mortality is least with surface planting and when rainfall is relatively less heavy in the week after planting. Suckers planted in August survived best, with a mean mortality of 25 per cent and those planted on the surface have the lowest mortality rate of 17.5 per cent (Pattanshetty *et al.*, 1972, 1974).

Month-wise planting trial with var. *Mysore* was taken up at Horticultural Research Station, Yercaud, located on the eastern side of the Western Ghats; commencing from June to November for 3 years, in order to assess the ideal planting time at elevations of 1300–1500 m above sea level. It was found that July planting gave the best establishment (87.92 per cent) followed by August, September, October and November plantings (77.9, 75.4, 63.7 and 61.6 per cent respectively). In the case of June planting the establishment was only 19.6 per cent. For better establishment there should be a minimum of 322 mm rainfall and the minimum and maximum temperature should not exceed 15.5–17.5 °C and 19.5–25.0 °C during the month (Nanjan *et al.*, 1981).

4.5   Planting

The general practice is to scoop a small depression in the filled soil and the seedling is placed in the centre of depression. Soil is then replaced taking care that the roots are distributed in their normal position and well pressed around the base of the clump. Deep planting should be avoided as it results in suppression of the growth of new shoots and may cause decaying of underground rhizomes. Seedlings are normally planted at an acute angle to the ground to prevent them from being blown over or broken by strong winds, which follow closely the planting season (Anonymous, 1952). A light root pruning is desirable but this should be restricted to longer roots, of 0.3 m or more. In the case of rhizome planting, the planting material can be kept in pits in a slanting position and rhizomes covered as done in the case of seedling planting.

Immediately after planting, the seedlings should be supported by stakes in order to prevent them from being damaged or blown over by wind and then the base is mulched with dry leaves. The cris-cross staking with two stakes is the best. Plants may be loosely tied to the stakes with dried banana sheath or jute threads for facilitating emergence and growth of aerial shoots.

After transplanting, care should be taken to offset the transplanting shock and to save the seedlings from heavy rains. Unhealthy plants are prone to disease incidence, and it is advisable to spray the entire seedling with 1 per cent Bordeaux mixture or any other suitable fungicides as a prophylactic measure. The newly planted area should be inspected periodically and gaps if any should be filled immediately, if the climate is favourable.

4.6   Planting of suckers

The system of propagation using suckers consists of splitting up of established clumps into sections consisting normally of at least one old and one young shoot (Fig. 4.3). Planting material of 20 cm long rhizome results in more shoots per clump, early bearing and large net returns than short rhizomes of 2.5 cm (Pillai, 1953; Pattanshetty, 1972a,b; Pattanshetty *et al.*, 1974). The section of rhizome is placed in a small depression in a pit that is already prepared and covered over with soil and mulch. The leafy shoots are placed almost parallel to the soil surface. New shoots arise from the

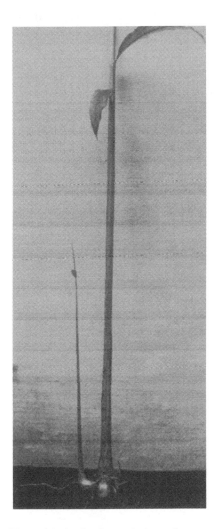

Figure 4.3 A planting unit in cardamom, consisting a portion of a rhizome, one old tiller and a developing shoot. Further tillering takes place from the rhizome after planting.

rhizome and constitute a clump. In the high ranges of Kerala straight planting of rhizome with stake is recommended. In Guatemala propagation of cardamom is invariably by suckers, where *katte* virus disease is not a threat (Anonymous, 1977). Three suckers per pit are used in order to induce more tillering in a short period. Fast growth and yielding are features in Guatemala mainly due to the well-distributed rainfall, rich fertile soil and good management. In India also under intensive care and irrigation, farmers have realized high yield per unit area by planting suckers from elite clones.

4.7 Gap filling

Successful initial establishment is an essential step in raising a good plantation of cardamom. On an average 5 per cent gaps are estimated in any cardamom plantation resulting in reduction of total production at the same rate. Use of healthy and disease-free seedlings or clumps for planting can reduce mortality of plants during establishment. It is quite possible that there can be a failure of monsoon during planting time. If the subsequent supporting rains are not received, the seedlings have to be irrigated at least once in a week. It is advisable to use healthy, sufficiently grown up seedlings or preferably clonal materials, for gap filling. The best time for gap filling is May–June, with commencement of monsoon. However, if proper care is given, it is possible to carry out gap filling in August–September also. Success of gap filling depends on aftercare till plants establish to the level of other plants. After planting, young plants should be kept on a regular schedule of cultural operations consisting of mulching, weed control, trashing, raking/digging, irrigation, shade regulation, manuring, filling up of gaps and plant protection measures for keeping the plants in a healthy and vigorous condition.

4.8   Mulching

Conservation of soil moisture in cardamom field plays an important role in realizing high productivity. The pre-monsoon showers have become erratic and not dependable in recent years. As a result often cardamom plants have to face drought for 5–6 months. Mulching is a practical solution for conserving soil moisture and has been acclaimed as the most important cultural operation for the overall improvement of soil and yield of cardamom (Zachariah, 1976). Advantages of mulching in a cardamom field are:

  (i) Soil moisture is conserved by minimizing the evaporation of water from the soil surface.
 (ii) During rains the soil does not get puddled up by the beating action of rain drops and hence the physical condition of the soil is not impaired.
(iii) Prevents runoff and erosion.
 (iv) Physical conditions like friability and structure of the soil are improved by enhanced biotic activity under mulch. Soil becomes more porous as the number of macropores is increased by biotic activity. This helps in the better percolation of water and moisture conservation.
  (v) Mulching keeps an even soil temperature and prevents diurnal variation in temperature of the surface soil.
 (vi) Fertility of the soil is increased. The soil becomes rich in organic matter and availability of nutrients is increased by the biotic flora under mulch.
(vii) Weed growth is suppressed.
(viii) Mulching helps in better root development and extraction of soil moisture from deeper layers (depth) of the soil.
 (ix) The mulch decomposes in course of time and improves soil fertility.

Soon after planting, the base is adequately mulched. Mulching is a simple operation in which dried leaves and other similar vegetational residues are used for the protection of soil. This operation should be attended before commencement of summer. The leaves shed by shade trees can be conveniently used for mulching bases of cardamom plants during November–December.

A study was conducted to assess the relative merits of locally available mulching material (viz. leaf mulch, paddy husk, phoenix leaves, coir dust and stratified leaf mulch) under uniform shade of coir matting and by using suckers of cv. P_1 combined with two levels of irrigation (viz. 75 per cent and 25 per cent available moisture). Results did not show any significant difference on sucker production due to irrigation (Table 4.2). However, among the mulches tried, leaf mulch (19.3) and Phoenix leaves (14.9) were on par and significantly superior to other mulches in sucker production (Raghothama, 1979).

Demulching is also equally important during May after the pre-monsoon showers for facilitating honeybee movement to obtain better pollination and capsule setting. The practice of uncovering the panicles shortly after the commencement of flowering improves fruit set in cardamom. The average number of capsules per panicle is 27.4 and 2.1 in case of exposed and covered panicles respectively in var. *Malabar* (Pattanshetty and Prasad, 1974). Removing the mulch accumulated in the centre of clump and thereby releasing panicles beneath would not only facilitate movement of honey bees but also provides better aeration and minimizes incidence of clump rot/rhizome rot diseases.

4.9   Weed control

Cardamom is a surface feeder, therefore, in the first year of planting, frequent weeding is necessary to avoid root competition between young cardamom seedlings and weeds. Weeding is done either over the whole area, or around the plants, (ring weeding), and weeds are used for mulching young plants. As many as 21 dicotyledonous weeds are identified in cardamom estates in Kodagu (Karnataka State). Out of these, *Strobilanthes ureceolaris* Gamb, is the most common. Weeds are controlled mainly by hand weeding, only rarely herbicides used. Two or three rounds of weeding are essential in the first year to remove the regenerating undergrowth. Generally first hand weeding is given in May–June, the second in August–September and the third in December–January. When weeding is done in May–June and August–September the weeded materials are heaped in the inter row spaces and are later used for mulching. In November–December weeding, the material is directly utilized for mulching. Normally slash weeding is carried out in cardamom plantations. Spraying gramaxone at the rate of 1.5 ml/l twice a year is also resorted to in some plantations though rarely, and this is found to be quite economical and convenient.

4.10   Other field operations

Other field operations practiced in cardamom fields are trashing, raking, digging and earthing up. Trashing consists of removing old and drying shoots of cardamom plants. From second year after planting, trashing is to be carried out every year. Trashing facilitates better sunlight penetration and aeration, there by promoting tiller initiation and growth as well as reduction in thrips and aphids population. It also helps in better pollination by honeybees (Korikanthimath and Venugopal, 1989). In the rainfed areas, trashing time is May, after the receipt of pre-monsoon showers. The trashed leaves and leafy stems may be heaped between the rows and allowed to decay or used for composting.

Towards the end of monsoon rains, a light raking or soil digging is done around plants at a radius of about 75 cm. The soil mulch thus formed around plants would help to conserve moisture for the ensuing dry period. This practice is particularly useful in low rainfall areas. Digging, not less than 25 cm deep once in 2 years, may be recommended to the whole area and immediate application of farmyard manure or organic manure such as bone meal, sterameal, groundnut cake, etc. that will produce a fairly good result. Digging can also be done in patches; however, it is necessary every year, particularly in clayey soil (Kuttappa, 1969).

Towards the end of the rainy season a thin layer of fresh fertile soil, rich in organic matter may be provided to the base of the clump, covering up to the collar region. The soil is taken by scraping between the rows or collecting soil from trenches/check pits. The thin layer of soil applied in the centre of clumps will not only keep them intact and cover the exposed roots but also will check the walking habit (radial growth) of cardamom (Korikanthimath and Venugopal, 1989). Care must be taken not to heap up the soil above collar region of the clump.

4.11   Replanting of cardamom

In cardamom, though a perennial, yield decline after a certain period is a regular feature. In order to maintain high productivity regular replanting of cardamom, every 8–10 years is necessary. One major reason for the low national average productivity in India is the existence of many old plantations having very low yielding potential (Korikanthimath *et al.*, 1989). Replanting with clonal material from superior, high-yielding varieties will maintain the productivity of the plot high.

Korikanthimath *et al.* (2000b) studied the economics of replanting. He replanted a cardamom field after 10 years, adopting the trench system of planting, and maintained them with all required inputs as per the HPT (High Production Technology) package. The planting materials used were 10-month-old seedlings from high yielding mother plants. The replanted field gave 155 kg dry cardamom in the second year. The following year an yield of 1775 kg dry cardamom/ha was realized. In subsequent 3 years the yield (dry in kg) realized were 385, 560 and 870, there by reaching an average yield of 749 kg/ha. Economic analysis carried out by Korikanthimath (2000) indicated a net return of Rs. 2,03,465/-per hectare, and the benefit cost ratio (BCR) is around 2.78. This study indicates that replanting of cardamom at a 10-year cycle is economically very advantageous.

5   PROPAGATION

Cardamom can be propagated by seeds and by vegetative means. Seedling population is variable because cardamom is a cross pollinated crop. Hence vegetative propagation is usually adopted in the case of elite clones. Vegetative propagation can be either through rhizome bits (suckers) or by micropropagation (tissue culture). Tissue culture propagation has gained a lot of importance in recent years and has been commercialized.

5.1   Propagation through seeds

In order to get quality seedlings, cardamom nursery has to be managed carefully and scientifically. This involves sowing seeds on raised beds, transplanting into primary and then to secondary nursery beds and finally into the field (Cherian, 1979; Kasi and Iyengar, 1961).

5.1.1   Seed selection

Seeds should be collected from high yielding vigorous plants, with well-formed compact panicles and well-ripened capsules free from infestation of pest and diseases.

Number of flowering branches formed on the panicles, percentage of fruit set and number of seeds per capsule should be given due consideration while selecting the mother plants for seed collection. (Anonymous, 1979; John, 1968; Ponnugangam, 1946; Siddaramaiah, 1967; Subbaiah, 1940). Apart from these desirable attributes, the mother clump should have more number of tillers (shoots) per plant, leaves with dark green colour and high percentage of fruit set. Colour of capsules should be dark green (Krishna, 1968). On an average, 1 kg fruits contain 900–1000 capsules with 10–15 seeds per capsule. Taking into consideration the percentage of germination, mortality due to diseases etc., on an average 1 kg of seed capsules are required to get about 5000 plantable seedlings.

5.1.2   *Preparation of seeds*

Seeds for sowing are collected from fully ripe capsules preferably from second to third round of harvest and are then either washed in water and sown immediately or mixed with wood ash and dried for 2 to 9 days at room temperature. The first method gives better results and is adopted widely. After picking, seed capsules should be immersed in water and gently pressed for separating seeds and washed well in cold water for removing mucilaginous coating of seeds. After draining water, seeds should be mixed with ash and surface dried in shade.

5.1.3   *Viability of seeds*

Storage of seeds results in loss of viability and delay in germination. In a trial, germination was found to be highest in case of fresh seeds, 59 and 50.6 per cent in vars. *Mysore* and *Malabar* respectively (Korikanthimath, 1982). Germination got reduced when there was a delay in sowing after storing the seeds for longer periods especially stored in air-tight containers. Seeds treated with organo-mercurials and stored in open bottles maintained viability up to a period of 4 months. Germination was highest (71.8 per cent) when sown in September (Pattanshetty and Prasad, 1973; Pattanshetty *et al.*, 1978). Korikanthimath (1982) observed that in a clone of var. *Malabar* there was a gradual decline in germination; 56.7, 51.0, 46.4, 34.1, 32.5 and 29.6 per cent when sown after 60 days in the case of seeds sown on 1st August, 15th August, 30th August, 14th September, 29th September and 14th October, respectively.

Cardamom seeds sown immediately after harvest in September germinate uniformly, early and satisfactorily and seedlings are ready for transplanting at the end of 10 months. If they are further retained in nursery beds for the next planting season either by proper thinning or by transplanting at wider spacing in secondary nursery beds, they develop rhizomes with large number of tillers, and are ideal for field planting. (Pattanshetti and Prasad, 1972). The ideal sowing season has been reported to be November–January for Kerala and Tamil Nadu and September–October for Karnataka (Anonymous, 1970, 1979).

5.1.4   *Pre-sowing treatment of seeds*

Cardamom seed possess a hard seed coat that delays its germination. Various studies have been undertaken on the effect of pre-sowing treatments of seeds to overcome this delay in germination. Treatment of freshly extracted seeds with concentrated nitric acid or hydrochloric acid for 5 minutes significantly improved the germination of cardamom seeds sown during November (Pattanshetty *et al.*, 1978; Prasad and Pattanshetty, 1974). Treating with nitric acid (20 per cent), acetic acid (25 per cent) or hydrochloric acid (50 per cent) for 10 minutes was found to be better than other treatments with 97.6 per cent, 98.6 per cent and 91.5 per cent germination respectively. Korikanthimath (1988) found that treatment of seeds with 20 per cent nitirc acid for 10 minutes is the best for enhancing germination in the case of var. *Malabar*. Apart from other factors, ambient temperature also plays a role in germination. The low winter temperature in the cardamom growing areas reduces germination as well as delays it (Krishnamurthy *et al.*, 1989). Gurumurthy and Hedge (1987) found that germination is significantly correlated with maximum and minimum temperatures prevalent in the area.

5.1.5   Nursery site

It is always advantageous to select nursery site on gentle slope, having an easy access to a perennial source of water. The nursery area should be cleared off all existing vegetation, stumps, roots, stones etc. Raised beds are prepared after cultivating the land to a depth of about 30–45 cm. Usually beds of 1 m width and convenient length and raised to a height of about 30 cm are prepared for sowing the seeds. A fine layer of humus-rich forest soil is spread over beds. The beds, when treated with formaldehyde solution (4 per cent) are found to control the damping off disease (Anonymous, 1985). After this treatment beds are covered with polythene sheets for a few days and seeds are to be sown 2 weeks after treatment. Before sowing, beds have to be flushed with water for removing any formaldehyde remaining.

5.1.6   Seed rate and sowing

The seed rate is 2–5 g per square metre of bed for raising 10 months old seedlings and 10 g for raising 18 months old seedlings (Anonymous, 1976, 1986). Seeds are sown in lines, usually not more than 1 cm deep. Rows are spaced at 15 cm apart and seeds are sown 1–2 cm apart within row. Deep sowing of seeds should be avoided for better and quicker germination. Seedbeds are to be dusted with Lindane at the rate of 60 g/5 m^2 to prevent termite attack. After sowing, beds are covered with a thin layer of sand or soil and pressed gently with wooden plank and thin mulch is given. Thereafter, beds are to be watered daily. Germination will commence in about 30 days and may continue for a month or two. The mulch materials are removed soon after the commencement of germination. The young seedlings are to be protected against exposure to sun and rain by providing shade over the seedbeds.

5.1.7   Mulching of nursery beds

Mulching of seedbeds influence germination (Abraham, 1958). Korikanthimath (1980) carried out studies on the effect of different mulches on germination by using locally available mulch materials like paddy straw, paddy husk, dry leaves of rose wood tree, saw dust, wild fern, coffee husk, goose berry (*Phyllanthes emblica*) leaves, sand, charcoal, polythene sheet and control (no mulching). The beds covered with paddy straw recorded highest germination (40 per cent) when seeds were sown in September and was on par with dry leaves of rose wood (37 per cent), goose berry leaves (37 per cent) and wild fern (38 per cent) (Table 4.2).

Table 4.2 Effect of different mulch materials on germination of cardamom seeds

Treatments	Germination (%)	Leaf spot disease % (45th day of sowing)
Paddy straw	40.8	12.0
Dry leaves of rose wood	37.4	16.2
Paddy husk	27.2	23.6
Saw dust	33.6	12.8
Wild fern	38.2	14.5
Coffee husk	21.6	25.6
Charcoal	35.2	19.2
Polythene sheet	1.8	–
Phyllanthes emblica leaf twigs	37.1	17.3
Sand	12.2	–
Control	13.3	–
SE/Plot	4.3	5.9
Gen. Mean	27.1	12.8
CV (%)	15.9	46.4
CD ($p = 0.05$)	6.2	8.6

There are also other reports that mulching with coconut coir dust, paddy straw or goose berry leaves enhances germination (Mayne, 1951; Sulikeri and Kololgi, 1978; Korikanthimath, 1983a).

5.2   Secondary nursery

In Kerala and Tamil Nadu regions the seedlings are transplanted to secondary nursery beds when they are about 6 months old, where as in Karnataka, the practice is sowing seeds in the primary nursery and thinning out excess seedlings and then allowing the remaining ones to grow right in the same place. Transplanting seedling to secondary nursery reduces nursery diseases. Korikanthimath (1982) has shown that following both primary and secondary nursery practices would be needed to get vigorous seedlings having 4–5 tillers with in a span of 10 months and with lesser incidence of pests and diseases.

On an average, 10 secondary beds are required for transplanting seedlings from one seedbed. Beds for transplanting are prepared in the same way as for primary nursery seedbeds. A mixture of powdered cow dung and wood ash is spread over the secondary beds before transplanting. In Karnataka, where seeds are sown during August–September, transplanting or thinning out takes place in November–January. In Kerala and Tamil Nadu states, seedlings from primary beds are transplanted to secondary nursery beds at a spacing of 20 × 20 cm during June–July. Rate of mortality was found to be higher when transplanting was done in second leaf stage (25.4 per cent) as against fifth leaf stage (1.1 per cent). Korikanthimath (1982) reported that the number of tillers produced per seedling was significantly more at the wider spacing of 30 × 30 cm (11.9) compared to the lesser spacing at 22.5 × 22.5 cm (9.2) or 15 × 15 cm (7.3). However, taking into consideration the larger area required and greater expenditure involved in raising nursery by transplanting seedlings at 5–6-leaf stage, a spacing of 15 × 15 cm is commonly recommended.

5.2.1   Manuring

Well-decomposed compost, cattle manure and fertile top forest soil are used for application to each bed at the rate of 8–10 kg (2.5 × 1.0 m) both in primary and secondary nursery. On average, 120 g nitrogen, 20 g phosphorous and 300 g potash, 50 g magnesium and 75 g calcium are removed from a bed planted with 100 seedlings. To promote uniform growth, 250 g mixture made of nine parts of NPK 17:17:17 and eight parts of zinc sulphate dissolved in 10 l of water may be sprayed once in 15–20 days, starting one month after transplanting (Anonymous, 1990). Regional Research Station, Mudigere, recommends NPK mixture at the rate of 160 g per bed one month after planting. This is to be increased by 160 g every month until a maximum of 960 g per bed is reached. The proportion of NPK is one part urea, two parts superphosphate and one part muriate of potash (Anonymous, 1979).

Korikanthimath (1982) observed that application of 45 g N, 30 g P_2O_5 and 60 g K_2O per bed of 2.5 × 1 m size in three equal splits at an interval of 45 days would result in better growth and higher number of tillers. First dose of fertilizer may be applied 30 days after transplanting in the secondary nursery. Application of diammonium phosphate (DAP) along with muriate of potash is found to be beneficial for tiller and root production (Anonymous, 1989).

5.2.2   Overhead shade

To protect the seedlings from sun, overhead shade has to be provided. Erecting a framework with wooden poles and sticks and then spreading over it nylon nursery (green house) mat or coconut leaves is usually used for this purpose. Nursery mat that can give 50 per cent shade is ideal, as growth and tiller production is found to be much better under such conditions. The shade nets have to be removed with the onset of monsoon.

5.2.3   Irrigation and drainage

The nursery beds should be irrigated twice a day immediately after planting up to 8–10 days, thereafter once a day up to 30 days. Once the seedlings establish and put forth new growth, watering may be resorted on alternate days till the monsoon starts. Flood and splash irrigations should be avoided as it may increase the problem of damping off and leaf spot diseases. Adequate drainage should be ensured to avoid stagnation of water, particularly in the low-lying areas during monsoon by providing central and lateral drains.

5.2.4   Weeding

Hand weeding has to be done once in 20–25 days to keep nursery beds free from weeds. The weed growth will be smothered once the seedlings attain sufficient growth.

5.2.5   Earthing up

The topsoil between the rows of cardamom seedlings would normally get washed away and deposited in pathways provided between nursery beds. Scraping of soil deposited from pathways and application in a thin layer up to collar region may be taken up 2 months after transplanting seedlings in secondary nursery. Application of the fertile soil collected from jungle along with cattle manure would be beneficial. Earthing up may be taken up immediately after split application of fertilizers.

5.2.6   *Rotation and fallow of nursery site*

It would be ideal to shift nursery sites once in 2–3 years to avoid build up of insects/pests in the area. Where such shifting is not possible due to non-availability of alternate site, it would be better to follow rotation of land with green manure crops (like *Crotalaria, Sesbania* etc.) and raising of cardamom seedlings. Green manure crops should be ploughed back and incorporated in soil once in 2 years and then cardamom seedlings may be raised. The practice of leaving part of the area fallow after deep digging/ploughing for a year would help in exposing of insects/pests to sun and bringing down the inoculum build up in the nursery site during the previous years.

By following cultural practices regularly, seedlings would be ready for transplanting in main field (plantation), 10 months after sowing seeds. Raising of seedlings in primary nursery and later transplanting them to secondary nursery is found to be more advantageous, as it facilitates better establishment and initiation of adequate number of suckers per plant.

5.2.7   *Raising of seedlings in paddy fields*

A survey conducted in Coorg, Hassan and Chickmagalur districts of Karnataka has shown that most small and marginal farmers raise cardamom nursery in paddy fields (wet lands) as water is easily available. Sufficient drainage has to be provided both in and around the nursery to avoid water stagnation during monsoon. Beds are separated by deep channels where in natural water is always available. As soils are frequently heavy, there is some risk of excessive moisture in the soil (Mayne, 1951). Such areas which normally possess sandy loam soils facilitate better root development and adequate growth of seedlings.

5.3   Dry nursery

Dry nursery is followed in plot (Malay) system of cardamom cultivation in Coorg district. The protection offered by the forest belt is so congenial that it is possible to raise cardamom nursery in small plots without providing overhead shade and without watering. As no watering is involved, the nursery is known as dry nursery. The nursery operations are very limited. The leaf litter is heaped up and dry seeds of cardamom are broadcasted after the initial showers in April (Mayne, 1951). Seeds are raked into soil and surface is covered with leaf mould. The branches of shade trees are cut to regulate shade. Hand weeding is carried out after germination. Before the end of monsoon, jungle soil is applied as a thin layer and beds are adequately mulched with leaf litter. Seedlings withstand drought while growing. Seedlings are planted in field after attaining sufficient growth, as is the case with other conventional method of raising seedlings.

5.4   Polybag nursery

Polythene bags of 20 × 20 cm size and thickness of 200–300 gauges with six–eight holes can be used for raising cardamom seedlings. Bags may be filled with nursery (pot) mixture in the ratio of 3:1:1 of jungle topsoil, farmyard manure and sand. The bags are arranged in rows of convenient length and breadth for easy management. Seedlings of 3–4-leaf stage can be transplanted in bags. Later, in between bags adequate space may be provided for better tillering (Fig. 4.4). The advantage of raising seedlings in polythene bags are: seedlings of uniform growth and tillering can be obtained, nursery period can be reduced and better establishment and growth can be ensured when transplanted to the main field.

A comparative trial using seedlings raised in polybags and in nursery beds conducted at Cardamom Research Centre, Appangala, showed 6 per cent mortality after 30 days in the plots planted with seedlings as against one per cent in case of polybag seedlings. Polybag seedlings may be conveniently used in homestead gardens and small estates, but, in large plantations it is rather difficult due to high cost of raising and transport.

5.4.1   Cost of raising seedlings

The cost of raising cardamom seedlings depends on duration of the nursery period and age of seedlings viz., 10 months (one season) or 18–22 months old (two seasons). For raising one lakh seedlings of 10 months old, an expenditure of Rs. 84,723 (US$ 1 = Rs. 46/-) is incurred in the departmental nursery of Spices Board in Madikeri, Karnataka. Normally increasing the number can bring down the cost per seedling. Presently the Spices Board supplies seedlings of 10 months old at the rate of Rs. 3 each and Rs. 4–4.50 for 18–22 months old seedlings.

5.4.2   Age of seedlings for field planting

Generally comparative success of establishment in the estate (plantation) and the cost of raising the seedlings should be given due consideration while recommending one season (10 months old) or two season (18–22 months) old seedlings. One-year-old seedlings can be successful for new plantings, while 2-year-old seedlings with a well-developed rhizome would be more suitable for gap filling. Seedlings of 18–22 months old are usually preferred for planting in Kerala and Tamil Nadu regions (Kasi and Iyengar, 1961).

5.4.3   Disadvantages of seed propagation

Because of the naturally cross-pollinated nature of cardamom plant, seedling populations will be highly heterogeneous, and the average yield from such populations will be generally low. A survey has indicated that approximately only 36 per cent of plants are good yielders in a plantation raised from seedlings (Krishnamurthy *et al.*, 1989). Another study of a population of 1490 plants from seedlings has shown that 45 per cent of the plants are poor yielders (less than 100 g of green capsules per plant) contributing only to 12.5 per cent of total yield; 36 per cent are medium yielders contributing to 40 per cent of total yield; while 15 per cent of population are good yielders that contributed to 32.1 per cent of total yield. In this last group of good yielders, plants giving high yield of 500–900 g were hardly 4 per cent; but they contributed about 15 per cent to the total yield. The average yield of this experimental population was only 170 g of green capsules per plant. The high degree of variability in yield and high percentage of poor yielders in the seedling population necessitates selection of elite clones and their vegetative propagation.

Figure 4.4 Polybag nursery under artificial shade. Each bag has 4–5 tillers, and they establish well and grow quickly when planted.

5.5 Vegetative propagation

Suckers from elite clones can be used for establishing plantations capable of high productivity. Methods have been devised for achieving high rate of sucker multiplication of selected high-yielding clumps (Anonymous, 1978). Plants raised from rhizomes come to bearing earlier than seedlings. Clonal propagation is thus advantageous in areas where *Katte* virus disease is not a menace. Clonal propagation is also followed in Guatemala (Anonymous, 1977). Vegetative propagation can be resorted to both by using tillers (suckers) and micropropagation by tissue culture (Kumar *et al.*, 1985). Small-scale growers especially in Tamil Nadu and Kerala commonly adopt propagation through suckers. Suckers should not be used in areas where *Katte* and other virus diseases (such as *Kokkekandu* and Nilgiri necrosis) are common.

One of the important constraints for achieving high yield is the non-availability of high-yielding selections to cardamom growers. Attempts have been made at various cardamom research stations to select high-yielding lines (Korikanthimath, 1998). In one such attempt 80 clones yielding more than 400 g of green capsules per clump were selected, each clump divided into four sets and the yield performance of 320 plants was studied for 6 years. It has been reported that 42 per cent of the good yielders contributed 43 per cent of the yield and 17 per cent of the very good yielders contributed 34 per cent of the yield, indicating that the average yield of the population can be improved significantly through the use of suckers (Parameshwar *et al.*, 1979). Such studies led to the development of clonal propagation techniques that are fairly rapid. Clonal multiplication through tissue culture has also been standardised (Kumar *et al.*, 1985)

5.5.1 *Rapid clonal nursery technique*

A quick method of proliferation of suckers was developed at the IISR (Cardamom Research Centre, Appangala, Kodagu, Karnataka) under controlled overhead shade to generate more number of planting units as well as high yield in a short time by resorting to high density planting in trenches at close spacing (Korikanthimath, 1992, 1999). The steps involved in this process are:

(i) High-yielding plants free from pest and diseases, with bold capsules are marked and part of the clump uprooted for clonal multiplication leaving the mother clump in its original site to induce subsequent suckers for further use;

(ii) Each planting unit consists of one grown-up sucker and a growing young shoot;

(iii) The planting units are placed at a spacing of 1.8 × 0.6 m in trenches, thus accommodating approximately 6800 plants per hectare of clonal nursery area;

(iv) On an average 32–42 suckers per planting unit will be produced by 12 months after planting. After one year it is possible to get 16–21 planting units from one clump (Fig. 4.5);

(v) In an area of 1 ha clonal nursery, 1–1.4 lakhs planting units can be produced after 12 months;

(vi) A crop of 190 g/plant dry cardamom (1759 kg/ha) was harvested within just 19 months of planting i.e. from planting to the final harvest (Korikanthimath, 1990, 1992). This system is easy and cheap and can be adopted by farmers.

6   SHADE MANAGEMENT

Cardamom is very sensitive to moisture stress and performs comparatively better in cool, shady environment. The shade canopy provides suitable environment by maintaining humidity and evaporation at suitable level (Abraham, 1965). Cardamom does not tolerate direct sunlight, at the same time under too much shade the metabolic activities of plants are retarded and they fail to grow well and yield. Hence removal of

Figure 4.5 Rapid clonal multiplication of cardamom in the nursery through close planting in trenches. Each clump when separated gives around 20 planting units per year.

excess shade is also essential so as to allow sufficient light penetration. Shade has to be regulated based on the lay of land, moisture retention etc. so as to get about 50 per cent filtered sunlight for proper growth and flowering. Following are the beneficial effects of shade in cardamom plantations:

(i) Provides good canopy and maintains cool temperature and protects soil from hot sun. Shade also checks surface evaporation of soil moisture thereby helping in retention of moisture for longer time; which is an important factor for realising good yield;

(ii) Protects plants from scorching effect of sun. Often in the border areas where cardamom plants are exposed to direct sun, severe scorching associated with an increase in pest and diseases have been noticed;

(iii) Checks high velocity of rain drops and minimizes mechanical damage to plants especially splitting of leaves;

(iv) The network of root system of shade trees prevents soil erosion and protects soil loss due to beating action of rainwater and thus improves physical properties of soil;

(v) Maintains adequate humidity and soil moisture that are essential for proper growth, flowering and setting of capsules;

(vi) Builds up sufficient humus and organic matter and fertility of soil by leaf litter that naturally gets accumulated. Leaves falling from shade trees provide enough mulch material;

(vii) Act as a wind break and minimize ill effects of gale and heavy wind;

(viii) One or the other shade trees flower round the year and thus act as an alternate source of nectar to honey bees which are the principal pollinating agents of cardamom. Forest areas ensure foraging capacity and availability of nectar for a long period;

(ix) Besides providing congenial microclimate for proper growth and performance, shade trees minimize weed growth which otherwise would grow luxuriantly in open areas and their control would have been a big problem.

6.1 Ideal shade trees

Usually all kinds of trees that are found in forests are maintained as shade trees. Trees belonging to 32 families of Angiosperms constitute the major tree flora in cardamom hills of Kerala, Karnataka and Tamil Nadu (Shankar, 1980). An ideal shade tree in cardamom plantation should have the following characteristics (George *et al.*, 1984):

(i) Wide canopy so that number of shade trees in an unit area is minimum;

(ii) No shedding of leaves during flowering phase of cardamom so that pollination is not affected;

(iii) They should be of medium size having evergreen nature retaining their foliage through out the year;

(iv) The trees should have small leaves and a well spread branching system;

(v) Root system should be deep to avoid competition for surface feeding;

(vi) Trees should be fast growing to provide necessary immediate shade;

(vii) They must have hard heartwood to resist wind.

It is desirable to maintain a mixed population of medium sized shade trees, that facilitate shade regulation, and to maintain more or less optimum conditions throughout the year.

The main considerations while selecting shade trees are adaptability to the climate, rate of growth and ease of establishment. Among the common trees *Balangi* (*Artocarpus fraxinifolius* Wt), *Nili* (*Bischofia javanica* Blum), Jack (*Artocarpus heterophyllus* Lamk), Red cedar, (*Cedrella toona* Roxb), *Karimaram* (*Diospyros ebenum* Koenig), *Karna* (*Vernonia monocis* C.B. Darke) are desirable as shade trees for cardamom (Rai, 1978; Abraham, 1957). *Maesopsis eminii,* an introduction from Africa, is a very good shade tree and is, (Korikanthimah, 1983). Another introduction is silky (silver) oak (*Grevellia robusta*) and this is now very popular as a shade cum support tree for black pepper. The wood is hard and is a useful cabinet wood.

Heterogeneity of shade tree species and their characteristics is the major constraint in conducting any experiments on shade requirements. Certain studies have been carried out for evaluating the existing shade trees and identifying the most suitable ones for cardamom (George *et al.,* 1984). In this study four important species viz., *Karimaram* or ebony (*Diospyrus ebenum*), *Elangi* or Spanish cherry (*Mimusops elangi*), *Nandi* or beatrack (*Lagerstroemia lanceolata*) and Jack (*Artocarpus heterophyllus*) have been evaluated. Results indicated that cardamom plants grown under *Karimaram* produced significantly more number of panicles, longer panicles and longer leaves and yielded 40–50 per cent more than those plants under other shade trees (Table 4.3).

Trees that carry crowded crown of canopy are undesirable as shade trees as they hardly allow filtered sunlight. *Erythrina lithosperma* and *Erythrina indica* (dadaps) are commonly planted by growers, especially when cardamom is planted in low-lying areas, but they are unsuitable as they compete for nutrients and soil moisture (shallow rooted) and act as alternate host for nematodes.

6.2   Shade requirements

Shade requirement varies from place to place depending on the lay of the land, soil type, rainfall pattern, crop combination etc. (Abraham, 1965; Korikanthimath, 1991). In Guatemala, which receive well distributed rainfall and has cool climate round the year, cardamom is grown practically in open areas with either no shade or having only very sparse shade (Anonymous, 1977). This is a major factor contributing to higher productivity in that country.

Table 4.3 Mean yield of cardamom per plant (g) under different shade trees

Shade trees	Year				Mean
	1975–76	*1976–77*	*1977–78*	*1978–79*	
Karimaram	112	81	81	207	121
Elangi	82	62	46	126	79
Jack	66	65	52	135	79
Nandi	109	65	42	135	89
Mean	92	68	55	151	92

Notes
SE for species = 57.5 CD ($p = 0.05$) = 18; SE for years = 49.4 CD ($p = 0.05$) = 15.
CD for any 2 years for any species = 31; CD for any two species for any years = 32.

Gaps in the shade canopy have almost always lead to leaf scorching under Indian conditions. It appears that the performance of cardamom plants under Indian conditions depends on their interaction with shade, sunlight and soil moisture (Aiyappa and Nanjappa, 1967). Sulikeri (1986) in a study on light intensity and soil moisture levels on growth and yield of cardamom, reported that under high density planting (9000 plants/ha) the yield figures were 1873 and 1928 kg/ha under medium (40–45 per cent) and high (65–70 per cent) light intensity respectively, as against 864 under low light intensity (15–20 per cent) (Table 4.4). Plants under medium light produced heavier capsules (75.6 g/100 capsules) compared to plants under high light intensity (72.3 g) and low light intensity (71 g). Harvest index (HI) under medium light intensity was 0.073 as against 0.066 under low light intensity and 0.037 under high light intensity. Thus it is necessary to regulate shade to provide adequate light especially during rainy season when the intensity of light is less. It is equally important to ensure that shade trees put up sufficient flush of leaves and provide adequate shade by the time summer sets in. The overhead canopy should, therefore be regulated once in a year during May–June (Fig. 4.6).

With the denudation of forests in the Western Ghats, the normal ecosystem is affected destabilizing the microclimate and rainfall in the cardamom growing tracts. With the onset of dry season the cool humid microclimate in the plantation changes rapidly as the hot wave of air from the hinterlands pass across the cardamom tracts without much hindrance due to deforestation all around the cardamom pockets. Consequently cardamom has to face inclement environment thereby resulting in poor growth and yield. Due to high velocity of wind, the rate of transpiration and evaporation will be increased. Moreover, plants suffer due to the physical pull of the high velocity of wind. The enhanced evaporation and transpiration deplete soil moisture rapidly (Abraham *et al.*, 1979).

Table 4.4 Dry yield of cardamom kg/ha as influenced by varying light intensity and soil moisture levels

Moisture levels	Light intensity			
	Low (5000–7000 lux)	Medium (15,000–17,000 lux)	High (25,000–27,000 lux)	Mean
Control	706.50	1217.25	645.75	856.50
Water at 25% ASM	945.90	2030.22	2521.53	1832.55
Water at 75% ASM	941.22	2373.03	2618.88	1977.51
Mean	864.54	1873.50	1928.52	1555.52

Effects	Significance	CD at 5%	CD at 1%
Main treatment	**	366.84	521.82
Sub treatment	**	221.49	298.26
Interactions			
(a) Two levels of treatment at a fixed level of main treatment	**	383.58	510.60
(b) Two levels of main treatments at a fixed of different levels of sub-treatments	**	492.21	685.35

Figure 4.6 Shade regulation is important for proper growth and flowering of cardamom. A young
cardamom field after shade regulation.

6.2.1   *Pest outbreaks in relation to shade*

The ecological upset, particularly the edaphic factors, has added new pest problems in
cardamom plantations. Many pests that were considered minor are assuming alarming
proportions in many cardamom-growing pockets. Root grub has emerged as a serious
pest in many areas (Gopakumar *et al.*, 1987). More population of root grub is seen
in exposed, warm and less shaded conditions. Similarly white flies are threatening
the existence of cardamom plantations in many locations. Outbreak of locusts
in Udubanchola taluk in Idukki district of Kerala is another example of ill effects
of changes in the microclimate (Joseph, 1986).

6.3   Bio-recycling

Among the plantation crops, no other crop experiences this benefit through the maintenance of tree growth *in situ*. Often, nutrients applied to soil have chances of leaching
down below the surface layer (Korikanthimath, 1987b). Trees absorb such nutrients and
trace elements from great depths and translocate them to leaves according to their
requirements. Falling leaves thus effect a recycling of these minerals to upper soil
humus layer. This is a normal and routine beneficial process taking place in forests.
Such a facility is being lost forever as a result of deforestation.

7   WATER REQUIREMENTS AND IRRIGATION MANAGEMENT

Cardamom is generally grown as a rainfed crop, and cardamom tracts of India experience a dry spell of about 5–6 months. Increased denudation of forests, deterioration in
forest ecology, coupled with erratic trends of rainfall, leads to aridity effects, adversely
affecting cardamom production (Ratnam and Korikanthimath, 1985). Even if there is

no reduction in total rainfall, failure of pre-monsoon and post-monsoon showers affects the crop adversely. During monsoon, post monsoon and winter months though there is sufficient moisture in the soil, plant growth is rather slow because of low ambient temperature. During summer months, if adequate moisture is available cardamom plant puts forth luxuriant growth.

Under normal conditions, panicles start emerging during January and continue to produce flowers from May onwards. Failure of post-monsoon rains and subsequent stress situation leads to flower and fruit drop, and under severe conditions drying up of panicle tips. Therefore, irrigation is necessary, almost from January to May.

In the above context, determination of optimum soil moisture for higher yield of cardamom needs no emphasis. Raghothama (1979) studied the effect of mulches and irrigation on the production of suckers in cardamom, using two levels of irrigation viz., at 25 per cent and 75 per cent available soil moisture (ASM). No control was included in this study. Although irrigation at 75 per cent ASM recorded higher values for all the growth parameters compared to irrigation at 25 per cent, there was no significant difference between the two levels. The differential effects of specific levels of irrigation were nullified by the presence of effective mulch in most of the sub-treatments and by the reduction in dry spell. These studies were further continued by Raju (1981) who made observations on initial yield in addition to growth parameters and reported that plants irrigated at 75 per cent ASM produced more bearing suckers and leaves, greater sucker height, fresh weight of plants and dry matter production compared to plants irrigated at 25 per cent ASM, but the differences were again non-significant. Further, it was reported that plants at the higher level of irrigation produced more and longer panicles, with more number of internodes, recorded higher fruit set, more capsules and higher weight of capsules compared to plants irrigated at lower level.

Sulikeri *et al.* (1978) reported that although cardamom requires high moisture level, it is very sensitive to high water table and consequent waterlogged situations. For better growth, drains should be opened at regular intervals so as to keep the water table 30 cm below the ground surface. Sulikeri (1986) conducted a trial on the effects of light intensity and soil moisture on growth and yield of cardamom with three soil moisture levels viz., control (no watering), watering at 25 per cent ASM and watering at 75 per cent ASM and three light levels viz., low light (5000–7000 lux), medium light (15,000–17,000 lux) and high light (25,000–27,000 lux). The results in the above experiment indicated that plants irrigated at 25 per cent ASM and at 75 per cent ASM produced 2.01 and 2.14 times more capsules than those without irrigation. The yield computed on hectare basis was 1832 kg and 1977 kg in plants irrigated at 25 and 75 per cent ASM respectively as against 856 kg per hectare in plants under no irrigation (Table 4.4). Irrigation at 75 per cent ASM combined with high light resulted in the highest yield of 2618 kg per hectare. The lowest yield of 645 kg per hectare was obtained under high light intensity and without irrigation. HI was highest at medium light combined with irrigation at 75 per cent ASM (0.075) and was least at high light without irrigation (0.032). The size of capsules also increased with higher moisture level.

7.1   Methods of irrigation

Among the various methods of irrigation viz., surface irrigation, sub-surface or trench irrigation and overhead irrigation or sprinkler irrigation, the last one is ideally suited in cardamom plantations.

7.1.1   Sprinkler irrigation

Overhead irrigation with sprinkling unit is found well suited for cardamom on account of its several advantages over other irrigation methods. Cardamom is grown on hill slopes with undulating topography and for such land, sprinkler irrigation can provide uniform water supply. The plantation can be irrigated without excessive loss of water due to surface run off or conveyance loss as the rate of application of water can be well regulated with sprinklers (John and Mathew, 1977; Saleem, 1978; Anonymous, 1985; Bambawale, 1980; Vasanth Kumar and Sheela, 1990). This will also avoid puddling, leaching and runoff that are common with other methods of irrigation. The humid atmosphere required for the successful growth and production of cardamom can be created by overhead sprinkling. Frequent light sprinkling can be done on soils of poor water-holding capacity. Irrigation equivalent to a rainfall of 4 cms, every 15 days would be quite sufficient for cardamom crop.

Installation of a sprinkler system should be processed after a careful survey of the area for an efficient and economical design. There should be a perennial source of water nearby. Sprinkler systems are designed to meet specific requirements, which may vary from one plantation to another depending on the lay of the land, area to be irrigated and source of water. The pumping site should be selected in a convenient place so as to cover the entire area with minimum number of pipes. Portable units are more economical to use, but operation costs are a little higher compared to the permanent system. The main line and laterals can be made portable so that they can be moved easily from one position to another.

Considering the high felt-need for a systematic study of the sprinkler irrigation system on cardamom, an investigation was conducted by Vasanthkumar and Sheela (1990) in Idukki district of Kerala state. Field experiments were conducted for two consecutive years (1982–83) and were laid out in split plot designs with two main plots viz., irrigated and non-irrigated (control) and three sub plots viz., three varieties of cardamom (*Malabar. Mysore* and *Vazhukka*). The results showed that the mean number of panicles produced per clump showed significant difference between the two main plots, irrigated and non-irrigated. Var. *Vazhukka* produced the highest number of panicles per clump, in the irrigated treatment (114.6) as against the non-irrigated treatment plots (90.8). Var. *Mysore* showed the lowest panicle production, in non-irrigated treatment plots (51.8). All the cultivars produced more flowers in plants that received sprinkler irrigation (3048, 1894 and 3756 respectively in *Malabar, Mysore* and *Vazhukka*) compared to the non-irrigated plants (1624, 1275 and 2068 respectively). The levels of irrigation did not significantly influence the fruit set per clump. The effect of sprinkler irrigation was in reducing the immature shedding of cardamom capsules, which was rather high in the non irrigated plots in all the three cultivars (26.4, 26.3 and 27.6 per cent respectively in *Malabar. Mysore* and *Vazhukka*). Capsule shedding was comparatively low in the irrigated plots (14.5, 14.9 and 11.7 per cent respectively in the three cultivars). Panicles of the irrigated plants showed faster growth, the effect being more pronounced in the var. *Vazhukka* (155.8 cm as against of 98.4 cm in the non-irrigated plants). Even in the var. *Mysore*, which normally produced shorter panicles, sprinkler irrigation increased the panicle length up to 73.8 cm whereas the non-irrigated plants produced shorter panicles (54.0 cm). The percentage of capsules that reached final maturity, was significantly influenced by sprinkler irrigation in all the three varieties, *Malabar. Mysore* and *Vazhukka* showed 61.3, 55.8 and 62.5 per cent respectively in the irrigated plots; the corresponding figures in the non irrigated plots were 50.6, 46.9 and 52.7 per cent.

Yield of capsules was almost double in the irrigated plots compared to the non-irrigated plants (Table 4.5). Essential oil content of capsules was more in the irrigated plants on dry weight basis. The mean values for the irrigated and non-irrigated plots were 10.2 and 8.2 per cent respectively, the highest was in the irrigated plot of var. *Mysore* (12.2 per cent) and the lowest in the non-irrigated plot of var. *Malabar* (7.1 per cent).

7.1.2 Drip irrigation

Drip irrigation is also becoming popular in cardamom plantations where water is scarce. It is a method of watering plants frequently with a volume of water just sufficient for the crop, there by minimizing losses such as deep percolation, runoff and evaporation. Water is applied to plant bases by using small diameter plastic lateral lines with devices called drippers (Fig. 4.7). This system applies water slowly in drops and keeps soil moisture within the desired range for plant growth. Flow of water through the drippers are so adjusted that two–three litres of water is discharged per hour. As water is applied directly to the plant basins and the interspaces are not irrigated, less quantity of water is used in this method. Application of 10–15 l of water per plant per day would be sufficient for cardamom. If cultivation is done on contour, this system can be practiced without any difficulty. Water from small farm ponds can be drawn in this system without pumping. If necessary, pumping can also be done from the lower down tanks, as the quantity of water required to each plant will be very less in this system. Drip irrigation has 80–95 per cent efficiency (Kurup, 1978).

Initial cost of drip irrigation equipment is considered to be its limitation for large-scale adoption. Cost of the unit per hectare depends mainly on the spacing. The investment for drip irrigation equipment for cardamom will be more in Karnataka where closer spacing is adopted than in Kerala and Tamil Nadu. Approximate cost for a drip irrigation unit would be around Rs. 25,000–Rs. 47,000 per hectare.

Figure 4.7 A young plantation maintained by drip irrigation.

Table 4.5 Effect of sprinkler irrigation on yield components and yield

Cultivars	Extension growth of panicles (cm) to the total flowers borne			Percentage of capsules matured			Fresh weight of capsules (g)			Dry weight of capsules (g)			Essential oil content of capsules (dwb)		
	I	NI	Mean	I	NI	Mean	I	NI	Mean	I	NI	Mean	I	NI	Mean
Malabar	101.3	78.5	89.9	61.3	50.6	56.0	1465	648	1061.5	353	154	253.5	8.7	7.1	7.9
Mysore	73.8	54.0	63.9	55.7	46.9	51.3	757	421	589.0	189	107	148.0	12.2	10.0	11.1
Vazhukka	155.8	98.4	127.1	62.5	52.7	57.6	1919	889	1404.0	446	213	329.5	9.6	7.4	8.5
Mean	110.3	76.97		59.87	50.11		1380.33	656.0			158		10.2	8.2	
'F' Test	**		*				**			**			**		
CD (P = 0.05)	19.59		5.32				3.8			67.40			0.47		
Irrigation	12.42		4.43				226			52.94			0.59		
Cultivars	17.56		6.27				320			74.86			0.83		

Notes

I – Irrigation; NI – No Irrigation; NS – Not significant; * – Significant at 5% level; ** – Significant at 1% level.

7.1.3 *Perfospray irrigation*

Perfospray is a type of irrigation in which water is sprayed under medium pressure. In this system, aluminium or PVC pipes are placed at 6–9 m interval and water is pumped to this system and after the desired time, it can be shifted to other locations in order to cover more areas. This is also a suitable mode of irrigation for cardamom (Sivanappan, 1985).

7.1.4 *Ground irrigation (contour furrows)*

This system can be followed in plantations that possess water source on the highest point. By taking advantage of water flow in the stream available at higher elevations, most portion of the plantation can be irrigated, through contour furrows on natural gradient. Contour furrows are opened to allow water to go along the furrow to irrigate plants, if necessary, small basins of 30–45 cm diameters may be made around cardamom plants. This does not require any expenditure towards pumping. Water can also be stored in tanks or ponds constructed on the slope for irrigation during summer months.

7.2   Time and frequency of irrigation

Using sprinkler set at an interval of 10–12 days to get the effect of 25 mm rain during January–May should provide irrigation. The moisture level should be maintained above 45–50 per cent water holding capacity of the soil. A stress period for about 45 days during December–January has been found to be quite beneficial for cardamom.

It would be ideal to commence irrigation during first week of February and continue at an interval of 12–15 days till the regular monsoon commences in the first week of June.

7.3   Water harvesting

Often the terrain of cardamom estates is undulating with moderate to steep slopes. Quite a number of small and fairly big streams pass through many of these areas. Runoff from the cardamom watersheds can be collected in farm ponds and check dams or underground water tapped through dug wells.

Harvested water can be stored in ponds and checkdams by minimizing the losses through seepage, evaporation and recycling. Apart from improving and stabilizing yields under rainfed cardamom cultivation, checkdams, farm ponds and dug wells reduce flood hazards and recharge ground water. Such devices in many cases will serve as percolation tanks that would substantially augment the ground water availability in the area.

8   CARDAMOM-BASED CROPPING SYSTEMS

Cardamom in the evergreen forests is a good example of forestry cum cash crop (plantation crop) management. Even from pure commercial forestry point of view some of these forests are uneconomical for management or utilization due to the heavy cost involved in building up suitable infrastructure for extraction and transportation of timber. Such forests, when brought under cardamom will become financially viable for management and benefits from both cardamom and tree culture can be derived. It is a harmonious system as it will not upset or adversely affect environment or protective quality of forest. Cardamom is perhaps the only plantation crop, which involves least disturbance to the existing forest trees as against partial felling of trees for raising coffee, pepper etc.

8.1   Economics and labour utilization pattern of sole forest vs forest cultivated with cardamom

It is worth comparing the additional benefits accruing from cardamom cultivation beneath shade trees. An evergreen forest in the Western Ghat region managed ideally (sole forestry system) can yield 10 m^3 of timber/hectare/annum. Our present yield is much below this level. At present the returns will be approximately 10 times more due to increase in timber cost. Moreover clearing of forests for non-forestry purposes is not permitted now. Computed over a period of 20 years, the net returns from this will be Rs. 7 lakhs and the job opportunities created 5 lakhs (Joseph, 1978). The same area when brought under cardamom for the same period will give a total net profit of Rs. 450 lakhs and employment opportunities created will be approximately 80 lakhs at a moderate estimate of price and yield. Hence the need is to focus attention towards efficient soil and land use. Besides this, mixed cropping of cardamom with other plantation crops (like coffee and pepper) having their canopies at different vertical heights and their roots foraging the soil at different depths and lateral distances, is practiced for increasing production of cardamom, and income from unit area. Such a cropping system, often called multitier cropping, is defined as a compatible combination of crops having varying morphological frame and rooting habits grown together and intercepts solar energy at varying heights and their roots forage the soil mass at different zones. The mixed crop and multitier crop combinations that are useful in maximizing yield and net returns from unit area are briefly discussed here.

8.2   Nutmeg–clove–cardamom mixed cropping system

Nutmeg, clove and cardamom tap solar radiation at different heights and so also mine soil moisture and nutrients at different vertical and horizontal spaces. Instead of planting other forest tree species, clove and nutmegs planted in vacant, open areas can provide sufficient shade to cardamom. This crop combination is ecologically feasible and economically sustainable. An example of such a cropping system is available at Burliar at an elevation of 3500 feet above MSL in Tamil Nadu state. This plantation was started as a nutmeg garden, later clove seedlings were planted in between nutmeg plants. Subsequently, cardamom seedlings were interplanted. Nutmeg and cloves have started bearing in 6–7 years after planting. Though both these plants require regulated shade in the early stages of growth, clove does not require so much shade once it starts yielding. Cloves and nutmegs can provide shade to cardamom in this plantation. Cardamom plants started yielding from third year and from 600 plants, the yield obtained was 150 kg dry capsules. Clove started yielding during 1978, and on an average, 1 kg of dried clove per plant was obtained during 1980. Subsequently the yield has gone up regularly, giving steady income. Nutmeg plants started yielding in 1979. At present there are about 100 nutmeg trees in bearing stage, and here also the yield increased as years passed and constituted a regular income. To satisfy the shade requirements of cardamom, in addition to nutmeg and clove, a few tall growing shade trees at regular intervals were retained. However, the problem of comparatively less shade was circumvented to a considerable extent by providing sprinkler irrigation throughout the plantation for all the three crops. As the area is exclusively utilized for cardamom, nutmeg and clove, the plantation requires heavy manuring and after care for getting steady yields. Due share of fertilizer has to be given to each crop as per recommendation.

8.3   Cardamom–arecanut mixed cropping

The long pre-bearing age of the main crop arecanut, small income from initial harvest, insecurity against pests and diseases, remoteness from markets and lack of transport were considered to be some of the reasons that might have prompted farmers to grow different crops in arecanut gardens (Abraham, 1956; Khader and Antony, 1968; Nagaraj, 1974). Till cocoa was introduced, cardamom was the main mixed crop cultivated beneath arecanut palms in Uttara Kannada district of Karnataka. Taking into consideration unit value of several mixed-crops tried in arecanut, cardamom has served as the most remunerative and quick yielding spice crop (Korikanthimath, 1990), at elevations between 700–1080 m, because at higher altitudes arecanut cultivation is not successful. However, under suitable agro-climatic conditions, this practice has been found to be highly remunerative. Normally, arecanut gardens with irrigation facilities during summer or areas that are low lying, flat land with adequate drainage facilities are selected for planting cardamom.

The studies conducted by Bhat and Leela (1968) and Bhat (1974) have shown that more than 80 per cent of the roots of arecanut are within a radius of 75 cm from base in palms spaced at 2.7 × 2.7 m. In cardamom, 80 per cent of its roots occur in a zone of 25 cm radius, 14 per cent in a zone of 25–50 cm radius and only 6 per cent in a zone of 50–75 cm. Vertically, cardamom roots penetrate only up to 40 cm. Though arecanut is also more or less a surface feeder, there will not be much competition between these two as far as the root system or fertilizer requirements are concerned since the nutrient requirement of cardamom is very much less than that of arecanut. Muralidharan (1980) reported that 32.7–47.8 per cent of incident light rays pass down through the canopy of a 14-year-old arecanut garden depending on the time of day. Normally in a pure arecanut crop spaced at 2.7 × 2.7 m, this light energy, which reaches the ground, is wasted. According to Balasimha (1989) approximately 27 per cent photosynthetically active radiation (PAR) passes through arecanut canopy. Cardamom plants in arecanut garden can advantageously utilize this energy.

In an arecanut–cardamom mixed crop field, arecanut should be planted in Northeast direction at spacing of 2.7 m between plants and rows. September–October is the ideal season for planting arecanut (Khader and Antony, 1968). Establishing of banana plants before planting arecanut seedlings at a spacing of 5 × 5 m will not only provide adequate shade to young arecanut seedlings but also fetches sizeable initial income to farmers.

After 7 years of planting arecanut i.e., ensuring that the arecanut palms put forth adequate shade, cardamom seedlings can be planted between two arecanut palms. In old plantations, cardamom is planted in alternate rows with a plant-to-plant spacing of 2 m. In new areas, cardamom is planted between rows of arecanut plants. About 1250–1500 plants are accommodated in a hectare of area garden.

Both arecanut and cardamom share common cultural operations like weeding, mulching, irrigation etc. thereby bringing down the total cost of cultivation. It is quite necessary to apply adequate quantities of plant nutrients right from the establishment of the garden to ensure early bearing and continued yield over a long period. Recommended doses of fertilizers have to be applied to each component crop to prevent inter-plant competition for nutrients. In Uttara Kannada district (Sirsi and adjoining areas) of Karnataka, regular replanting of cardamom is taken up once in 5–7 years due to heavy incidence of *katte* disease. Here, the planters pick 2–3 crops of cardamom and then take up replanting. The high production technology demonstration trials conducted on mixed cropping of arecanut and cardamom have given good growth and yield of both the crops; as high as 625 kg dry cardamom and 3750 kg of dry arecanut/ha/year with irrigation and an average of 375 kg cardamom and 2250 kg arecanut (dry)/ha under unirrigated conditions (Korikanthimath, 1989). Korikanthmath *et al.* (2000) studied the microclimatic and photosynthetic characteristics in arecanut–cardamom mixed cropping system. In this study arecanut recorded higher photosynthetic rate, transpiration rate, carboxylation efficiency and water use efficiency at gas exchange level. Stomatal conductance as well as intercellular CO_2 concentration were higher in cardamom than in arecanut.

8.4   Arecanut – tree spices and cardamom crop combination

A spice cafeteria maintained in certain arecanut gardens is available at Belur village in Shimoga district of Karnataka. Cardamom was planted between the arecanut palms, followed by introduction of nutmeg at a spacing of 10×10 m. Clove plants were also planted both inside as well as on the borders of arecanut gardens. The nutmeg trees have grown up to a height of 8–10 m and now stabilized their yield. Performance of clove is found to be more encouraging in the border areas than the ones planted amidst the arecanut palms. On an average each nutmeg plant is yielding about 500–750 fruits. The clove trees have yielded 750–1250 g per tree. The yield of cardamom is around 275–400 kg (dry) capsules per hectare.

The crop combination of arecanut as a top canopy that provides shade for nutmeg, cloves and cardamom is found to be useful in increasing the unit area income. Cardamom is phased out in course of time when both nutmegs and cloves grow up and put out sufficient top canopy.

8.5   Coconut–pepper–cardamom and coffee combination

Examples of crop combination of coconut, pepper, cardamom and coffee combinations are available at Koppa taluk in Karnataka. In this crop combination, coconut acts as first tier, followed by pepper, which is trained on to the trunk of coconut trees as second tier. Cardamom and coffee (robusta) are also planted between the coconut rows. All these crops are complementary to each other, and help in maximizing return from unit area.

8.6   Cardamom–coffee mixed cropping

Cardamom can be successfully mixcropped with coffee. In this cropping pattern, coffee – both arabica and robusta – constitute the first tier and cardamom the second tier. Often black pepper, trained on the shade trees, forms another component, and the shade tree canopy forms the top tier. Many experimental studies have gone into the coffee–cardamom cropping system, such as the one given below.

Arabica coffee (S.795) planted in 1976 at a spacing of 1.8 × 1.8 m (triangular) was selected for studying the feasibility of mixed cropping of cardamom under rainfed condition. Here, every third row of coffee was uprooted in 1981–82 and subsequently cardamom was planted by providing a spacing of 5.4 × 0.9 m (2058 plants/ha). The fertilizers due to cardamom and coffee were applied separately. Plant protection measures and cultural operations like weeding, mulching, shade regulation, irrigation etc., were followed commonly both for cardamom and coffee. An average yield of 259 and 308 kg/ha dry cardamoms and coffee respectively was obtained during 1983–86 (mean of 3 years). This crop combination of cardamom with arabica coffee resulted in a net profit of Rs. 30,305/ha (Table 4.6).

In another case, Robusta coffee (Ferdinia) planted at 2.7 × 2.7 m in 1947 was interplanted with cardamom after removing alternate rows of coffee in 1985 to accommodate var. *Malabar* (Cl.37), between two rows of coffee (spaced at 5.4 × 2.7 m) at a spacing of 1.8 × 1.2 m. An average yield of 1907 kg/ha dry coffee (mean of four years) and 950 kg/ha of dry cardamom capsules were obtained (Korikanthimath, 1989). This clearly shows the high production potential of the coffee–cardamom mix crop combination. Srinivasan *et al.* (1992) reported that the monetary benefit can be enhanced by more than 30 per cent by adopting cardamom–coffee cropping system planted at 1:1 ratio in alternate rows either along or across the slopes. Thus the combination of coffee–cardamom offer prosperity to planters through enhanced net returns, favourable cost benefit ratio and partial guarantee a against market glut of a single commodity in monoculture. Additional employment potential, efficient land and water use, effective weed control etc. are some other advantages of multistoried cropping systems over monoculture.

Inclusion of either pepper alone as in Kerala or a combination of coffee and arecanut as in Karnataka in the multiple cropping system with cardamom is highly remunerative and can give higher benefit–cost ratios. In Karnataka a benefit cost ratio of 3.53 was recorded in a mixed cropping system of cardamom and coconut (Korikanthimath *et al.*, 1998a). With the cardamom–coffee cropping system, the benefit cost ratio was 1.94 (with single hedge system with arabica coffee) and 4.25 (in the case of double hedge system with Robusta coffee (Korikanthimath *et al.*, 1998b).

Table 4.6 Economics of cardamom mixed cropped with Arabica coffee (rain fed)

Year	Yield (dry) (kg/ha)		Gross income (Rs./ha)			Total expdr. (Rs./ha) (both for cardamom and coffee)	Net profit (Rs./ha)
	Cardamom	Coffee	Cardamom	Coffee	Total		
1983–84	122.5	437.5	24,010	5486	29,496	18,750	10,746
1984–85	300.0	175.0	54,600	2450	57,050	19,550	37,500
1985–86	355.0	312.5	59,640	4531	64,171	21,500	42,671
Average	259.0	308.0	46,083	4156	50,239	19,933	30,305

Note
1 US$ = Rs. 46.50.

8.7   Rhizosphere changes due to mixed cropping

Microbiological investigations on cardamom mixed cropping systems were carried out (Korikanthimath unpublished) and the results indicated significant increase in microbiological activity under mixed cropping compared to monocropping. Under the arecanut–cardamom mixed crop, the population of bacterial count in the rhizosphere of cardamom registered an increase of 93 per cent (at 0–15 cm depth) and 29 per cent (at 15–30 cm depth) compared to the cardamom monocrop. In the case of fungal population an increase of 32, 59 and 61 per cent were recorded at 0–15, 16–30 and 31–45 cm depths over the values obtained for monocrop. Actinomycetes increased by 66 per cent at 0–15 cm depth, but decreased at deeper layers.

Under the robusta coffee and cardamom mixed crop the population of bacteria registered and increase (41 per cent) at 0–15 cm depth only, while fungal population increased by 69 per cent under mixed crop at 0–15 cm, but did not show any significant increase in deeper layers. Mixed cropping did not alter the actinomycetes population in this case.

In the arabica coffee–pepper–cardamom mixed cropping system, Korikanthimath *et al.* (2000b) did not get constant results on microbial population. However, in comparison with the respective monocrops, intercropping registered an increase of 71, 98 and 52 per cent respectively in the rhizospheres of arabica coffee, cardamom and pepper. Similar increase in the population of fungi was also recorded in comparison with respective monocrop. However the above workers did not make any effort in identifying the microorganisms present in the rhizosphere.

9   CONCLUSION

The high production technology of cardamom evolved at IISR Cardamom Research Centre at Madikeri, has been demonstrated in many farmers' fields. It has been amply proven that remarkable increase in yield could be achieved through an integration of factors such as high quality planting material, proper agrotechnology, judicious nutrient management and control of pests and diseases. There are many production constraints that eventually influence the production and productivity. Constraint alleviation also becomes an integral part of the high production technology, especially the management of heat and drought stress and control of pests and diseases. The emerging scenario is a system approach where cardamom forms one component of a cropping system tailored to give a higher, sustainable productivity.

REFERENCES

Abraham, P. (1956) Spices as intercrops in coconut and arecanut gardens. *Arecanut J.*, 7, 56–58.
Abraham, P. (1957) 'Karuna' is excellent as shade tree for cardamom. *Indian Farming*, 7(9), 14–16.
Abraham, P. (1958) New knowledge for cardamom growers. *Indian Farming*, 8(2), 34–38.
Abraham, P. (1965) The cardamom in India. *ICAR Publication*, New Delhi, pp. 1–46.
Abraham, P. and Tulasidas, G. (1958) South Indian cardamom and their agricultural value. *Indian Council of Agricultural Research Bulletin*, 79, 1–27.
Abraham, V.A., Gopinathan, K.V., Padmanabhan, V. and Saranappa (1979) Deforestation and change of micro-macro climate conditions. *Cardamom J.*, 12(18), 3–7.

Aiyappa, K.M. and Nanjappa, P.P. (1967) Highlights on cardamom research – its problems and prospects. *Cardamom Industry*, Cardamom Board, Cochin, Kerala.

Anonymous (1952) *Elettaria cardamomum* Maton (Zingiberaceae). *The Wealth of India*, Vol. 3, CSIR, New Delhi.

Anonymous (1970) Cardamom culture. Part I. *Cardamom News*, 4(6), 2–3.

Anonymous (1976) Cardamom in Karnataka. UAS *Tech. Series* No. 14. University of Agricultural Sciences. Hebbal, Bangalore, p. 5.

Anonymous (1977) Report of the Indian cardamom delegation to Guatemala. *Cardamom*, 6, 3–9.

Anonymous (1978) New clone developed. *Farmers Friend*, 3(2), 4–5.

Anonymous (1979) Propagation of cardamom (Nursery practices of cardamom). University of Agricultural Sciences, Regional Research Station, Mudigere, p. 5.

Anonymous (1982) Condiments and spices. In *Handbook of Agriculture*, ICAR, New Delhi.

Anonymous (1985) Cardamom package of practices, Pamphlet No. 28. Central Plantation Crops Research Institute, Kasaragod, p. 4.

Anonymous (1986) Cardamom cultivation. Cardamom Board, Cochin, p. 11.

Anonymous (1989) Cardamom Package of Practices, National Research Centre for Spices, Calicut, Kerala.

Anonymous (1990) Cardamom nursery management Extension Folder No. 1. Spices Board. Indian Cardamom Research Institute, Myladumpara, p. 7.

Balasimha, D. (1989) Light penetration patterns through arecanut canopy and leaf physiological characteristics of intercrops. *J. Plantation Crops*, (suppl.), 61–67.

Bambawale, P.M. (1980) Economics of sprinkler irrigation system. *Cardamom* 12(12), 21–27.

Bhat, K.S. (1974) Intensified inter/mixed cropping in areca garden – the need of the day. *Arecanut and Spices Bull.*, 5, 67–69.

Bhat, K.S. and Leela, M. (1968) Cultural requirement of arecanut. *Indian Farming*, 18(4), 8–9.

Charles, J.K. (1986) Productivity in cardamom – an insight. *Cardamom J.*, 19(12), 13–20.

Cherian, A. (1977) Environmental ecology – an important factor in cardamom cultivation. *Cardamom J.*, 9(1), 9–11.

Cherian, A. (1979) Produce better cardamom seedlings for ensuring high yield. *Cardamom J.*, 11(1), 3–7.

George, C.K. (1990) Production and export of cardamom in Guatemala. *Spice India*, 3(9), 2–6.

George, M.V., Mohammed Syed, A.A. and Korikanthimath, V.S. (1984) *Diospyros ebenum* Koenig, an ideal shade tree for cardamom. *J. Plantation Crops*, 12(2), 160–163.

Gopakumar, B., Kumaresan, D. and Varadarasan, S. (1987) Management strategy of root grub in cardamom. *Cardamom J.*, 20(10), 5.

Gurumurthy, B.R. and Hegde, M.V. (1987) Effect of temperature on germination of cardamom (*Elettaria cardamomum* (L) Maton). *J. Plantation Crops*, 15, 5–8.

John, J.M. and Mathew, P.G. (1977) Sprinkler irrigation in cardamom – a sure step to increased production. *Cardamom*, 9(6), 17–21.

John, M. (1968) Hints for raising cardamom nurseries. *Cardamom News*, 2(6), 4–5.

Joseph, K.J. (1978) Multicropping concept in Forestry. In E.V. Nelliat *et al.* (eds) *Proceedings of First Annual Symposium on Plantation Crops (PLACROSYM-1)*. CPCRI, Kasaragod, pp. 441–443.

Joseph, K.J. (1986) Spotted locusts (*Aularches milians* L) and its integrated management. *Cardamom J.*, 19(17), 5.

Kasi, R.S. and Iyengar, K.G. (1961) Cardamom propagation. *Spices Bull.*, 1(3), 10–11.

Khader, K.B.A. and Antony, K.J. (1968) Intercropping: A paying proposition for areca growers – what crops to grow. *Indian Farming*, 18(4), 14–15.

Korikanthimath, V.S. (1980) Nursery studies in mulches for use in cardamom primary nursery. *Annual Report*, Central Plantation Crops Research Institute, Kasaragod, Kerala.

Korikanthimath, V.S. (1982) Nursery studies in cardamom. *Annual Report*, Central Plantation Crops Research Institute, Kasaragod, Kerala.

Korikanthimath, V.S. (1983a) Nursery studies in cardamom. *Annual Report*, Central Plantation Crops Research Institute, Kasaragod, Kerala.

Korikanthimath, V.S. (1983b) Practical planting and shade management in cardamom. *Planters Chronicle*, UPASI, pp. 405–406.

Korikanthimath, V.S. (1987a) Studies on analysis of rainfall and its impact on cardamom. *Proceedings of the National Seminar on Agrometerology of Plantation Crops*, Regional Agricultural Research Station, Kerala Agricultural University, Pilicode, Kerala, 12–13, March, 1987.

Korikanthimath, V.S. (1987b) Systems of planting-cum-fertiliser levels in cardamom under rainfed condition. *Ann. Rep.*, National Research Centre for Spices, 1989, pp. 30–31.

Korikanthimath, V.S. (1987c) Impact of drought on Cardamom. *Cardamom*, 20, 5–12.

Korikanthimath, V.S. and Venugopal, M.N. (1989) High Production Technology in Cardamom. *Tech. Bull.*, National Research Centre for Spices, Calicut, Kerala, India.

Korikanthimath, V.S., Venugopal, M.N. and Naidu, R. (1989) Cardamom production – a success story. *Spices India*, 119, 19–24.

Korikanthimath, V.S. (1989) Systems of planting-cum-fertilizer levels in cardamom under rainfed condition. *Ann. Rep.*, National Research Centre for Spices, 1988–89, Calicut, Kerala, pp. 24–26.

Korikanthimath, V.S. (1990) Efficient management of natural resources for cardamom (*Elettaria cardamomum* Maton) production. Presented at *International Symposium on Natural Resources Management for a Sustainable Agriculture. Feb. 6–10, 1990. Indian Society of Agronomy*, New Delhi.

Korikanthimath, V.S. (1991) Shade management for high productivity in cardamom (*Elettaria cardamomum* Maton). *Spice India*, 4(2), 15–21.

Korikanthimath, V.S. (1992) Large scale multiplication of cardamom – the NRCS experiment. In Y.R. Sarma, S. Devasahayam and M. Anandraj (eds) *Black pepper and Cardamom – Problems and Prospects*, Indian Society for Spices, Calicut, pp. 65–67.

Korikanthimath, V.S. (1999) Rapid clonal multiplication of elite cardamom selections for generating planting material, yield upgradation and its economics. *J. Plantation Crops*, 27, 45–53.

Korikanthimath, V.S. and Mulge, R. (1998) Pre-sowing treatment to enhance germination in cardamom. *Karnataka J. Agric. Sci.*, 11, 540–542.

Korikanthimath, V.S., Kiresur, V., Hiremath, G.M., Hegde, R., Mulge, R. and Hosmani, M.M. (1998a) Economics of mixed cropping of coconut with cardamom. *Crop Research*, 15(2&3), 188–195.

Korikanthimath, V.S., Mulge, R., Hegde, R. and Hosmani, M.M. (1998b) Crop combinations and yield pattern in coffee mixed cropped with cardamom. *J. Plantation Crops*, 26(1), 41–49.

Korikanthimath, V.S., Ankegowda, S.J., Yadkumar, N., Hegde, R. and Hosmani, M.M. (2000a) Microclimatic and photosynthetic characteristics in arecanut and cardamom mixed cropping systems. *J. Spices and Aromatic Crops*, 9, 61–63.

Korikanthimath, V.S., Hegde R., Rao, G. and Gayathri, A. (2000b) Investigations on cardamom based cropping systems. Report of the ad-hoc scheme (2000), IISR, Calicut.

Krishna, K.V.S. (1968) Cultivation of cardamom-selection of plants suitable for seed purpose. *Cardamom News*, 2(9), 1–2.

Krishna, K.V.S. (1997) Cardamom Plantations in Papua New Guinea. *Spice India*, 10(7), 23–24.

Krishnamurthy, K., Khan, M.M., Avadhani, K.K., Venkatesh, J., Siddaramaiah, A.S., Chakravarthy, A.K. and Gurumurthy, B.R. (1989) *Three Decades of Cardamom Research at Regional Research Station, Mudigere (1958–1985)*, University of Agric. Sci., Bangalore.

Kumar, K.B., Prakash Kumar, P., Balachandran, S.M. and Iyer, R.D. (1985) Development of clonal plantlets from immature panicles of cardamom. *J. Plantation Crops*, 13(1), 31–34.

Kurup, K.R. (1978) Trickle irrigation. *Cardamom*, 10(6), 11–16.

Kuttappa, K.M. (1969) Cardamom – digging. *Cardamom News*, 2(7), 3–5.

Mayne, W.W. (1951) *Report on Cardamom Cultivation in South India*. Bulletin 50, ICAR Publication, New Delhi, pp. 1–53.

Mini Raj, N. and Murugan, M. (2000) Ecological decline of cardamom hills – an analysis. In *Spices and Aromatic Plants*, ISS, IISR, Calicut, pp. 172–176.

Mohanchandran, K. (1984) Planting for plantations – a study on cardamom. *Cardamom J.*, 17(11), 5–8.

Muralidharan, A. (1980) *Biomass Productivity. Plant Interactions and Economics of Inter-cropping in Arecanut.* Ph.D Thesis, University of Agricultural Sciences, Bangalore, India. p. 271.

Nair, K.N. (1989) *Ecology and Economics in Cardamom Development.* Centre for Developmental Studies, Thiruvananthapuram.

Nanjan, K., Muthuswami, S., Thangarajan, T. and Sundarajan (1981) Time of planting of cardamom at Shevaroy hills. *Cardamom*, 13(1), 13–15.

Parameshwar, N.S., Haralappa, H.S. and Mahesh Gowda, H.P. (1979) Cardamom yield can be increased by clonal propagation. *Current Research*, 8, 150–151.

Pattanshetty, H.V., Nusrath, R., Sulikeri, G.S. and Prasad, A.B.N. (1972) Effects of season-cum-depth of planting on the mortality of vegetatively propagated cardamom suckers (Rhizome sets of *Elettaria cardamomum* Maton). *Mysore J. Agril. Sci.*, 6(4), 413–420.

Pattanshetty, H.V. and Prasad, A.B.N. (1972) Early August is the most suitable time for planting cardamom in Mudigere area. *Current Res.*, 1(9), 60–61.

Pattanshetty, H.V. and Prasad, A.B.N. (1973) September is the most suitable month for sowing cardamom seeds. *Current Res.*, 2(5), 26.

Pattanshetty, H.V. and Prasad, A.B.N. (1974) Exposing the cardamom panicles from a layer of leaf mulch to open pollination by bees and thereby improving the fruit set. *Current Res.*, 3(8), 90.

Pattanshetty, H.V., Nusrath, R., Sulikeri, G.S. and Prasad, A.B.N. (1974) Suitable season and depth of planting cardamom suckers. *Current Res.*, 3(8), 84.

Pattanshetty, H.V., Rafeeq, M. and Prasad, A.B.N. (1978) Influence of the length and type of storage on the germination of cardamom seeds. In E.V. Nelliat (ed.) *Proceedings of the First Annual Symposium on Plantation Crops*, Indian Society for Plantation Crops, Central Plantation Crops Research Institute, Kasaragod, pp. 267–274.

Pillai, V.G. (1953) Using long rhizomes for cardamom planting. *Indian Farming*, 3(9), 12–13.

Ponnugangam, V.S. (1946) Management of cardamom seedlings in the nursery. *Planters Chronicle*, 41(6), 117–119.

Prasad, A.B.N., Pattanshetty, H.V. and Kololgi, S.D. (1994) Seed treatment to improve and hasten the germination of cardamom seeds sown during winter. *Cardamom News*, 6(8), 6.

Raghothama, K.G. (1979) *Effect of Mulches and Irrigation on the Production of Suckers in Cardamom* (*Elettaria cardamomum* Maton) M.Sc. (Agri.) Thesis, University of Agricultural Sciences, Bangalore.

Rai, S.N. (1978) Nursery and planting of some tropical evergreen and semievergreen species. *Tech. Bull.*, Karnataka Forest Dept., p. 49.

Raju, B. (1981) *Effects of Mulches and Irrigation on the Growth and Initial Yield of Cardamom* (*Elettaria cardamomum* Maton). M.Sc. (Agri.) Thesis, University of Agricultural Sciences, Bangalore.

Ratnam, B.P. and Korikanthimath, V.S. (1985) Frequency and probability of dry spells at Mercara. *Geobios*, (12), 224–227.

Saleem, C.P. (1978) Sprinkler irrigation doubles cardamom yield. *Cardamom*, 10(6), 17–20.

Shankar, B.D. (1980) Shade trees and their role in cardamom cultivation. *Cardamom*, 12(3), 3–11.

Siddaramaiah (1967) Hints on raising of cardamom nursery in Mysore state. *Cardamom News*, 1(6), 4.

Sivanappan, P.K. (1985) Soil conservation and water management for cardamom. *Cardamom*, 18(6), 3–8.

Srinivasan, K., Sudarshan, M.R., Nair, C.K. and Naidu, R. (1992) Study on the feasibility of cardamom. *J. Plantation Crops*, 20(suppl.), 53–54.

Subbaiah, M.S. (1940) Cardamom cultivation in the Bodi hills. *Madras Agricultural J.*, 28(10), 279–288.

Subbarao, G. and Korikanthimath, V.S. (1983) The influence of rain fall on the yield of cardamom (*Elettaria cardamomum* Maton) in Coorg district. *J. Plantation Crops*, 11(1), 68–69.

Sulikeri, G.S. and Kololgi, S.D. (1978) *Phyllanthes emblica* L. leaves – a suitable mulch for cardamom nursery beds. *Current Res.*, 7(1), 3–4.

Sulikeri, G.S., Pattanshetty, H.V. and Kololgi, S.D. (1978) Influence of the level of water table on the growth and development of cardamom (*Elettaria cardamomum* Maton). In *Proceedings of the First Annual Symposium on Plantation Crops*, Kottayam, 216–219. In E.V. Nelliat (ed.) Indian Society for Plantation Crops, CPCRI, Kasaragod p. 531.

Sulikeri, G.S. (1986) *Effects of Light Intensity and Soil Moisture Levels on Growth and Yield of Cardamom (Elettaria cardamomum* Maton). Ph.D. (Hort.) Thesis, University of Agricultural Sciences, Bangalore.

Vasanthkumar, K. and Sheela, V.C. (1970) Sprinkler irrigation in cardamom plantations. *Cardamom*, 23(3), 5–9.

Zachariah, P.K. (1976) Mulching in cardamom plantations. *Cardamom News*, 8(1), 3–9.

5   Nutrition of cardamom

V. Krishnakumar and S.N. Potty

1   INTRODUCTION

Conventionally, in India, cardamom was grown as an undergrowth in dense evergreen forests of Western Ghats of South India, without tillage or nutrient application. Later with the development of intensive agriculture, tilling of soil became a routine practice, and gradually soil got depleted, especially because of cutting of forest trees to reduce shade. Besides, due to heavy rainfall in these areas and undulating topography, soil erosion and leaching of nutrients become inevitable leading to reduction in native soil fertility. Continuous cropping in the same area also leads to rapid depletion of nutrients leading to poor growth and yield. These factors necessitate the application of balanced doses of nutrients for realizing sustainable crop production in cardamom plantations. Judicious agro-management practices and use of high-yielding varieties are important factors influencing productivity vis-à-vis unit cost of cultivation of any crop. Among various management techniques, nutritional management is of great importance even though it is not by itself sufficient enough for improving productivity of a crop. Maintaining soil fertility status at optimum levels should be one of the prime concerns of any cardamom planter.

2   CARDAMOM-GROWING SOILS

Soil functions as a medium for plant growth by way of supplying essential nutrients and water apart from support. The availability of plant nutrients at the appropriate time in soil plays a significant role in influencing crop production and productivity of the land. It is, therefore, very important that all these nutrients are supplied to plants in sufficient quantities at appropriate stages of plant growth.

In India, soils of major cardamom-growing areas come under the order Alfisols, formed under alternate wet and dry conditions, and the suborder ustalfs derived from schists, granite and gneiss and are lateritic in nature (Sadanandan *et al.*, 1990). Soils most favourable for growth and development of cardamom are red lateritic loam with layers of organic debris present in evergreen forests, although it grows on a variety of soils with only a shallow zone of humus accumulation. In general cardamom-growing soils are fairly deep having good drainage. The clay fraction is predominantly kaolinitic and hence there is some fixation of potassium in these soils. The cardamom-growing soils of Karnataka are mostly clay loam (Kulkarni *et al.*, 1971).

In Guatemala, cardamom growing areas have rich forest soils. Towards the northern region, it is grown in the newly cleared forestlands, the soil having dolomitic limestones, underlined with typical tropical clay. In the southern regions, soil is sandy clay loam with volcanic ash deposits. Soils in the south are more fertile than those of the north because of the presence of volcanic ash (George, 1990).

2.1   Soil pH

Analysis of soil samples from different cardamom-growing areas of Kerala have shown that majority of them are acidic, the pH being in the range of 5–5.5 (Zachariah, 1975). Nair et al. (1975) reported that pH of soils in three important cardamom-growing districts in Kerala viz., Idukki, Wynad and Palghat, ranged from 4.7 to 6.15, 4.75 to 5.2 and 5.5 to 5.8 respectively and there was a slight variation in the pH with depth of soil, the surface layer having higher pH in most of the cases. Ranganathan and Natesan (1985) reported that pH of cardamom-growing soils of Karnataka was higher (5.8–6.5) than that of Kerala soils. Soils of cardamom-growing areas of Tamil Nadu have pH ranging from 4.7 to 7 (Vadiraj et al., 1998).

2.2   Cation exchange capacity

The major contribution to soil cation exchange capacity (CEC) comes from the high organic matter content of soils. Sadanandan et al. (1990) noticed that CEC of soils varied from 8.6 to 58.5 c mol (p+)/kg. The CEC of Coorg soils in Karnataka is higher than that of Idukki soils in Kerala. In general, with increase in altitude the CEC values also increase. They attributed the reason to the active form of humus present in high proportion to total organic matter. It was also observed that CEC was positively correlated with soil organic matter.

2.3   Organic carbon

The organic carbon content of soil is found to be low at lower altitudes and increases gradually with increase in elevation. Kulkarni et al. (1971) reported a mean organic carbon content of 5.9 per cent in various cardamom-growing soils of Karnataka. Nair et al. (1978), based on extensive studies, reported the organic carbon content to be 3.3 per cent, 3.6 per cent and 4.6 per cent in top soils of Palghat, Idukki and Wynad districts of Kerala, respectively. They also observed that organic carbon content decreased with depth of soils. Srinivasan (1984) found a significantly positive correlation between organic carbon and total as well as available nitrogen. Rate of decomposition of organic matter in cardamom fields is much slower than that of other fields cultivating plantation crops such as tea at the same elevation because of forest tree association and consequent lower mean annual temperature (Ranganathan and Natesan, 1985).

2.4   Soil phosphorus

Soil test results of cardamom-growing areas have indicated that majority of soils are low to medium in available phosphorus (less than 5–12.5 kg/ha). The percentage of samples falling in this category was 83, 68 and 63 for Karnataka, Kerala and Tamil Nadu, respectively (Anonymous, 1998). Cardamom soils also contain large quantities of iron and aluminum oxides and they strongly fix soluble phosphates into insoluble ones. From laboratory studies, Srinivasan and Mary (1981) found that the overall fixation of phosphorus in acid soils of cardamom ranged from 55 per cent at 1 h to 88 per cent after 90 days of application in the lower level (25 kg P_2O_5/ha) and from 43 to 85 per cent in the higher level (50 kg P_2O_5/ha) of application. However, under field conditions, the thick mulch on the ground may help to prevent immediate fixation of soluble phosphates and the high organic matter content facilitates solubilization of phosphates (Ranganathan and Natesan, 1985). Availability of phosphorus was found to decrease sharply with soil depth (Nair *et al.*, 1978; Srinivasan *et al.*, 1986).

2.5   Soil potassium

Available potassium in cardamom-cultivating soil is high in majority of cases studied; the percentage of soils rated as high being 78 in Kerala, 71 in Tamil Nadu and 53 in Karnataka (Anonymous, 1998). Although the available potassium appears to be high as judged by the traditional soil test methods, Ranganathan and Natesan (1985) opined that it is intrinsically low as the soil tests estimate only the exchangeable form of potassium which is highly mobile and it could either be taken up by the plants or lost by leaching. Though these researchers have reported that there is no fixation of potassium in cardamom soils due to the predominance of kaolinitic clay fraction, studies by Srinivasan (1990) showed that the potassium-fixing capacity of cardamom soils ranged from 16.9 to 32.1 per cent for soils of Kerala region, from 11.9 to 21.3 per cent for soils of Karnataka and from 19.0 to 23.3 per cent for soils of Tamil Nadu region. Availability of potassium also decreased with depth (Nair *et al.*, 1978; Srinivasan *et al.*, 1986). Nair *et al.* (1997) are of opinion that K recommendations based on NH_4OAC extractable K are unreliable, as it does not indicate the real K status of the soil. They suggested that determination of K-buffer power of the soils and its integration into routine soil test K data would result in better predictability of soil K availability. They found that capsule yield of cardamom is related to the K-buffer power of the soils and not NH_4OAC extractable K.

On an average the cardamom soils from Coorg in Karnataka contain much less NH_4OAC and HNO_3 extractable K compared to soils of Idukki District in Kerala, though the other soil properties are comparable. Despite the lower K, the Coorg soils produced nearly twice as much cardamom yield compared to Idukki soils. This is probably because in soils having low K buffering, the K concentration at the root surface may decline rapidly, while in a soil with high K buffering capacity this does not happen. Thus actual K uptake by roots depends on both K concentration and K-buffering power. It is suggested that a reorientation of K fertilization schedule based on K-buffer power of soils might lead to much higher yield realization than what is now realized based on the present recommendation.

2.6   Secondary and micronutrients

Srinivasan *et al.* (2000) studied the sulphur status of cardamom-growing soils of South India. Results of analysis of 100 soil samples collected from Kerala (60), Karnataka (28) and Tamil Nadu (12) revealed that sulphur content ranged from traces to 36 ppm in Kerala, traces to 27.5 ppm in Karnataka and 15–36 ppm in Tamil Nadu. By taking a level of 10 ppm as the critical limit, it was observed that

43.3 per cent of Kerala samples and 50 per cent of Karnataka samples showed deficiency of this nutrient. However, in Tamil Nadu soils there is no deficiency for this nutrient.

Srinivasan *et al.* (1993b), who investigated the micronutrient status of cardamom-growing soils of South India, indicated that the status of available iron ranged from 14.6 to 65.8 ppm, manganese from 1.3 to 44.8 ppm, copper from 0.66 to 32.2 ppm, zinc from 0.01 to 2.71 ppm, boron from 0.05 to 3.70 ppm and molybdenum from 0.01 to 11.10 ppm. Based on critical limits prescribed for various micronutrients, it was observed that 68 per cent of areas sampled were deficient in available zinc, 49 per cent in available boron, 28 per cent in available molybdenum and 9 per cent in available manganese. None of the soils recorded deficiency of available iron and copper. There was no deficiency for available manganese in Karnataka and for available molybdenum in Tamil Nadu. They also noticed a significantly negative correlation between available iron content and pH, whereas, the correlation was positive between available zinc as well as copper. Among the micronutrients, only available manganese showed a positive significant correlation with organic carbon.

3   NUTRIENT DEFICIENCY SYMPTOMS

Deshpande and Kulkarni (1973) used sand culture studies to find out deficiency symptoms of various plant nutrients on cardamom seedlings. Deficiency symptoms described by these workers are given below.

3.1   Major nutrients

Deficiency symptoms for nitrogen first appeared in older leaves mainly in the form of reduction in leaf size. There was also reduction in sucker production, and newly formed suckers were found to dry up after some time. In the case of phosphorus, symptoms appeared after 4–5 months of experimentation. Small patches of purplish spots appeared on the leaves followed by premature dropping of older leaves. Stunting and reduction in number of suckers were also observed. Deficiency symptoms for potassium appeared on older leaves. There was reduction in growth of shoots and roots and plants showed browning of leaf tips, which extended downward. Later, the whole leaf turned dark brown in colour. Further sucker production was completely absent and the plants died within 2 weeks after deficiency symptoms were first appeared.

3.2   Secondary nutrients

In case of calcium, deficiency symptoms first appeared on young leaves after 75 days. There was reduction in growth of shoots and roots and further growth of aerial shoot ceased. There was thickening of aerial stem giving a bulb-like appearance. Scattered yellow spots appeared on leaves and margins that turned brown with golden yellow band underneath. Reduction in inter-nodal length was seen in case of magnesium deficiency and plants showed broom-like appearance. Twisting of top leaves was observed with tip drying. Later, the whole leaf became pale yellow leaving the mid rib green. White papery spots appeared on leaf lamina, which was the most commonly noticed deficiency symptom in the nursery. Inhibition of sucker production was also seen. Sulphur deficiency appeared on young leaves and the growing leaf became whitish in colour followed by death of leaves starting from margins.

3.3   Micronutrients

The symptoms of zinc deficiency in cardamom plantations are poor growth, narrow curled leaves, interveinal chlorosis and sometimes rosetting (Anonymous, 1979). Boron deficiency expressed on seedlings as reduction in leaf size and cracking of leaf lamina.

4   SHADE TREES AND SOIL FERTILITY

Cardamom is cultivated under shade trees and the soils in general have high fertility status due to addition of leaf litter and recycling (Zachariah, 1978). Nair *et al.* (1980) observed that shade trees bring to the soil surface the nutrients they take up from lower layers by way of leaf shed and thus they help to maintain a high fertility status and soil pH in surface soils. Even though on an average about 5.5 t/ha of organic material as leaf litter, weeds and pruned plant parts are recycled in a cardamom plantation in a year, the nutrients are mainly in the organic form and are available to the crop only by the process of mineralization. As the rate of mineralization is always low, the nutrients that become available to the crop will be able to sustain only average growth with average production. (Srinivasan *et al.*, 1993a) (See Chapter 4 also).

5   PLANT NUTRIENT COMPOSITION AND UPTAKE

5.1   Tissue composition

The tissue composition of cardamom plant as a whole and after partitioning into different parts has been studied and well documented by various workers (Ratnavele, 1968; Srivastava *et al.*, 1968; Kulkarni *et al.*, 1971; Raghothama, 1979; Venkatesh, 1980; Pattanshetty, 1980; Sulikeri, 1986). Among the second and fifth leaf from tip, the content of phosphorus and potassium were more in second leaf than the others, whereas, the reverse was the case with calcium (Ratnavele, 1968). The aerial stem contained more potassium than the leaves. Kulkarni *et al.* (1971), from analysis of different plant parts at various stages of growth of 10 year-old plants, reported that while nitrogen, phosphorus and calcium contents of leaves increased from young to mature stages, there were generally reduction in potassium and magnesium contents as the plants grow. At bearing stage, while major nutrients decreased in the leaf tissue, an increase was observed in the case of calcium and magnesium. The same trend was noticed in aerial stem for various nutrients with increase in age of plants. Rhizomes contain lower levels of major nutrients than roots, whereas, roots contained more of secondary nutrients. In general the content of nitrogen was highest in leaves followed by shoots, rhizomes and roots (Raghothama, 1979; Venkatesh, 1980; Pattanshetty, 1980), while, potassium content was highest in shoots followed by rhizome and leaves. However, no definite trend could be observed with respect to phosphorus from these studies. Sulikeri (1986) reported highest nitrogen content in leaves followed by rhizome, shoot and roots during the pre-bearing stage. By harvest period, the nutrient content increased in all plant parts except in the case of leaves that showed a decrease. Phosphorus content did not vary much in different parts during the pre-bearing period, whereas, at harvest, capsules contained more of this nutrient than other plant parts. Highest content of potassium was seen in rhizomes followed by shoots, leaves and roots during the pre-bearing period. At harvest stage also the same trend was noticed except the fact that capsules and panicles contained more potassium than leaves and roots. At harvest, potassium content decreased in all the plant parts compared to the pre-bearing stage except in rhizome where its concentration increased. Content of calcium was highest in leaves at pre-bearing stage and at harvest stage capsules recorded the highest values. The content of magnesium was the highest in leaves and the lowest in shoots for both the periods of observation.

5.2   Nutrient uptake

The uptake studies by Kulkarni *et al.* (1971) revealed that for production of 1 kg of cardamom capsules, 0.122 kg N, 0.014 kg P and 0.2 kg K are removed by the plant. For the cardamom plant as a whole, the ratio of N, P, K, Ca and Mg in which these nutrients are present was worked out as 6:1:12:3:0.8. A similar ratio was also arrived at by Venkatesh (1980). Highest uptake of N (53 per cent) and Ca (47.5 per cent) was shown by leaves followed by shoots, whereas, in case of P, K and Mg the reverse was the order observed. Uptake by rhizomes and roots followed in the descending order (Pattanshetty, 1980). Among the major nutrients, uptake of N by leaves was higher than P and K during pre-bearing stage compared to the harvest stage. However, there was considerable reduction in N uptake by leaves at harvest stage than at pre-bearing stage, whereas, N uptake by rhizomes and shoots showed an upward trend at harvest stage. In case of P and K, the highest uptake was at harvest stage by shoots and there was reduction in the uptake of N by leaves. Rhizomes occupied the second position in uptake of K at harvest stage indicating the importance of this nutrient in crop production. Among secondary nutrients, the highest uptake of Ca was seen in shoots at both the stages, whereas, for Mg it was the leaves which showed highest values. Uptake of Ca by shoots at harvest stage was more than that of pre-bearing stage; on the other hand the reverse was the trend in case of Mg (Sulikeri, 1986).

The nutrient requirement of cardamom shows the need for liberal application of K, N and P in that order. Importance of K in the nutrition of cardamom is also highlighted from studies of various workers that soils giving high yields in Karnataka (Kulkarni *et al.*, 1971) and Kerala (Vadiraj *et al.*, 1998) have high fertility ratings for potassium.

6   FERTILIZER REQUIREMENTS

6.1   Fertilizer scheduling

As stated earlier, cardamom is being grown in the rich fertile soils of forest eco-system. Up to the middle of fifties, it was cultivated without any manuring or limited to organic manures, if at all applied. As importance of this spice crop was increasingly felt in the national and international market scenario, its cultivation was taken up in a more systematic and scientific manner and many of the planters started applying chemical fertilizers. de Geus (1973) suggested a fertilizer dose of 45:45:45 kg N, P_2O_5 and K_2O/ha for Kerala, 67:34:100 kg N, P_2O_5 and K_2O/ha for Karnataka with half N in organic form and 45:34:45 kg N, P_2O_5 and K_2O/ha for Tamil Nadu. However, considering the low requirement of cardamom and the high status of N and K in cardamom-growing soils, a maintenance dose of 30:60:30 kg N, P_2O_5 and K_2O/ha was recommended for healthy and vigorous growth of plants (Zachariah, 1978). Based on further studies and the factors affecting the availability of nutrients in the soil, a fertilizer dose of 75:75:150 kg N, P_2O_5 and K_2O/ha was recommended under rainfed situation for a normal crop of 100 kg dry capsules/ha. If the yield is more, the fertilizer doses are to be increased proportionately. Additional fertilizer dose of 0.65 kg N, P_2O_5 and 1.30 kg K_2O/ha is to be applied for increase in yield of every 2.5 kg dry capsules over normal yield (Anonymous, 1976; Kologi, 1977). Results of recent studies (Srinivasan *et al.*, 1998) indicated that significant yield increase and savings in fertilizer requirements would be possible by way of soil-cum-foliar application method. In Karnataka, N, P_2O_5 and K_2O @37.5:37.5:75 kg/ha through soil and 2.5 per cent urea + 0.75 per cent single super phosphate + 1.0 per cent muriate of potash through foliage gave 64 kg yield increase/ha over no fertilizer application. The yield increase recorded was 43 kg/ha under Tamil Nadu conditions, where the dosages required were 20:40:20 kg N, P_2O_5 and K_2O/ha through soil and 3 per cent urea + 1 per cent single super phosphate + 2 per cent muriate of potash through foliage. In irrigated plantations, the general fertilizer recommendation is to apply N, P_2O_5 and K_2O at the rate of 125:125:250 kg/ha to soil in three splits (Anonymous, 1997). Kumar *et al.* (2000) also suggested application of NPK @ 75:75:150 kg/ha under Karnataka conditions for increasing yield of the crop. Urea was found to be the best source for N compared to ammonium sulphate (Deshpande *et al.*, 1971). For cardamom-growing soils, being rich in organic matter and acidic in nature, mussoriephos was found to be the ideal fertilizer source for P (Nair and Zachariah, 1975). When plant population is increased to 5000/ha under trench method of planting, fertilizers are to be applied at the rate of 120:120:240 kg N, P_2O_5 and K_2O/ha (Korikanthimath, 1986), which was subsequently revised as 150:75:300 kg N, P_2O_5 and K_2O/ha (Korikanthimath *et al.*, 1998a). Under low rainfall situations prevailing in the Lower Pulney tracts of Tamil Nadu, a fertilizer dose of 40:80:40 kg N, P_2O_5 and K_2O/ha is recommended (Natarajan and Srinivasan, 1989). Later on Srinivasan and Bidappa (1990) worked out nutrient models for P and K based on adsorption isotherms. The total phosphorus requirement (TPR) of soil was arrived at as TPR (kg/ha) $= (60 - x) \times 3.24$ where 60 is the desired soil fertility level for P, x is the soil test value for P and 3.24 is a constant derived from the model taking into account of bulk density, weight of soil/ha etc. The Total Potassium Requirement worked out was TPR (kg/ha) $= (300 - x) \times 2.64$, where 300 is the desired soil fertility level for K, x is the soil test value for K and 2.64 is the constant. The models aim at maintaining soil P and K levels at 60 and 300 kg/ha as these levels are considered to be optimum for cardamom. Ranganathan and Natesan (1985) reported beneficial effect of application of zinc sulphate to nursery plants by incorporating it along with fertilizer mixture. Sivadasan *et al.* (1991) found that application of 500 ppm zinc sulphate to yielding plants on foliage not only enhanced growth and yield but also quality of produce. Deshpande *et al.* (1971) found the beneficial effect of liming of cardamom soils in correcting acidity and enhancing the rate of nitrification resulting in better growth. However, subsequent observations (Nair *et al.*, 1988; Anonymous, 1997) indicated that liming is not required as a routine practice in cardamom plantations. Fertilizer schedule is given in Table 5.1.

When application of fertilizer through soil is resorted to, it would be necessary to give one-third of the recommended dose (75:75:150 kg/ha of NPK) during the first year of plant growth both under rainfed and irrigated conditions. During the second year of plant growth, the dose may be increased to one-half of the recommended dose (75:75:150 and 125:125:250 kg NPK/ha for rainfed and irrigated conditions respectively) and fertilizers at full dose may be applied from third year of plant growth onwards (Anonymous, 1993).

All the above fertilizer recommendations are based on the conventional analytical techniques and not based on more recent concepts such as the nutrient buffer power. Application of this concept to K requirements led Nair *et al.* (1997) to suggest that a thorough reorientation of the nutrient schedule in cardamom is essential.

Organic manures are considered essential for improving physical characteristics of soil, apart from their nutritional values, and they are indispensable for cardamom irrespective of whether fertilizers are applied or not. Application of organic manures such as neem cake (1–2 kg/plant) or farmyard manure or cow dung compost at the rate of 5 kg/plant may be made once in a year in May–June along with mussoriephos and muriate of potash (Anonymous, 1997). Thimmarayappa *et al.* (2000) suggested an integrated nutrient management to meet the 25 per cent of requirement of nitrogen

Table 5.1 Fertilizer schedule for cardamom

Region	Soil application	Soil-cum-foliar application	Time of application	
			Soil	Foliar
Karnataka	NPK 75:75: 150 kg/ha	NPK 37.5:37.5: 75 kg/ha	May–June, August–September	September, November, January
		Urea (2.5%) Single super phosphate (0.75%) Muriate of potash (1.0%)		
Kerala	NPK 75:75: 150 kg/ha	NPK 37.5: 37.5:75 kg/ha	May–June, August–September	September, November, January
		Urea (2.5%) Single super phosphate (0.75%) Muriate of potash (1.0%)		
Tamil Nadu	NPK 40:80: 40 kg/ha	NPK 20:40: 20 kg/ha Urea (3.0%) Single super phosphate (1.0%) Muriate of potash (2.0%)	September, November	June, August, November–December

Note
Instead of straight fertilizers, 2 per cent each of di-ammonium phosphate and muriate of potash can also be used.

through farmyard manure and the balance 75 per cent through inorganic nitrogen source for sustained production of cardamom over a period of time.

6.2 DRIS norms

Sadanandan *et al.* (2000) made studies on foliar nutrient diagnostic norms for optimizing cardamom production. They analyzed different nutrients in the youngest mature leaf (fifth pair) from terminal of newly emerged panicle-bearing tillers along with crop yield and developed the DRIS (Diagnostic Recommendation Integration System) norms. Leaf nutrient composition was categorized into deficient, low, optimum and high. As per the norms developed, leaf having 1.26–2.81 per cent N, 0.1–0.2 per cent P, 1.1–3.4 per cent K, 0.51–1.38 per cent Ca, 0.18–0.31 per cent Mg, 135–370 ppm Fe, 261–480 ppm Mn, 20–45 ppm Zn, 10–46 ppm Cu and 0.28–0.84 per cent Mo is considered optimum for producing cardamom yield ranging from 131 to 625 kg/ha. From the study, it was revealed that for getting a high yield level of 378 kg/ha, the indicator leaf of cardamom should have a mean N, P, K, Ca and Mg status of 2.04, 0.15, 2.25, 0.95 and 0.25 per cent respectively. The optimum status for micronutrients for realizing high yield is 253 ppm Fe, 371 ppm Mn, 33 ppm Zn, 28 ppm Cu and 0.56 ppm Mo. They found that the requirement of K is the highest (2.25 per cent) among all the nutrients for producing better yield. The diagnostic chart denoting the status viz., deficient, low, optimum, high and excessive for major leaf nutrients developed by the workers is given in Fig. 5.1.

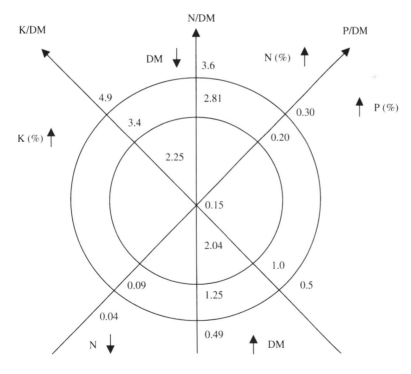

DRIS norms chart for cardamom (leaf).

Figure 5.1 Foliar diagnostic norms for optimizing cardamom production.

6.3   Method and time of application

Root spread of cardamom, which is the feeding zone, should be taken into consideration while fertilizer application is made. Lateral spread of roots of a full bearing 8-year-old plant was found to have 80 per cent of its roots in a zone of 25 cm radius, 14 per cent in a zone of 25–50 cm and only 6 per cent of them are seen in 50–75 cm zone. Vertically cardamom roots penetrates only up to 40 cm (Khader and Sayed, 1977). Further studies conducted at the Indian Cardamom Research Institute (Nair, 1988) also showed that roots of cardamom are confined to a shallow depth. Nearly 70 per cent of roots were seen in the top 5 cm depth. Horizontally only 10 per cent of the roots forage an area 120 cm away from the clump. Therefore for the maximum efficiency of applied fertilizers it would be necessary to apply them at a radius within 50 cm and being a surface feeder, deep placement of fertilizers is not advisable. Before application of fertilizers, if the panicles are spreading on the ground, they are to be kept coiled encircling the base of the plant and the mulch removed. Fertilizers may be applied around the plant base in a circular band of width 15 cm, leaving about 30 cm from plant base and incorporated in the soil by using a hand fork (Fig. 5.2a,b). Mulching should be followed immediately after incorporating fertilizers. Panicles may then be released and spread on the ground for *Malabar* types to facilitate honeybee movement for better pollination and setting of the capsules.

Cardamom growth is influenced by seasonal conditions, especially rainfall pattern. Vegetative buds emerge from the bases of tillers almost throughout the year. However, majority of vegetative buds are produced after the rainy period (Sudharshan *et al.*, 1988). It was also observed that the linear growth of tiller increases with the onset of southwest monsoon and growth rate slows down with cessation of rain. Peak flowering and fruit set period coincide in cardamom and nearly 70–90 per cent of flower production was recorded between May and August. Hence for efficient utilization of fertilizers, time of application is very important. Applications of fertilizers in May and later in September are found to be the best (Pattanshetty and Nusrath, 1973). However, under irrigated conditions, tiller initiation and panicle initiation are continuous processes and hence more split applications are beneficial.

Cardamom is cultivated under shade tress, and hence complete utilization of applied fertilizers by cardamom may not take place because of root interference. For yielding cardamom plants, soil-cum-foliar application will be an effective method. Soil application of fertilizers in two rounds during May–June and August–September and subsequent applications through foliage during September, November and January are recommended (Srinivasan *et al.*, 1998).

George (1990) reported that in Guatemala, almost all growers fertilize cardamom. There are different fertilizer mixtures available there. One such fertilizer mixture is 15:15:15, which is marketed under brand name Barco Vikingo. Another mixture which contains calcium and magnesium in addition to NPK is 18:6:12:4:0.2 with the brand name of Agrovet SA. Fertilizer application starts from the second year of planting with 2 oz per plant. This is gradually increased to 3 oz in the third year and 4 ounces from the fourth year onwards. Some planters add 1 oz of urea over and above this normal fertilizer application. The recommended dose is applied 2–4 times a year (George, 1990).

When cardamom is cultivated with controlled, artificial shade, a fertilizer dose of 100:25:100 kg NPK/ha is recommended (Korikanthimath *et al.*, 1998b).

(a)

(b)

Figure 5.2 Fertilizer application as practiced in the major cardamom growing tracts of Idukki district of Kerala: (a) fertilizer mixture is applied around the clump about 30 cm away from the base; (b) fertilizer is incorporated into the soil by hand forking.

7 CONCLUSION

Though cardamom is a perennial crop, its growth behaviour resembles more to a biennial crop in the sense that vegetative phase (tillers) emerging in one year turns into reproductive phase during the second year and produce panicles, flowers and capsules. Cardamom, being cultivated as an undergrowth with shade trees, competition for inputs among them makes nutritional management an important practice in realizing optimal yield of the crop. Escalation in cost of fertilizers makes necessary their use in an efficient and economic manner. Even though there exists greater demand for different spices produced through organic cultivation practices, the demand for organic cardamom is not much at present. However, production of cardamom through low input sustainable agriculture incorporating integrated nutrient management system involving use of various kinds of organic manures and bio-fertilizers should be aimed at in the present day context of preservation of natural ecosystem and environmental protection.

REFERENCES

Anonymous (1976) *Cardamom in Karnataka.* University of Agricultural Sciences, Bangalore, pp. 8–10.

Anonymous (1979) *Cardamom Culture and Package of Practices.* Pamphlet No. 1/79. Cardamom Board, Cochin, p. 24.

Anonymous (1993) *Nutritional Management for Cardamom – Technical Guide for Cardamom Planters.* Spices Board, Cochin.

Anonymous (1997) *Cardamom – Package of Practices.* Spices Board, Kochi, p. 19.

Anonymous (1998) Final Report of ICAR ad-hoc scheme "Evaluation of crop response to application of micronutrients to small cardamom (*Elettaria cardamomum* Maton)", p. 10.

Deshpande, R.S., Kulkarni, S.D., Viswanath, S. and Suryanarayana Reddy, B.G. (1971) Influence of lime and nitrogenous fertilizers on soil biology. *Mysore J. Agric. Sci.*, 5(1), 77–81.

Deshpande, R.S. and Kulkarni, D.S. (1973) Deficiency symptoms in cardamom (*Elettaria cardamomum*). *Mysore J. Agric. Sci.*, 7, 246–249.

de Gues, J.G. (1973) *Fertilizer Guide for the Tropics and Subtropics*, (2nd edn), Zurich Centre d'Etude de l'Azote.

George, C.K. (1990) *Report of the Visit of the Delegation for Mutual Understanding in Cardamom Trade with Guatemala*, 1990, p. 12.

Khader, K.B.A. and Sayed, A.A.M. (1977) Fertilizing cardamom – its importance. *Cardamom*, 9(1–2), 13–14.

Kologi, S.D. (1977) Cardamom Research at Mudigere. In *Towards Higher Yield in Cardamom*, CPCRI, Kasaragod, pp. 28–29.

Korikanthimath, V.S. (1986) Systems of planting-cum-fertilizer levels in cardamom under rain-fed condition. *National Research Centre for Spices. Annual Report 1986*, pp. 30–31.

Korikanthimath, V.S., Hegde, R., Mulge, R. and Hosmani, M.M. (1998a) Growth and yield parameters of cardamom (*Elettaria cardamomum* Maton) as influenced by nutrition and planting density. *J. Spices and Aromatic Crops*, 7, 39–42.

Korikanthimath, V.S., Hegde, R. and Hosmani, M.M. (1998b) Influence of yield and yield parameters of cardamom grown under controlled shade. *J. Med. Aromatic Plant Sci.*, 20, 700–702.

Kulkarni, D.S., Kulkarni, S.V., Suryanarayana Reddy, B.G. and Pattanshetty, H.V. (1971) Nutrient uptake by cardamom (*Elettaria cardamomum* Maton). *Proc. Int. Symp. Soil Fert. Evaluation*, New Delhi, Vol. 1, pp. 293–296.

Kumar, M.D., Santhaveerabhadraiah, S.M. and Ravishankar, C.R. (2000) Effect of fertilizer levels on the yield of small cardamom under natural shade. In *Spices and Aromatic Plants*, Indian Soc. Spices, IISR, Calicut, pp. 179–180.

Nair, C.K. and Zachariah, P.K. (1975) Suitability of Mussoriephos as a phosphatic fertilizer for cardamom. *Cardamom News*, 7(6), 23–24.

Nair, C.K., Srinivasan, K. and Zachariah, P.K. (1978) Distribution of major nutrients in the different layers of cardamom soils. *Proc. 1st Annual Symp. Plantation Crops*, pp. 148–156.

Nair, C.K., Srinivasan, K. and Zachariah, P.K. (1980) Soil reaction in cardamom growing soils in relations to the base status. *Proc. Seminar on Diseases of Plantation Crops and Manuring of Plantation Crops.* Tamil Nadu Agricultural University, Madurai, pp. 50–53.

Nair, C.K. (1988) Phosphatic fertilizers for small cardamom. *Proc. Seminar on the use of Rock Phosphate in West Coast Soils*. UAS, Bangalore and PPCL, p. 79.

Nair, C.K, Srinivasan, K. and Sivadasan, C.R. (1988) Soil acidity in cardamom plantations. *Spice India*. 1(6), 40–41.

Nair, K.P.P., Sadanandan, A.K., Hamza, S. and Abraham, J. (1997) The importance of Potasium buffer power in the growth and yield of cardamom. *J. Plant Nutrition*, 20(7–8), 987–997.

Natarajan, P. and Srinivasan, K. (1989) Effect of varying levels of N, P and K on yield attributes and yield of cardamom (*Elettaria cardamomum* Maton). *South Indian Horticulture*, 37, 97–100.

Pattanshetty, H.V. and Nusrath, R. (1973) May and September are most optimum time for fertilizer application to cardamom. *Current Res.*, 2(7), 47–48.

Pattanshetty, H.V. (1980) *Selections and Cloning for High Productivity in Cardamom (Elettartia cardamomum (L) Maton var.Minor.Watt.)* Ph.D (Hort.) Thesis, *Uni. Agri. Sci.*, Bangalore.

Raghothama, K.G. (1979) *Effect of Mulches and Irrigation on the Production of Suckers in Cardamom (Elettaria cardamomum Maton)*. M.Sc (Agri.) Thesis, *Uni. Agri. Sci.*, Bangalore.

Ranganathan, V. and Natesan, S. (1985) Cardamom soils and manuring. *Planters Chronicle*. 80(7), 233–236.

Ratnavele, M.V.S. (1968) Value of soil and plant analysis. *Cardamom News*, 2(6), 1–4.

Sadanandan, A.K., Korikanthimath, V.S. and Hamza, S. (1990) Potassium in soils of cardamom (*Elettaria cardamomum* Maton) plantations. *Seminar on Potassium in Plantation Crops. Uni. Agri. Sci.*, Bangalore.

Sadanandan, A.K., Hamza, S. and Srinivasan, V. (2000) Foliar nutrient diagnostic norms for optimising cardamom production. In *Spices and Aromatic Plants*, Indian Soc. for Spices, IISR, Calicut, pp. 101–104.

Sivadasan, C.R., Nair, C.K., Srinivasan, K. and Mathews, K. (1991) Zinc requirement of cardamom (*Elettaria cardamomum* Maton). *J. Plantation Crops*, 18(suppl.), 171–173.

Srinivasan, K. (1984) Organic carbon, total and available nitrogen in relation to cardamom soils *J. Institution of Chemists (India)*, 56, 15–16.

Srinivasan, K. (1990) Potassium status and Potassium fixing capacity of cardamom soils *J. Institution of Chemists (India)*, 62, 245–246.

Srinivasan, K. and Mary, M.V. (1981) Studies on fixation of Phosphorous in acid soils of cardamom. *J. Institution of Chemists (India)*, 53, 145–146.

Srinivasan, K., Roy, A. K., Sankaranarayanan, S. and Zachariah, P.K. (1986) Radial distribution of available nutrients in cardamom soils. *J. Institution of Chemists (India)*, 58, 61.

Srinivasan, K. and Bidappa, C.C. (1990) Evaluation of nutrient requirement of cardamom by desorption isotherm. *J. Indian Soci. Soil Sci.*, 38, 166–168.

Srinivasan, K., Rama Rao, K.V.V. and Naidu, R. (1993a) Organic matter addition and its nutrient contribution in cardamom plantations. *J. Tropical Agri.*, 31, 119–121.

Srinivasan, K., Vadiraj, B.A., Krishnakumar, V. and Naidu, R. (1993b) Micronutrient status of cardamom growing soils of South India. *J. Plantation Crops*, 21(suppl.), 67–74.

Srinivasan, K., Vadiraj, B.A., Krishnakumar, V. and Potty, S.N. (2000) Available sulphur status of cardamom growing soils of South India (in press).

Srinivasan, K., Krishnakumar, V. and Potty, S.N. (1998) Evaluation of fertilizer application methods on growth and yield of small cardamom (*Elettaria cardamomum* Maton). In N. Muralidharan and R. Rajkumar (eds) *Recent Advances in Plantation Crops Research*. Allied Pub., New Delhi, pp. 199–202.

Srivastava, K.C., Bopaiah, M.G. and Ganapathy, M.M. (1968) Soil and plant analysis as a guide to fertilization in cardamom. *Cardamom News*, 2(3), 4–6.

Sudharshan, M.R., Kuruvilla, K.M. and Madhusoodanan, K.J. (1988) Periodicity phenomenon in cardamom and its importance in plantation management. *Spice India*, 1(6), 14–21.

Sulikeri, G.S. (1986) *Effect of Light Intensity and Soil Moisture on Growth and Yield of Cardamom (Elettaria cardamomum Maton)*. Ph.D (Hort). Thesis, *Uni. Agri. Sci.*, Bangalore.

Thimmarayappa, M., Shivashankar, K.T. and Shanthaveerabhadraiah, S.M. (2000) Effect of organic manure and inorganic fertilizers on growth, yield attributes and yield of cardamom (*Elettaria cardamomum* Maton). *J. Spices and Aromatic Crops*, 9, 57–59.

Vadiraj, B.A., Srinivasan, K. and Potty, S.N. (1998) Major nutrient status of cardamom soils of Kerala. *J. Plantation Crops*, 26(2), 159–161.

Vadiraj, B.A., Srinivasan, K. and Potty, S.N. (1998) Fertility status of cardamom soils of Tamil Nadu. *South Indian Hort.*, 46, 122–125.

Venkatesh, J. (1980) *Comparison of the efficiency of soil application and soil-cum-foliar application of NPK nutrients on proliferation of suckers in cardamom*. M.Sc (Hort.) Thesis, *Uni. Agri. Sci.*, Bangalore.

Zachariah, P.K. (1975) The fertility status of cardamom growing soil. *Cardamom News*, 8 (3), 5–8.

Zachariah, P.K. (1978) Fertilizer management for cardamom. In *Proc. 1st Ann. Symp. on Plantation Crops*, Kottayam, pp. 141–156.

6 Viral diseases of cardamom

M.N. Venugopal

1 INTRODUCTION

Crop losses caused by widespread occurrence of cardamom mosaic virus (car-MV-*Katte*) is a major production constraint for cardamom in India. Occurrence of cardamom necrosis virus (car-NV-Nilgiri necrosis virus) and cardamom vein clearing disease (car-VCV-*Kokke kandu*) in some endemic zones are also matters of concern to the cardamom industry. Diseases of cardamom were earlier reviewed by Chattopadhyay (1967), Agnihothrudu (1987), Naidu and Thomas (1994) and Venugopal (1995). Based on severity of occurrence and crop losses, four serious viral diseases of cardamom and their integrated management are discussed in this chapter.

2 MOSAIC OR *KATTE* DISEASE (CAR-MV)

Mosaic disease is locally known as *katte*, meaning a disorder, and is known as marble disease in Anamalais, Tamil Nadu (India) (Varma and Capoor, 1953).

2.1 Distribution

Earliest reference on *katte* dates back to 1900 by Mollison. In South India, *katte* is widely distributed in all cardamom growing tracts, incidences ranging from 0.01 to 99 per cent (Mayne, 1951; Venugopal and Naidu, 1981). Cardamom in Guatemala remained free of virus disease until the 1970s. In 1975, a disease with virus-like symptoms was observed in some parts and within 5 years the disease has spread to all nearby plantations of the southern-pacific coastal region, which produces 60 per cent of cardamom in Guatemala (Gonsalves *et al.*, 1986). Recent surveys conducted in cardamom areas of India for car-MV and car-VCV have revealed the prevalence of car-MV in most of the cardamom plantations in Karnataka, India (Govindaraju *et al.*, 1994).

2.2 Crop loss

Loss in yield due to the disease depends on growth stage at the time of infection and duration that the plants have been subjected to infection. If plants are infected at the seedling stage or early pre-bearing stage the loss will be almost total (Samraj, 1970; CPCRI, 1980). Infection on bearing stage results in gradual decline in productivity

(CPCRI, 1984). Crop losses of 10–60 per cent, 26–91 per cent and 82–92 per cent were reported under cardamom–areca mixed cropping in first, second and third years of production respectively (Varma, 1962). Similarly in monocrop situations, infection of bearing plants led to yield reduction of 38 per cent during the same year, 62 per cent in the second year and 68.7 per cent in the third year of infection (Venugopal and Naidu, 1987). Total decline of plants occurs in 3–5 years of infection.

2.3 Symptomatology

The first visible symptom appears on the youngest leaf of affected tiller as slender chlorotic flecks measuring 2–5 mm in length. Later these flecks develop into pale green discontinuous stripes (Fig. 6.1). These stripes run parallel to veins from midrib to leaf margins. All subsequently emerging leaves show characteristic mosaic symptoms with stripes of green tissue almost evenly distributed over the entire lamina (Uppal *et al.*, 1945). Often mosaic-type of mottling is seen on leaf sheaths and young leaf shoots. Variation in field symptoms are seen in different cardamom growing tracts of South India and on inoculation of different virus isolates on a common host (Venugopal and Naidu, 1985). Plants of all stages are susceptible to virus infection; it is systemic in nature, and gradually spreads to all tillers in a clump. In advanced stages, the affected plants produce shorter and slender tillers with only a few short panicles, and the plants degenerate gradually.

Figure 6.1 Cardamom leaves showing symptoms of *katte* virus infestation. The leaf on the left shows severe symptoms, the right leaf shows the early stage.

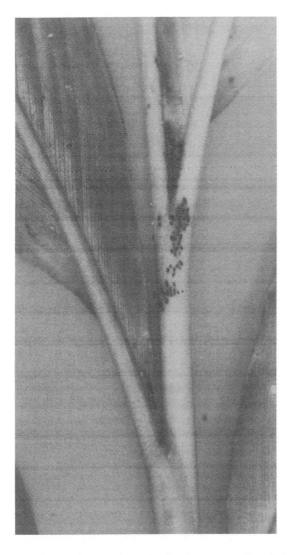

Figure 6.2 Cardamom shoot tip showing aggregation of the *katte* transmitting aphid, *Pentalonia nigronervosa* f. *caladi*.

2.4   Transmission

The virus is not transmitted through seed, soil, root to root contact and through manual operations (Thomas, 1938; Rao, 1977a,b). The only method of dissemination of the virus is by means of banana aphid (*Pentalonia nigronervosa* Coq.) and also through infected rhizomes (Fig. 6.2). The first experimental transmission of *katte* virus in India was obtained with banana aphid (Uppal *et al.*, 1945). So far 13 aphid species *(Aphis craccivora* Koch., *A. gossypii* Glover, *A. nerii* B. de. F., *A. rumicis* L., *Brachycaudus helichrysi* L., *Greenidia artocarpi* W., *Macrosiphum pisi* Kalt., *M. rosaeformis* Das, *M. sonchi* L., *Schizaphis cyperi* van der Goot, *S. graminum* Rondm., *Pentalonia nigronervosa* f. *typica* and *P. nigronervosa* f. *caladii* van der Goot) were reported to transmit *katte* virus (Rao and Naidu, 1974).

2.5   Spread of the disease

2.5.1   *Sources of infection*

Infected clones and apparently healthy clones, seedlings raised in the vicinity of infected plantations, volunteers collected from severely-infected plantations and a few infected zingiberaceous hosts (*Amomum* sp.) are the sources of infection. In a contiguous area, infected plantations are the reservoirs of virus sources (Varma, 1962; Naidu and Venugopal, 1987, 1989).

2.5.2   *Primary spread*

In plantations, primary spread occurs at random due to activity of viruliferous alate forms of the vector. Under field conditions, in plantations located 400–600 m from concentrated virus sources, the percentage of primary infection varied from 0.07 to 5.19 (Venugopal *et al.*, 1997a). The frequency of random spread directly depends upon access to virus sources.

2.5.3   *Secondary spread*

After the primary spread, secondary spread of disease is mainly internal and the rate of spread is very low (Deshpande *et al.*, 1972; Naidu and Venugopal, 1989). Centrifugal influx was found around primary foci due to spread by the activity of apterate adults. In plantations, the disease is concentrated within 40 m radius with occasional random spread up to 90 m distance. Gradient of infection is steep within 40 m from initial foci and it flattens thereafter. In Guatemala, rate of disease spread is very fast and natural infection may reach 83 per cent within 6 months of planting. A similar situation occurs in areca–cardamom mixed cropped areas in South India.

2.5.4   *Disease incubation period*

In field, incubation period of *katte* virus varies from 20 to 114 days during different months and the expression is directly influenced by growth of plants. Usually young seedlings at 3–4 leaves stage express the symptoms within 15–20 days of inoculation, whereas grown up plants take 30–40 days for symptom expression during active growing period and 90–120 days during winter months (Naidu and Venugopal, 1989). Senile leaf sheaths, which are natural breeding sites of the vector, are poor inoculum sources compared to young actively growing shoots (Naidu and Venugopal, 1989).

2.5.5   *Virus–vector relationship*

It was earlier thought that the aphids found on banana and cardamom were same, but later it was found that *P. nigronervosa* f. *typica* breeds on *Musa* and related genera, while *P. nigronervosa* f. *caladii* breeds on cardamom, *Colocasia* and *Caladium* (Siddappaji and Reddy, 1972). In cardamom plantations, the aphid populations are seen throughout the year with one or two peak periods during November–May and the population is drastically reduced during the monsoon season. All four nymphal instars and adult are capable of transmitting the disease (Rajan, 1981). Bimodal transmission was examined by using two distinct virus strains with respect to acquisition, latent period and persistence. Naidu *et al.* (1985) established the non-persistent nature of the *katte* virus.

2.5.6 Host range of the virus

Several plants belonging to Zingiberaceae viz., *Amomum cannecarpum, A. involucratum, A. subulatum, Alpinia neutans, A. mutica, Curcuma neilgherrensis, Zingiber cernuum* and a member of Marantaceae (*Maranta arundinacea,* West Indian arrow root) were found infected in lab inoculation tests (Rao and Naidu, 1973; Viswanath and Siddaramaiah, 1974; Yaraguntaiah, 1979; Siddaramaiah *et al.*, 1986).

2.6 Etiology

The evidence for the viral nature of the disease was first established in 1945 with the successful transmission of virus through the banana aphid *P. nigronervosa* Coq. (Uppal *et al.*, 1945). Studies in Guatemala and India have shown the association of flexuous rod-shaped virus particles measuring 650 nm in length and 10–12 nm in diameter in dip and purified preparations (Naidu *et al.*, 1985; Gonsalves *et al.*, 1986; Usha and Thomas, personal communication). Purified preparations of six strains also revealed homogenous flexuous particles (Fig. 6.3). Presence of inclusion bodies was also reported from leaf tissues of car-MV infected plants. Based on morphology of virus particles and presence of characteristic pinwheel-shaped inclusion bodies, it was suggested to include car-MV in 'poty virus' group (Naidu *et al.*, 1985). In Guatemala, mosaic-affected cardamom leaves revealed pinwheel type inclusion bodies, which is a common feature in other poty viruses. Leaf dip extracts showed particles of 660 nm length and those of purified preparation showed 700–720 nm long particles.

Serological affinity of car-MV of Guatemala with some poty viruses was demonstrated through indirect ELISA. Four viruses viz., Zucchini yellow mosaic, papaya ringspot types w and p, cow pea aphid borne mosaic virus and a severe strain of bean common mosaic virus consistently gave positive reaction in indirect ELISA. Presence of inclusion bodies, particle morphology and serological affinity of car-MV have confirmed the inclusion of it in the poty virus group (Dimitman *et al.*, 1984; Gonsalves *et al.*, 1986). Sequence analysis of the coding regions for coat protein and the 3'-untranslated region of the Yeslur isolate (from Saklespur, Karnataka) placed *katte* virus as a new member of the genus *Madura* virus of Potyviridae (Jacob and Usha, 2001). Considerable genetic diversity was noted among various isolates (Jacob *et al.*, in press).

Some consider *katte* as a complex disease caused by more than one component, or viruses (Rao, 1977a). So far the studies conducted in India and Guatemala does not support the complex nature of *katte* disease.

2.6.1 Strains of car-MV

Presence of three natural strains was first reported on the basis of symptomatology on the main host and cross-protection studies (Rao, 1977a). Further, occurrence of different natural strains was reported from the studies using 68 representative isolates of all cardamom growing zones of India. Three important biological criteria namely symptoms on the main host, transmission through *P. nigronervosa* f. *caladii* and reaction on the set of zingiberaceous differentials consisting of *Elettaria cardamomum* Maton var. *Malabar*, *Alpinia mutica*, *Amomum microstephanum* and *A. cannaecarpum* were used to identify the strains (Naidu *et al.*, 1985).

3   CARDAMOM VEIN CLEARING DISEASE OR *KOKKE KANDU* (CAR-VCV)

This disease is a new threat to cardamom in some endemic pockets in all the main cardamom growing districts of Karnataka, India. Surveys conducted in 1991–93 (Venugopal and Govindaraju, 1993; Govindaraju *et al.*, 1994) indicated the prevalence of car-VCV ranging from 0.1 to 80 per cent in plantations and nurseries. Because of its characteristic symptom it is locally referred as *Kokke Kandu*, meaning hook-like tiller.

3.1   Importance

In all the five cardamom-growing areas in Karnataka viz., Kodagu, Hassan, Chickmagalur, Shimoga and North Canara, the disease is present. In 381 plantations surveyed, widespread incidence of car-MV, car-VCV and mixed infections were seen in 375 plantations with incidence ranging from 0.1 to 82 per cent. A survey in 39 nurseries in hotspots also revealed the incidence of car-VCV.

3.1.1   Crop loss

Affected plants decline rapidly, yield reduction is to the extent of 62–84 per cent in the first year of peak crop (NRCS, 1994). Under mixed crop situations with arecanut as main crop, yield losses varied from 68–94 per cent in plants with different stages of infection (IISR, 1995). The affected plants become stunted and perish within 1–2 years of infection. Thousands of hectares of cardamom plantations in the Hongadahalla zone in Hassan district and arecanut-based mixed crop zone in North Canara district (both in the state of Karnataka, India) have become uneconomical due to the infection of mosaic and *Kokke Kandu*.

3.2   Symptoms

First, symptomatic leaves reveal characteristic continuous or discontinuous intraveinal clearing, stunting, rosetting, loosening of leaf sheath and shredding of leaves. Leafy stems exhibit clear mottling in all seasons. Clear light-green patches with three shallow grooves are seen on immature capsules. Cracking of fruits and partial sterility of seeds are other associated symptoms. In summer, the newly infected plants reveal only faint discontinuous vein clearing symptoms. Plants of all stages, right from seedling to bearing stage, show these symptoms. New leaves get entangled in the older leaves and form hook-like tiller (Fig. 6.3a,b), hence locally called as *Kokke-Kandu*.

3.3   Transmission and etiology

Car-VCV is not transmitted through seed, soil, leaves, roots, mechanical contact and farm implements. Mechanical transmission on set of differentials through combination of buffers, antioxidants, additives and abrasives was also not successful (Anand *et al.*,

Figure 6.3 Symptoms of cardamom vein clearing disease: (a) leaves showing vein clearing symptoms; (b) a shoot tip showing the symptoms, such as rosetting vein clearing and shredding of leaves.

1998; Venugopal *et al.*, unpublished). The disease is transmitted through cardamom aphid, *Pentalonia nigronervosa* f. *caladii* in a semi-persistent manner (IISR, 1996) or persistent manner (Anand *et al.*, 1998). Incubation period range from 22–128 days and a single viruliferous aphid can transmit virus to plants of all stages.

Exact etiology of associated virus is not yet established. In enzyme-linked immunosorbent assay (ELISA), antigen from infected host parts reacted positively with antibodies raised against poty viruses like peanut mottle virus, sugar cane stripe virus and Indian and Guatemalan car-MV isolates (Venugopal *et al.*, 1997b). These results indicate that car-VCV is possibly a member of poty virus group.

3.4   Disease spread

Like car-MV, initial spread occurs randomly in distant blocks due to the activity of incoming alate viruliferous vectors. Random spread was reported in new plantations located up to 2000 m away from infected plantations. Frequency of primary spread is directly dependent on the distance from foci of infection (NRCS, 1994; IISR, 1995, 1996).

Secondary spread within the infected plantations are both centrifugal and random. Alate forms of the aphid are responsible for random spread and apterate forms for the centrifugal spread. In infected plantations, the rate of spread varied from 1.3 to 8.5 per cent per year. Disease spread depends on the distance and level of incidence in the foci of infection. Gradient is steep, concentrated near the sources of virus inoculum (about 100 m) and it is shallow in the next 100 m (IISR, 1996).

4   CARDAMOM NECROSIS DISEASE (NILGIRI NECROSIS DISEASE)

This disease was first noticed in severe form in Nilgiris, Tamil Nadu, India hence the name Nilgiri Necrosis Virus (NNV). Recent surveys revealed new pockets in Kerala, Tamil Nadu and some spots in Karnataka. These are located in Nilgiris, Anamalais, Cardamom hills and Biligiri Rangan hills.

4.1   Importance

Random surveys in South India revealed low incidence of 0.1 to 1 per cent (CPCRI, 1985). Only in an isolated case in Valparai, Tamil Nadu incidence up to 13 per cent was recorded. Later surveys in South India indicated that the disease is prevalent in some of the cardamom-growing regions of Tamil Nadu as well, with an incidence ranging from 7.7 to 80 per cent (Sridhar *et al.*, 1990). In Lower Pulneys, out of 24 plantations surveyed, one plantation in Thadiankudisai showed 76 per cent incidence. In Valparai, 7.7 to 15.07 per cent incidence was recorded. Highest incidence of 80 per cent was recorded in Conoor of Nilgiris. Some estates in Munnar and Thondimalai areas of Idukki (Kerala), 4.6 and 1.46 per cent incidence respectively were recorded. Unlike *katte*, the infected plants decline rapidly and become stunted and unproductive.

4.2   Symptomatology and loss

Symptoms are seen on young leaves as whitish–yellowish, continuous or broken streaks, proceeding from midrib to leaf margins. In advanced stages of infection these streaks turn reddish-brown. Often leaf shredding is noticed along these streaks. Leaves are reduced in size with distorted margins. Plants infected early, produce only a few panicles and capsules, and in advanced stages of infection, tillers are highly stunted and fail to bear panicles. All the types of cardamom cultivars are susceptible to the disease (Sridhar, 1988).

Plants in the early infection stage recorded less reduction in yield when compared to the plants in advanced stage of infection (Sridhar *et al.*, 1991). One year studies carried out in diseased plantation indicated 55 per cent reduction in yield in early infected plants and total yield loss in late infected plants (Sridhar *et al.*, 1991).

4.3   Transmission

Seed, soil, sap and mechanical means do not transmit the disease. It is transmitted through planting of infected rhizomes. Aphids, *Pentalonia nigronervosa* f. *caladi*, thrips, *Sciothrips cardamomi* and white flies, *Dialeurodes cardamomi* were tested for their ability to transmit the disease. No insect transmission of the disease from infected to healthy were recorded so far (Sridhar, 1988).

4.4   Etiology and epidemiology

Association of flexuous particles of 570–700 nm long and 10–12 nm broad were seen in dip preparations of NNV infected leaf tissue and it belongs to Carla virus group (Naidu and Thomas, 1994).

Infected rhizomes/seedlings raised from diseased nurseries are the primary sources of inoculum. Plotting of new infections at regular intervals in a diseased plantation revealed that the spread of the disease is mainly internal and new infections occur in a centrifugal fashion from the source of inoculum. Most of the infections occurred within 10–15 m radius from the source of inoculum and the number of new infections decreased as the distance increased (Sridhar, 1988). The pattern of spread is similar to that of *katte* disease. The rate of spread of the disease is rather low, being 3.3 per cent for the period of one year. Occurrence of a few outbreaks around the infection foci is an indication that the disease can be successfully managed by periodical rouging of infected plants.

4.5   Infectious variegation virus

This disease was first noticed in Vandiperiyar area in Kerala in a severe form. Later it was also noticed in Coorg, Hassan and North Canara in Karnataka. Disease incidence of 15 per cent was recorded in Vandiperiyar zones. Infected plants show typical variegated symptoms on leaf with characteristic slender to broad radiating stripes of light and dark green on the lamina. Distortion of leaves, tillers and stunting are other common symptoms. Infected plants become unproductive within the same year of infection. Only 2 per cent transmission was obtained through the aphid *P. nigronervosa* f. *caladii*. Rouging resulted in near total elimination of disease in all the three test plantations.

5   INTEGRATED MANAGEMENT OF VIRAL DISEASES OF CARDAMOM

5.1   Production and use of virus-free planting material

Because of many infrastructure constraints like suitable nursery site, water resources, accessibility, availability of labour force and security, seedlings are raised repeatedly in the same nursery site, more commonly in the area adjacent to plantation, which invariably are infected. Seedlings require 10–18 months to attain plantable stage. This prolonged exposure to virus access through viruliferous aphids in the vicinity of concentrated virus sources results in infection at nursery stage itself. Further, secondary spread in nursery through aphids results in the spread of virus to many plants. As high as 28 per cent car-MV (Venugopal *et al.*, 1997a) and 73.33 per cent car-VCV (Govindaraju *et al.*, 1994) occur in the nursery stage itself. None of the three virus diseases (car-MV, car-VCV and car-NV) are seed transmitted, and lack long distance spread beyond 2000 m. Hence raising nursery in isolated places is necessary to produce healthy seedlings. For car-MV, isolation of 200 m from virus sources is adequate and for car-VCV, isolation of more than 200 m is necessary.

Apparently healthy high yielding plants are normally sub-cloned and planted for gap filling and raising plantation (Varma, 1962). As the infected plants take 23–168 days to express car-MV symptoms (Naidu and Venugopal, 1987) and 22–128 days to express car-VCV symptoms (Venugopal *et al.*, unpublished) it is not advisable to use clones from infected gardens. Like nurseries, clonal nurseries also have to be raised in isolated sites. In micropropagation, starting material has to be checked for virus-free status.

5.1.1   Avoidance of volunteers

Volunteers that sprout from remnants of infected materials are the potential primary sources of spread (Naidu and Venugopal, 1987). Self-sown seedlings in the infected plantations are exposed to virus access for 2–8 months. As high as 28 per cent infection was recorded in the nurseries raised from volunteers. Removal of infected volunteers in replanted area and total avoidance of volunteers for nursery activity in hotspots are most important pre-requisites for producing virus-free planting material.

5.1.2   Movement of planting material

In India, surveys conducted in 1981 (Venugopal and Naidu) and in 1994 (Govindaraju *et al.*) have shown that within the infected zone/plantation there are many disease-free pockets. Further, car-VCV is confined to certain endemic pockets only. In Guatemala also large area is free from mosaic infection (Dimitman, 1981). Creating awareness and preventing the movement of planting material have to be followed to check introduction/reintroduction of viruses.

5.2   Vector management

5.2.1   Chemical control

Because of non-persistent nature of car-MV and semi-persistent nature of car-VCV, chemical control measures are not that effective to reduce secondary spread. The insecticides at recommended concentrations do not kill aphids rapidly enough to prevent probing. Further, persistence of aphid vector throughout the year makes vector control measures almost impracticable. Thirty-four insecticides were evaluated to determine their effect on transmission and acquisition of *katte* virus under laboratory conditions. Transmission results showed that none of the insecticides tested were effective in checking acquisition and transmission of the virus even on the day of insecticide application. Being non-persistent (Rao, 1977a,b; Naidu *et al.*, 1985), car-MV can be transmitted within short periods of probing and feeding. Mere probing is sufficient for transmission of virus. This may be the reason for ineffectiveness of insecticides in checking secondary spread (Rajan *et al.*, 1989). The cardamom aphid, *P. nigronervosa* f. *caladii*, is photophobic and is found in colonies of 30–50, comprising nymphs, alate and apterate adults. These colonies are found in between the leafy stems and loose-leaf sheaths especially of old, partly dried or damaged parts. Occasionally the colonies are found on the leaf spindles, young suckers and panicles. Because of their concealed placement in the older parts, the possibilities of direct access to contact insecticides and indirect contact to systemic insecticides are less. As a result of insecticide treatment, the colony might have been disturbed and their hyperactivity, probing and intermittent migration in search of suitable hosts, may be responsible for ineffectiveness of some treatment like Phorate (granules), Carbofuran (granules) and Phosphamidon.

5.2.2 *Removal of breeding sites*

The photophobic vector breeds in senile, concealed parts of the host (Rajan, 1981). Periodical removal of old parts of the rhizomatous crop is effective in reducing the aphid population and the spread of car-MV (Rajan *et al.*, 1989). Other natural hosts like *Colocasia* sp., *Caladium* sp. (Siddappaji and Reddy, 1972; Rajan, 1981) etc., which are common weeds in the swampy areas of cardamom plantations, have to be removed periodically to check multiplication of the aphid, in addition to vector control measures.

5.2.3 *Use of biopesticides*

Extracts of many botanicals were found to be effective in reducing the breeding potential of aphid vector. Neem products significantly reduced the population of aphids on cardamom leaves even at 0.1 per cent concentration and were lethal to aphids at higher concentrations (Mathew *et al.*, 1997, 1999a,b). Aqueous extracts of *Acorus calamus* L. (dried rhizome), *Annona squamosa* L. (seeds) and *Lawsonia inermis* L. (leaves) reduced the settling percentage of aphids on leaves. Vapours of *A. calamus* are highly toxic to aphids and cause total mortality. Essential oil of turmeric (*Curcuma longa* L.) was also found to be repellant against the aphid (Saju *et al.*, 1998).

Entomogenous fungi like *Beauvaria bassiana* (Bals-Criv) Vuill, *Verticillium chlamydosporium* Goddard and *Paecilomyces lilacinus* (Thom.) Samson were promising in suppressing aphid population without causing hyperactivity (Mathew *et al.*, 1998).

5.2.4 *Resistant sources*

All the 168 germplasm accessions comprising *Mysore, Malabar* and *Vazhukka* types tested at Indian Institute of Spices Research (Cardamom Research Centre, Appangala, Karnataka, India), are susceptible to car-MV. Twenty-one elite accessions, distinct morphotypes like compound panicle types are also susceptible (Subba Rao and Naidu, 1981). Seventy natural disease escapes showed field resistance to car-MV in sick plots. They are also yielding better than the local cultivar (IISR, 1996, 1997). Screening trial consisting of 24 elite accessions of cardamom against car-VCV in sick plots showed that the accession 893 (of IISR, CRC, Appangala) is less susceptible compared to all other test accessions.

5.2.5   *Removal of virus sources*

In management of plant viruses, phytosanitation involves the detection and elimination of virus sources present within and outside plantations and the efficiency of phytosanitaion in management of viral diseases is centered around this operation. Attempts to control *katte* disease spread began as soon as researchers recognized the role of virus and control has been mainly based on sanitation or removal of virus source. Rouging is reported to be very effective in minimizing the spread and enhancing economic life of plantations (Varma and Capoor, 1958; Varma, 1962; Capoor, 1967, 1969; George, 1967, 1971; Deshpande *et al.*, 1972; Naidu and Venugopal, 1982). However, disease intensity and distribution of disease within a plantation are the prime factors influencing the efficacy of rouging. It is more appropriate to adopt rouging as an effective means of containing disease spread if the disease is less than 10 per cent (Naidu and Venugopal, 1982; Naidu *et al.*, 1985). Generally in plantations, the disease is seen concentrated in patches with random spread in certain spots. In such concentrated spots, survey to detect fresh infection and rouging may be undertaken at shorter intervals to minimize the chances of secondary spread (Naidu and Venugopal, 1982). This may be continued for 3–4 months till the new outbreaks are reduced and thereafter survey intervals can be increased to a few months. Through sustained timely efforts, new infections can be reduced to 2–3 per cent per annum, although it is impossible to eradicate the disease in a plantation because of predominance of small holdings and multiple chances of reintroduction. In contiguous cardamom holdings isolated attempts are not adequate to contain the disease economically. In such areas community approach through total removal of all the plants followed by re-planting and proper surveillance are shown to be more feasible (Varma, 1962). In varied field situations (like: (i) new plantation in isolated place, (ii) new plantation in hotspots, (iii) plantation with unidirectional virus source, (iv) plantation with multidirectional virus source, (v) plantation located between two infected plantations and (vi) plantation in continuous belt), different approaches involving total removal and replanting, selective rouging and gap filling, phased replanting etc. were shown to be effective in reducing secondary spread.

In car-MV and car-VCV infected areas, there are independent and also mixed infections of both viruses. In such areas comprehensive efforts involving use of healthy seedlings, periodical survey through trained disease surveillance gang, prompt removal and destruction of infected plants were reported to be effective in containing both viral diseases (Saju *et al.*, 1997).

5.2.6   *Early detection*

Inoculants take 23–120 days to express the visible symptoms. Early detection plays a very important role in elimination of virus sources. Attempts were made to test the polyclonal antiserum produced against car-MV before symptom expression as well as in the viruliferous aphid vector *P. nigronervosa* f. *caladii* for quick detection through ELISA (Saigopal *et al.*, 1992). Various host plant parts and young and mature seeds of infected and healthy plants were examined. Positive reaction for the presence of viral antigen was observed before symptom expression in all the host plant parts except in the mature seeds of infected plants. Virus concentration was more in the roots than the other host parts. Testing of viruliferous and non-viruliferous aphid vectors showed that the viral antigen could be detected in viruliferous aphids. The usefulness of quick detection technique through ELISA was further confirmed by indexing the primary cultures after *in vitro* multiplication. ELISA tests can be used for rapid field diagnosis of mosaic infection (Roberto, 1982) and in Guatemala, ELISA is being used extensively in the virus control programme (Gonsalves *et al.*, 1986).

Integration of several methods like: (i) strategies to produce healthy seedlings in isolated place, (ii) efforts to reduce vector population, (iii) use of virus resistant lines and (iv) removal of foci of infection are required to manage virus diseases in the field. Establishment of plant disease clinics in potential cardamom zones also helps to create awareness and to impart training to plantation community to manage viral diseases. In India, an attempt was made to contain car-MV through *katte* clinic programme (Naidu and Venugopal, 1982). There was encouraging response from the growers to this and in a period of 8 months, 60 plantations extending to 393 ha area distributed in 30 villages in Coorg district were covered. In India, the erstwhile Government of Bombay tried to eradicate the *katte* disease in North Canara district by providing technical assistance at Government cost (Varma, 1962). Similarly Cardamom Board also took up *katte* eradication programme in contiguous blocks by giving technical and financial assistance for rouging. These programmes have created awareness about identification and management techniques to contain spread of viral diseases.

5.3   Developing *katte* resistant cardamom

The cardamom research centre under the IISR carried out intensive survey of hotspot areas of *katte* virus disease that led to the collection of 134 disease escapes. This collection included var. *Mysore* (4), var. *Vazhukka* (29) and var. *Malabar* (105). A clonal nursery was established from these disease escapes, and they were subjected to screening in the green house using virulent virus isolates with the aid of the vector, 67 collections took infection and were discarded. The escapes from the screening tests were planted in sick plots and screening was carried out for 6 years. Most tester lines included in the trial took infection within 2 years, but 23 of them remained totally free from the symptoms. Four lines (NKE-11, NKE-16, NKE-22 and NKE-71) expressed faint granular symptoms in actively growing months, but such symptoms vanished in the subsequent months (Venugopal, 1999).

The NKE lines that passed the above screening were planted in hotspot areas. Nineteen lines remained free from disease out of which 17 had satisfactory agronomic characters. Venugopal *et al.* (1999) studied the breeding potential of the vector *Pentalonia nigronervosa* on these 17 lines and compared with those of a local susceptible check. The aphids colonized and multiplied on all the accessions, thereby indicating that the resistance of the 17 accessions is not due to deterrence to vector but due to some other host factors.

These 17 lines were further tested after interplanting with known susceptible checks, and they remained free of disease. Repeated inoculation did not produce any disease symptoms in these lines, thereby confirming the virus-resistant nature of these lines (Venugopal, 1999).

At present nothing is known about the mechanism of resistance in the above resistant lines. Different strains of car-MV have been reported from different virus-infected zones. Some zingiberaceous plants like *Alpinia mutica*, which was found resistant against Kodagu, Wynad, Hassan and Chickmagalur isolates showed higher suscepti-bility to Nelliampathy isolates. Hence further studies are required to assess the per-formance of the above virus-resistant clones against other distinct virulent strains from different cardamom-growing areas.

6   CONCLUSION

Viral diseases are responsible for rapid degeneration of production potential in early stages of plantation establishment and cause constant threats to sustained cardamom production. Very little information is available on the characterization of viruses. Systematic efforts are required to characterize the virus and to identify the virus reservoirs within and outside the crop to reduce the risk of reinfection in new plantations.

Rouging has been reported as viable and economical strategy to contain virus diseases of cardamom. However, continued persistence of infection and recurrence from the sources outside and within the crop is a matter of great concern to cardamom growers, commodity development and promotion agencies. Upgradation of disease management strategies depends on early diagnosis of virus infection in plants at the incubation stage itself and virus carriers. Though there have been attempts made in Guatemala, in India, application of biochemical and immunological techniques for early detection are not being practiced. Successful establishment of plantation and its sustained productivity is highly dependent on production of healthy planting material. Sensitive techniques are yet to be employed in diagnosis of virus infection and in mass multiplication programmes of location specific, high-yielding lines. Similarly indexing of fast depleting diverse genetic resources is the need of the hour to conserve them appropriately.

Many lines identified from disease escapes have shown field resistance to car-MV. Proper understanding of genetics and mechanisms of resistance will help in the long run to utilize them in crop improvement programmes through conventional and biotechnological approaches.

REFERENCES

Agnihothrudu (1987) Diseases of small and large cardamom. *Rev. Trop. Plant. Path.*, 4, 127–147.
Anand, T., Govindaraju, C., Sudharshan, M.R. and Srinivasulu, P. (1998) Epidemiology of vein-clearing virus of small cardamom. *Indian J. Virology*, 14, 105–109.
Buchanan, F. (1807) *A Journey From Madras Through the Countries of Mysore, Kanada and Malabar*, Vol. 3, p. 225.
Capoor, S.P. (1967) Katte disease of cardamom. Seminar on Cardamom – Cardamom Board, Cochin, p. 25.
Capoor, S.P. (1969) Katte disease of cardamom. *Cardamom News*, 3, 2–5.
Chattopadhyay, S.B. (1967) Diseases of plants yielding drugs, dyes and spices. *Indian Council of Agriculture Research*, New Delhi, p. 100.
CPCRI (1980) Annual Report. Central Plantation Crops Research Institute, Kasaragode, Kerala, India, pp. 121–122.
CPCRI (1984) Annual Report. Central Plantation Crops Research Institute, Kasaragode, Kerala, India, p. 27.
CPCRI (1985) Annual Report. Central Plantation Crops Research Institute, Kasaragode, Kerala, India, pp. 118–122.

Deshpande, R.S., Siddappaji, C. and Viswanath, S. (1972) Epidemiological studies on *katte* disease of cardamom (*Elettaria cardamomum* Maton). *Mysore J. Agric. Sci.*, 6, 4–9.

Dimitman, J.E. (1981) An aphid transmitted virus of cardamom in Guatemala (Abstr.). *Phytopathology*, 71, 104–105.

Dimitman, J.E., Flores, A. and Nickoloff, J.A. (1984) Cardamom mosaic – a member of the potyvirus group in Gautemala (abstr.). *Phytopathology*, 74, 844.

George, K.V. (1967) *Katte* disease threatens the future of cardamom plantation industry. *Cardamom News*, 1, 1–3.

George, K.V. (1971) Research on cardamom – some suggestions. *Cardamom News*, 5, 2–3.

Gonsalves, D., Trujillo, E. and Hoch, H.C. (1986) Purification and some properties of a virus associated with cardamom mosaic, a new member of the potyvirus group. *Plant Disease*, 70, 65–69.

Govindaraju, C., Venugopal, M.N. and Sudharsan, M.R. (1994) An appraisal of 'Kokke kandu'- a new viral disease of cardamom and *Katte* (Mosaic) disease in Karnataka. *J. Plantation Crops*, 22, 57–59.

IISR (1995) Annual report for 1994–95. Indian Institute of Spices Research, Calicut, Kerala, India, pp. 5–6.

IISR (1996) Annual report for 1995–96. Indian Institute of Spices Research, Calicut, Kerala, India, pp. 66–67.

IISR (1997) Annual report for 1996–97. Indian Institute of Spices Research, Calicut, Kerala, India, p. 63.

Jacob, T. and Usha, R. (2001) 3'-terminal sequence analysis of the RNA genome of the Indian isolate of cardamom mosaic virus, a new member of the genus *Madura* virus of Potyviridae. *Virus Genes*, 23, 81–88.

Jacob, T., Jebasingh, Venugopal, M.N. and Usha, R. (2002) High genetic diversity in the coat protein and the 3'-untranslated regions among the geographical isolates of cardamom mosaic virus from South India (in press).

Mathew, M.J., Saju, K.A. and Venugopal, M.N. (1997) Management of *Pentalonia nigronervosa* f. *caladii* Van der Goot, vector of Cardamom Mosaic Virus (Katte) and Cardamom Vein Clearing Virus (Kokke kandu) through eco-friendly vector control measures (abstract). In *Symposium on Economically Important Plant Diseases*. December 18–20th, 1997, Bangalore, p. 59.

Mathew, M.J., Saju, K.A. and Venugopal, M.N. (1998) Efficacy of entomogenous fungi on biological suppression of *Pentalonia nigronervosa* f. *caladii* Van der Goot of cardamom (*Elettaria cardamomum* Maton). *J. Spices and Aromatic Crops*, 7, 43–46.

Mathew, M.J., Saju, K.A. and Venugopal, M.N. (1999a) Effect of neem products on behaviour and mortality of cardamom aphid, *Pentalonia nigronervosa* f. *caladii* Van der Goot. In *Proc. National Symposium on Pest management in Horticultural Ecosystems: Environmental implications and thrusts*, IIHR, Bangalore (in press).

Mathew, M.J., Venugopal, M.N. and Saju, K.A. (1999b) Field evaluation of certain biopesticides in comparison with monocrotophos against cardamom aphid, vector of *katte* and *kokke kandu* diseases of cardamom (abstract). In *National Symposium on Biological Control of Insects in Agriculture, Forestry, Medicine and Veterinary Science*, 21–22 January, 1999, Bharathiar University, Coimbatore, p. 87.

Mayne, W.W. (1951) Report on cardamom cultivation in South India. *ICAR Bull.*, 50. Indian Council of Agriculture Research, New Delhi, p. 62.

Mollison, J.W. (1900) Cultivation of the betel palm, cardamom and pepper in the Kanara district of Bombay presidency with notes on the manures used in the spice gardens of that tract. *Bull No. 20, Dept. Land Rec. Agri.*, Bombay.

Naidu, R. and Thomas, J. (1994) Viral diseases of cardamom. In K.L. Chadha and P. Rethinam (eds) *Advances in Horticulture*, Vol. 10 – *Plantation and Spice Crops*, Part 2. Malhotra Publishing house, New Delhi, pp. 1101–1111.

Naidu, R. and Venugopal, M.N. (1982) Management of 'Katte' disease of small cardamom. In *Proc. PLACROSYM V*, Indian Society for Plantation Crops, Kasaragod, India, pp. 563–571.

Naidu, R. and Venugopal, M.N. (1987) Epidemiology of *katte* virus of small cardamom. 1 Disease incubation period and role of different host parts as a source of inoculum in relation to disease spread. In *Proc. PLCROSYM VI* 1984, Indian Society for Plantation Crops, Kasaragod, India, pp. 121–127.

Naidu, R. and Venugopal, M.N. (1989) Epidemiology of *katte* virus disease of small cardamom II. Foci of primary disease entry, patterns and gradients of disease entry and spread. *J. Plantation Crops*, 16(suppl.), 267–271.

Naidu, R., Venugopal, M.N. and Rajan, P. (1985) Investigations on strainal variation, epidemiology and characterization of 'katte' virus agent of small cardamom. Final Report of Research Project, Central Plantation Crops Research Institute, Kasaragod, Kerala, India.

NRCS (1994) Annual Report for 1993–94. National Research Centre for Spices, Calicut, Kerala, India, pp. 4–5.

Rajan, P. (1981) Biology of *Pentalonia nigronervosa* f. *caladii* Van der Goot, vector of *katte* disease of cardamom. *J. Plantation Crops*, 9, 34–41.

Rajan, P., Naidu, R. and Venugopal, M.N. (1989) Effect of insecticides on transmission and acquisition of *katte* virus of small cardamom and their use in relation to disease spread and vector control. *J. Plantation Crops*, 16(suppl.), 261–266.

Rao, D.G. (1977a) *Katte* disease of cardamom and its control. *Indian J. Hort.*, 34, 184–187.

Rao, D.G. (1977b) *Katte* disease of small cardamom. *J. Plantation Crops*, 5, 23–27.

Rao, D.G. and Naidu, R. (1973) Studies on *Katte* or mosaic disease of small cardamom. *J. Plantation Crops*, 1(suppl.), 129–136.

Rao, D.G. and Naidu, R. (1974) Additional vectors of *Katte* disease of small cardamom. *Indian J. Hort.*, 31, 380–381.

R.R.S., Mudigere (1990) Technical Bulletin-2. Three Decades of Cardamom Research (1958–88), Regional Research Station, Mudigere, Karnataka, India.

Roberto, C.J. (1982) The enzyme-linked immunosorbent assay (ELISA) in the diagnosis of cardamom mosaic virus. *Rev. Cafetaera*, 219, 19–23.

Saju, K.A., Venugopal, M.N. and Mathew, M.J. (1997) Effect of phytosanitation on recurrence of Vein Clearing Virus (kokke kandu) disease of small cardamom (abstract). In *National Symposium on Resurgence of Vector Borne Viral Diseases.* 1–3 August 1997, Gujarath Agricultural University, Anand, p. 15.

Saju, K.A., Venugopal, M.N. and Mathew, M.J. (1998) Antifungal and insect repellent activities of essential oil of turmeric (*Curcuma longa* L.). *Curr. Sci.*, 75, 660–663.

Saigopal, D.V.R., Naidu, R. and Joseph, T. (1992) Early detection of 'Katte' disease of small cardamom through Enzyme Linked Immunosorbent Assay (ELISA). *J. Plantation Crops*, 20 (suppl.), 73–75.

Samraj, J. (1970) Mosaic disease of cardamom. *Cardamom News*, 4, 12.

Siddappaji, C. and Reddy, D.N.R.N. (1972) A note on the occurrennce of the aphid, *Pentalonia nigronervosa* f. *caladii* Van der Goot (Aphididae: Hemiptera) on cardamom (*Elettaria cardamomum* Maton). *Mysore J. Agric. Sci.*, 6, 192–195.

Siddaramaiah, A.L., Chandrashekar, S.C., Balachandra, C.K. and Pattanshetti, V. (1986) Arrowroot (*Maranta arundinacea* L.) is a new collateral host for *Katte* disease of cardamom (*Elettaria cardamomum* Maton). *Curr. Sci.*, 55, 18.

Sridhar, V.V. (1988) *Studies on Nilgiri Necrosis, a new virus disease of cardamom.* Ph.D Thesis, Tamil Nadu Agricultural University, Coimbatore.

Sridhar, V.V., Muthuswamy, M. and Naidu, R. (1990) Survey on the occurrence of Nilgiri necrosis – a new virus disease of cardamom. *South Indian Hort.*, 38, 163–165.

Sridhar, V.V., Muthuswamy, M. and Naidu, R. (1991) Yield loss assessment in Nilgiri Necrosis infected cardamom. *South Indian Hort.*, 39, 169–170.

Subba Rao, G. and Naidu, R. (1981) Breeding cardamom for *Katte* disease resistance. In *Proc. PLACROSYM IV,* 1980. Indian Society for Plantation Crops, Kasaragode, Kerala, India, pp. 320–324.

Thomas, K.M. (1938) Detailed Administrative Report of the Government Mycologist, Madras, 1937–1938.

Uppal, B.N., Varma, P.M. and Capoor, S.P. (1945) A mosaic disease of cardamom. *Curr. Sci.*, 14, 208–209.

Varma, P.M. (1959) Information on *Katte* disease of small cardamom for cultivators. *Arecanut J.*, 10, 7–16.

Varma, P.M. (1962) Control of *Katte* or mosaic disease of cardamom in North Kanara. *Arecanut J.*, 13, 79–88.

Varma, P.M. and Capoor, S.P. (1953) Marble disease of cardamom. *Indian Farming*, 3, 22–23.

Varma, P.M. and Capoor, S.P. (1958) Mosaic disease of cardamom and its transmission by the banana aphid *Pentalonia nigronervosa* Coq. *Indian J. Agric. Sci.*, 28, 97–107.

Venugopal, M.N. (1995) Viral diseases of cardamom (*Elettaria cardamomum* Maton) and their management. *J. Spices and Aromatic Crops*, 4, 32–39.

Venugopal, M.N. and Govindaraju, C. (1993) Cardamom vein clearing virus – a new threat to cardamom. *Spice India*, 5(2), 6–7.

Venugopal, M.N. and Naidu, R. (1981) Geographical distribution of *Katte* disease of small cardamom in India (abstr.). In *Third International Symposium on Plant Pathology*, December 14–18, New Delhi, pp. 221–222.

Venugopal, M.N. and Naidu, R. (1987) Effect of natural infection of *Katte* mosaic disease on yield of cardamom – a case study. In *Proc. PLACROSYM – VI*, Oxford & IBH, New Delhi, pp. 115–119.

Venugopal, M.N., Saju, K.A. and Mathew, M.J. (1997a) Primary spread of cardamom mosaic virus under different field situations (abstract). In *National Symposium on Resurgence of Vector Borne Viral Diseases*, 1–3 August 1997, Gujarath Agricultural University, Anand, p. 15.

Venugopal, M.N., Saju, K.A. and Mathew, M.J. (1997b) Transmission and serological relationship of cardamom vein clearing disease (Kokke kandu) (abstract). In *Symposium on Economically Important Plant Diseases*. December 18–20th, 1997, Bangalore, p. 60.

Venugopal, M.N., (1999) Natural disease escapes as source of resistance against cardamom mosaic virus causing *katte* disease of cardamom (*Elettaria cardamomum* Maton). *J. Spices & Aromatic Crops*, 8, 145–151.

Venugopal, M.N., Mathew, M.J and Anandaraj, M. (1999) Breeding potential of *Pentalonia nigronosa f caladii* Van der goot (Horopotera: Aplindidae) on *katte* escaped cardamom (*Elettaria cardamomum* Maton). *J. Spices & Aromatic Crops*, 8, 189–191.

Viswanath, S. and Siddaramaiah, A.L. (1974) *Alpinia neutans*, a new host of Katte disease of cardamom. *Curr. Res.*, 3, 96.

Yaraguntaiah, R.C. (1979) *Curcuma neilgherensis* WT – a new host of Katte disease of cardamom. In *Proc. PLACROSYM II*, 1979 Central Plantation Crops Research Institute, Kasaragode, Kerala, India, pp. 313–315.

7 Diseases of cardamom (fungal, bacterial and nematode diseases)

Joseph Thomas and R. Suseela Bhai

Cardamom is affected by a number of diseases caused by various pathogenic fungi, bacteria and nematodes, in main plantations as well as in nurseries. As many as 25 fungal, bacterial and nematodal diseases have been reported till date. Based on severity, spread and extent of damage, these are grouped as major and minor diseases occurring in the main plantations and in nurseries. Among them, four major diseases in plantations and two major diseases in nurseries seriously affect the plant and cause considerable crop damage. Major diseases such as the rots, leaf blights and nematode infestation are often wide spread and lead to crop losses while minor diseases generally affect the foliage and occur in minor proportions. Diseases alone can cause up to 50 per cent crop loss if not managed properly.

1 MAJOR DISEASES

Diseases such as the Capsule rot ('Azhukal') and the Rhizome rot are comparatively severe and affect crop production while the widespread leaf blight and nematode infections, lead to weakening of plants, and subsequent reduction in productivity. The major diseases occurring in plantations are listed in Table 7.1.

1.1 Capsule rot (*Azhukal* disease)

Capsule rot, popularly known as *Azhukal* in the local (Malayalam) language means rotting, is perhaps the most serious fungal disease of cardamom. Menon *et al.* (1972) reported it for the first time from plantations of Idukki district in Kerala State.

1.1.1 *History and distribution*

Initially, rotting symptoms were observed on the fruits or the capsules only and accordingly it was named as capsule rot. Later the disease symptoms have been observed in several other plant parts. It is still a major problem affecting cardamom cultivation in Idukki and Wynad districts of Kerala and Anamalai hills in Tamil Nadu (Thomas *et al.*, 1989). The disease makes its appearance after the onset of southwest monsoon rains. However, capsule rot is not observed in low rainfall areas in Tamil Nadu. Cardamom plantations in Karnataka state receive a great deal of monsoon rains but this disease has not yet been found to occur in this geographical locality.

Table 7.1 Major fungal and nematode diseases of cardamom in plantations

Diseases	Parts affected	Causal agents
Capsule rot (*Azhukal*)	Capsules, leaves, panicles and young tillers	*Phytophthora meadii P. nicotianae* var. *nicotianae*
Rhizome rot (Clump rot)	Rhizomes, tillers and roots	*Pythium vexans, Rhizoctonia solani. Fusarium oxysporum*
'Chenthal' (Leaf blight)	Leaves	*Colletotrichum gloeosporioides*
Root knot nematode	Roots, leaves	*Meloidogyne incognita*

1.1.2 Symptoms and damage

Disease symptoms develop mainly on the capsules, young leaves, panicles and tender shoots. The first visible symptom appears as discoloured water-soaked lesions on young leaves or capsules. These lesions enlarge and the affected portions decay. Infection takes place on capsules or tender leaves simultaneously or first on capsules followed by foliar infection (Thomas *et al.*, 1991a). When foliage infection occurs, water-soaked lesions appear on leaf tips or leaf margins which later enlarge and adjacent lesions coalesce to form large patches. Immature unopened leaves fail to unfurl following infection. As the disease advances, the lesion areas turn necrotic, the leaves decay and shrivel and finally they give a shredded appearance (Fig. 7.1a). Infected capsules show water soaked discoloured areas; they turn brownish and later such capsules decay and drop off (Fig. 7.1b). Such rotten capsules emit a foul smell. Capsules of all ages are susceptible to infection. However, young capsules are seriously affected by the disease.

During favourable climatic conditions the disease is aggravated and infection extends to panicles and tender shoots also. In extreme cases the whole panicle or the whole psuedo-stem decays completely. In such cases the rotting extends to underground rhizomes also. The root system of such plants gets decayed and the entire plant collapses. Nair (1979) described similar symptoms and observed that the disease severity is uniform in the three major cardamom types, viz. *Mysore, Vazhukka* and *Malabar*. Nambiar and Sarma (1976), who studied the disease earlier, have reported a crop loss of 30 per cent. However later it has been shown that as high as 40 per cent crop loss can occur in severely disease-affected plantations (Anonymous, 1989a).

1.1.3 Causal organism

Menon *et al.* (1972) first reported *Phytophthora* sp. as the causal organism. Thankamma and Pillai (1973) identified it as *P. nicotianae* Brede de Haan var. *nicotianae* Waterhouse and as *P. palmivora* Butler (Radha and Joseph, 1974). Nambiar and Sarma (1976) reported the association of *Pythium vexans* and a *Fusarium* sp. along with *Phytophthora* sp. However later studies by Nair (1979) showed that *P. nicotianae* var. *nicotianae* is the causative organism which could be successfully isolated from all infected plant parts. *P. meadii* Mc Rae has also been widely observed as causing *Azhukal* disease (Anonymous, 1986). Host-range studies show that *P. palmivora* from coconut and rubber is infective to cardamom (Radha and Joseph, 1974), *P. palmivora* from cardamom is infective to cocoa, coconut, arecanut, black pepper and rubber (Manomohanan and Abi Cheeran,

Figure 7.1 *Azhukal* disease symptoms: (a) rotting and shredding of leaves; (b) rotting of capsules; (c) *Phytophthora meadii* in carrot agar medium; (d) Sporangia of *P. meadii*.

1984), and *P. meadii* from cardamom is infective to cocoa, black pepper and citrus (Sastry and Hedge, 1987, 1989). Nair (1979) observed that wild Colocasia plants in cardamom plantations serve as collateral hosts for *P. nicotianae* var. *nicotianae*.

Seven different isolates of *P. meadii* from various localities causing infection on capsules, leafy stems, rhizomes and leaves of cardamom have been characterized based on their culture characters, sporangial morphology, sexual behaviour and pathogenic

Table 7.2 Grouping of *P. meadii* isolates based on temperature requirements for optimal growth and sporangial characters

Isolates		Opt. temp. °C	Length μm	Breadth μm	L/B ratio	Pedicel length μm
*I**	*II***					
	159	20	45.96 (24.8–66.9)	24.96 (19.8–29.7)	1.83 (1.1–2.4)	32.86 (19.9–26.8)
209	–	20	34.72 (19.8–42.1)	20.01 (12.4–22.3)	1.75 (1.3–2.2)	6.51 (2.4–12.4)
–	239	24–28	40.82 (32.2–47.1)	22.01 (17.3–22.3)	1.82 (1.4–2.0)	35.22 (12.4–71.4)
–	240	24–28	46.55 (27.2–62.0)	22.51 (17.3–29.7)	2.09 (1.5–2.3)	40.92 (24.8–94.4)
241	–	24–28	33.57 (24.8–49.6)	19.84 (14.8–24.8)	1.80 (11.3–2.2)	20.90 (19.9–37.2)
–	244	32	33.57 (32.2–54.5)	19.84 (17.3–32.2)	1.80 (1.3–2.2)	20.90 (0.0–2.4)
252	–	32	34.72 (29.2–39.6)	17.69 (14.8–22.3)	1.85 (1.3–2.3)	22.87 (4.9–39.6)

Notes
* Sporangial length 33.57–34.72 μm; Sporangial breadth 17.69–20.01 μm.
** Sporangial length 40.82–46.55 μm; Sporangial breadth 20.01–24.96 μm.

virulence (Anonymous, 1989a). These seven isolates fall in two groups in their require-ment for optimum temperature for growth and mean sporangial dimensions (Table 7.2). In single cultures no oospores are formed but when paired with A1 mating type, five of them readily formed sex organs and oospores confirming that most of these isolates belong to the A2 mating type. The type species of *P. meadii* from cardamom readily grows on carrot agar and sporulates; the sporangia are caducous, ellipsoid, papillate and with short to medium pedicels (Fig. 7.1c,d). Although these seven isolates morphologically differ slightly, all of them were found to be pathologically virulent types. The pathogen, *P. nicotianae* var. *nicotianae*, survives in the soil and plant debris in the form of chlamydospores and in moist soil upto 48 weeks (Nair, 1979). However in the case of *P. meadii* no chlamydospore formation has been observed either in moist field soils or under laboratory conditions. Dantanarayana *et al.* (1984) also have reported the inability of *P. meadii* from rubber to form chlamydospores.

1.1.4 Epidemiology

Nair (1979) studied the epidemiology of *Azhukal* disease. He observed that high disease incidence is correlated to high and continuous rainfall during the monsoon seasons.

The number of *Phytophthora* propagules increases in soil and results in heavy disease incidence coinciding with high soil moisture levels (34.3–37.6 per cent), low tempera-tures (20.4–21.3 °C), high relative humidity (83–90.6 per cent) and high rain fall (320–400 mm) during the months of June to August (Nair and Menon, 1980). Presence of high level of soil inoculum, thick shade in the plantation, close spacing, high soil moisture, water logging together with favourable weather conditions such as high relative humidity, continuous rainfall and low temperature predispose the plants to *P. meadii* infection. Nair (1979) also found that the density of *Phytophthora* population reduces with increasing distance from plant base and from soil surface.

1.1.5   Disease management

As the outbreak of disease is during the monsoon season, disease management measures have to be initiated sufficiently in advance i.e. before the primary infection starts. During earlier years various types of fungicides have been extensively used for controlling the disease. Spraying and drenching of copper fungicides such as one per cent Bordeaux mixture and 0.2 per cent Copper oxychloride (Menon *et al.*, 1973; Nambiar and Sarma, 1974; Nair, 1979; Nair *et al.*, 1982) was recommended as the control measure. Inhibition of the fungus under *in vitro* conditions was reported following treatments with organomercurials (Wilson *et al.*, 1974). Nair (1979) observed 86 per cent reduction in soil population levels of *Phytophthora* when drenched with one per cent Bordeaux mixture or 100 ppm Dexon (Bay, 5072). Alagianagalingam and Kandaswamy (1981) observed that the disease could be controlled by spraying the plants with 0.2 per cent Dexon (Bay-5072) at the rate of 4 kg/ha.

Although a number of fungicides have been reported to control the disease, often disease control in the field has been a challenging experience. The factors responsible for the constraints in achieving satisfactory disease control include lack of phytosanitation, effective and timely application schedules, high cost and unavailability of fungicides and the continuous rain that makes any fungicidal application ineffective.

Thomas *et al.* (1989, 1991a) evaluated a number of contact and systemic fungicides under field conditions and concluded that two to three rounds of sprays including one round of prophylactic spray with one per cent Bordeaux mixture or 0.3 per cent Aliette (Fosetyl-Aluminium) after proper phytosanitation effectively controlled the spread of the disease.

1.1.6   Biological control

Bio-agents play an important role in an eco-friendly system of disease management to fight against plant pathogens in a totally safe manner avoiding the use of expensive and hazardous chemical fungicides. Inhibition of *P. meadii* under laboratory conditions and disease suppression in cardamom nurseries have been studied by Thomas *et al.* (1991b) using *Trichoderma viride, T. harzianum, Laetisaria arvalis* and *Bacillus subtilis*. Suseela Bhai *et al.* (1993) achieved field control of *Azhukal* disease using *Trichoderma viride* and *T. harzianum* and have further developed a simple carrier cum multiplication medium for *Trichoderma* application in the field (Suseela Bhai *et al.*, 1994, 1997). *Trichoderma viride* and *T. harzianum* isolates harbouring native cardamom soils have been screened and effective strains for high biocontrol potential have been developed (Dhanapal and Thomas, 1996). Today, field control of *Azhukal* disease of cardamom has become effective, environmentally safe and cost effective due to the bio-control potential of *Trichoderma* sp.

1.2   Rhizome rot

Rhizome rot, also known as Clump rot, is a common disease occurring in cardamom plantations during the monsoon period. The disease was first reported by Park (1937).

Subba Rao (1938) described it as clump rot disease. The disease is widely distributed through out cardamom plantations in Kerala and Karnataka states and in heavy rain fall areas of Tamil Nadu such as the Anamalai hills.

1.2.1   Disease symptoms

The disease makes its appearance during south-west monsoon period by about middle of June. The first visible symptom is the development of pale yellow colour in the foliage and premature death of older leaves. These leaves show wilting symptoms. The collar portion of the aerial shoots becomes brittle and the tiller breaks of at slight disturbance. Symptoms of rotting develop at the collar region, which becomes soft and brown coloured. At this stage the affected aerial shoots fall off emitting a foul smell. Mayne (1942) reported the incidence of the disease in cardamom hills of Kerala. The tender shoots or the young tillers also turn brown coloured and rot completely. As the disease advances, all the affected aerial shoots fall off from the base. The panicles and young shoots attached to this also are affected by rot. The rotting extends to the rhizomes and roots also. Falling off shoots resulting from rhizome rot infection becomes severe during July–August months. In severely affected areas as much as 20 per cent disease incidence was recorded.

1.2.2   Causal organism

Subba Rao (1938) observed that cardamom rhizome rot is caused by *Rhizoctonia solani* Kuhn., and it was associated with a nematode. Ramakrishnan (1949) reported *Pythium vexans* de Barry as the causal organism. Thomas and Vijayan (1994) reported that *Fusarium oxysporum* is also occasionally found to cause rhizome and root rot infections.

1.2.3   Disease management

The disease is usually observed in areas previously affected by rhizome rot disease. Therefore phytosanitation plays a major role in disease management. Presence of inoculum in the soil and plant debris, overcrowding of plants, and thick shade are congenial conditions for disease development. Therefore, any disease management schedule has to be followed with these points in mind. Application of superphosphate at the rate of 300–400 g per plant has been recommended for controlling clump rot in plantations (Anonymous, 1955).

Soil drenching with one per cent Bordeaux mixture or 0.25 per cent Copper oxychloride or neem oil cake at the rate of 500 g per plant followed with one round pre monsoon and two rounds post monsoon soil drenching with 0.25 per cent copper oxychloride at one month interval has been reported to be very effective for controlling the disease (Thomas and Vijayan, 1994).

1.2.4   Biological control

Recent attempts in rhizome-rot control are by the use of *Trichoderma* sp. *Trichoderma viride* and *T. harzianum* were reported to reduce rhizome rot incidence in plantations (Thomas *et al.*, 1991b). A formulation of *T. harzianum* in a carrier medium consisting of farmyard manure and coffee husk mixture has been developed for field application in the integrated disease management system for control of rot diseases of cardamom (Thomas *et al.*, 1997).

1.3   Chenthal (leaf-blight)

A leaf-blight disease popularly known as *Chenthal* has been reported in cardamom plantations (George *et al.*, 1976) from Idukki district of Kerala state. Since then, the occurrence of the disease has been observed in many plantations. The disease spread is faster in partially deforested areas and less shaded plantations. Though it was reported as a minor disease of limited spread, presently the situation is alarming as the disease is spreading to newer areas and is becoming a major problem.

1.3.1   Symptoms and damage

Chenthal makes its appearance mostly during the post-monsoon period and becomes severe during summer months. Symptoms develop on the foliage as water soaked rectangular lesions, which later elongate to form parallelly arranged streaks. The length of these streaks varies from a few millimeters up to 5 cm. The lesion areas become yellowish-brown to orange-red in colour and often the central portions become necrotic. Usually the two youngest leaves are not attacked by the disease. As the disease advances more and more lesions develop on older leaves, adjacent lesions coalesce and these areas begin to dry up. Severely infected plants show a burnt appearance. George and Jayasankar (1979) reported reduction in plant height, panicle length and crop loss due to failure in panicle formation in severely affected plants. However, Govindaraju *et al.* (1996) studied the symptomatology in detail and found that *Chenthal* infection affects only the leaves and not the plant height, panicle emergence or crop yield.

1.3.2   Causal organism

Chenthal was originally reported as a bacterial disease caused by *Corynebacterium* sp. (George and Jayasankar, 1977). They also recommended penicillin spray for controlling the disease. As later workers could neither isolate *Corynebacterium* sp. nor could control the disease with penicillin sprays, the bacterial etiology was suspected and the cause of the disease remained obscure for more than a decade. Govindaraju *et al.* (1996) conducted detailed investigations on symptomatology, etiology and management strategies of *Chenthal* and have clearly shown beyond doubt that the causal organism is the fungus *Colletotrichum gloeosporioides* (Penz.) Penz and Sacc. The fungus closely resembles *C. gloeosporioides* causing anthracnose disease of capsule reported by Suseela Bhai *et al.* (1988). Both the leaf and capsule isolates showed similar cultural and morphological characters and were cross-infective to capsules and leaves and *vice versa*. However, these two isolates exhibited considerable differences in their period of occurrence, type of symptoms, distribution and spread of the disease.

1.3.3   Disease management

As the disease was considered to be caused by *Corynebacterium* sp. penicillin spray was suggested as a control measure for the disease (George and Jayasankar, 1977). But it was not effective and was not followed by planters. Govindaraju *et al.* (1996) reported that three sprays at monthly intervals with carbendazim (Bavistin, 0.3 per cent) or Mancozeb (0.3 per cent) or Copper oxychloride (0.25 per cent) effectively controlled *chenthal* disease spread in the field.

1.4 Nematode diseases

Nematode infestation in cardamom is a major problem often amounting to heavy crop losses. Although, as many as 20 genera of various plant parasitic nematodes have been reported from cardamom soils (Ali, 1983), only the root knot nematode *Meloidogyne incognita* causes severe damage to the crop. Root knot nematodes are widely observed in almost all cardamom plantations and nurseries (Ramana and Eapen, 1992) while the lesion nematode *Pratylenchus coffeae* and the burrowing nematode *Radopholus similis* are noticed in mixed plantations.

1.4.1 *Symptoms and damage*

Root knot nematode causes aerial symptoms such as stunting of plants, reduced tillering, rosetting and narrowing of leaves, yellow banding on leaf blades and drying of leaf tips or leaf margins. The flowering is normally delayed. Immature fruit dropping results in yield reduction (Anonymous, 1972, 1989b). Underground symptoms develop on the roots of infected plants in the form of pronounced root galling. Tender root tips show spherical–ovoid swellings. Severe infestation can result in crop losses up to 80 per cent (Ramana and Eapen, 1992). Nematode population is high in cardamom soils during post-monsoon period (September–January). Heavy shade in plantations, moist soil and warm humid weather are predisposing factors for nematode multiplication.

Nematode infestation is a chronic problem in cardamom nurseries where the same site is repeatedly used for raising seedlings. Nematode-infested soils affect seed germination and result in severe and heavy galling in the root system, marginal yellowing and drying of leaves, stunting and reduced tillering. The leaves become narrow and the leaf tips show upward curling.

1.4.2 *Nematode control*

As infected seedlings serve as the source of inoculum, transplanting of affected seedlings should be avoided. Pre-treatment of infested nursery beds with methyl bromide at the rate of 500 g/10 M^2 or soil drenching with 2 per cent formalin is usually recommended. Solarization of nursery beds is reported to reduce nematode populations in the soil.

Application of nematicides such as aldicarb, carbofuran or phorate at the rate of 5 g ai per plant twice a year was recommended for controlling nematodes in plantations (Ali, 1984). Recently, a bio-control schedule was developed with *Trichoderma* and *Paecilomyces lilacinus* for control of damping off and nematode damage in cardamom nurseries (Eapen and Venugopal, 1995).

2 MINOR DISEASES

A number of minor diseases affecting leaves, capsules and aerial stems occur sporadically in cardamom plantations. While some of these are frequently observed in all areas, others are

Table 7.3 Minor fungal and bacterial diseases in cardamom plantations

Diseases	Parts affected	Casual agents
Leaf blotch	Leaves	*Phaeodactylium alpiniae*
Phytophthora Leaf blight	Leaves	*Phytophthora meadii*
Leaf rust	Leaves	*Phakospora elettariae (Uredo elettariae)*
Leaf spots	Leaves	*Sphaceloma cardamomi, Cercospora zingiberi, Glomerella singulata Phaeotrichoconis crotalariae Ceriospora elettariae*
Sooty mould	Leaves	*Trichosporiopsis* sp.
Stem lodging	Pseudostem (Tillers)	*Fusarium oxysporum*
Anthracnose	Capsules	*Colletotrichum gloeosporioides*
Capsule tip rot	Capsules	*Rhizoctonia solani*
Fusarium capsule rot	Capsules	*Fusarium moniliformae*
Capsule canker (Vythiri spot)	Capsules	*Bacterium* (?)
Capsule ring spot	Capsules	*Marasmius* sp.
Bacterial rot	Rhizomes	*Erwinia chrysanthimi*

restricted to certain specific localities. These include various types of leaf spots and capsule spots, stem infections etc. caused predominantly by fungal pathogens (Table 7.3).

2.1   Leaf blotch

Agnihothrudu (1968) reported a foliar disease in cardamom characterized by typical blotch symptoms on leaves. The disease appears during monsoon season i.e. from June to August months, normally under heavily shaded conditions. Thick shade, continuous rainfall and high atmospheric humidity predispose plants to infection. Leaf blotch was thought to be a minor disease. However, recently it was found spreading in high proportions in certain localities.

2.1.1   *Symptoms*

Symptomatology of leaf blotch disease has been studied in detail (Nair, 1979). During rainy period, round, ovoid or irregular water-soaked lesions develop on middle leaves, usually near the leaf tips or at the mid rib areas. These areas enlarge in size, become dark brown with a necrotic centre. In moist weather, a thick, gray coloured fungal growth is seen on the under side of these blotched areas (Fig. 7.2a). The periphery of lesion shows a dark band of water-soaked zone as the lesions spread. The lesion areas enlarge and characteristic dark and pale brown zonations develop in the blotched areas. However, the lesion spread is limited in size following a dry period.

2.1.2   *Causal organism*

Leaf blotch is a fungal disease caused by *Phaeodactylium venkatesanum* (Agnihothrudu, 1969). Later this fungus was identified as *P. alpiniae* (Sawada) (Ellis, 1971).

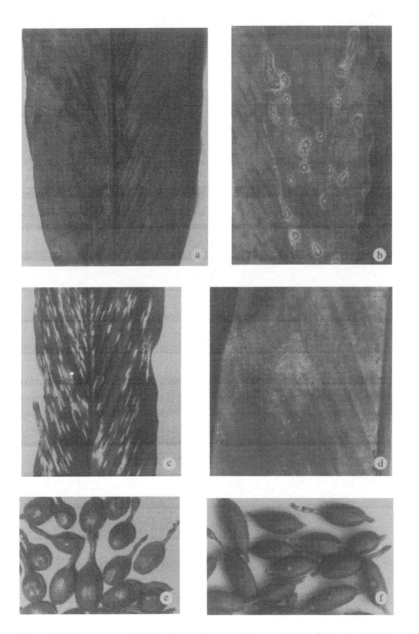

Figure 7.2 Symptoms of leaf and capsule spot diseases: (a) leaf blotch; (b) *Glomerella* leaf spot; (c) *Cercospora* leaf spot; (d) leaf rust; (e) capsule canker; (f) anthracnose.

The pathogen grows profusely on the underside of leaves and also grows abundantly on potato dextrose agar medium. Hyphae are hyaline, smooth, partially submerged, 6–10 μm thick, dichotomously or often trichotomously branched with conidia formed at their tips. Conidia are solitary, hyaline with three transverse septa, smooth, elliptical with tapered basal end and broad apices. Conidia measure 15–25 μm × 4–7 μm. The pathogen infects and produces typical symptoms on *Alpinia* sp., *Amomum* sp. and *Hedychium* sp.

2.1.3   Disease management

Nair (1979) has observed that the fungus was completely inhibited under *in vitro* conditions by Bordeaux mixture (1 per cent), Bavistin (0.1 per cent), or Hinosan (0.15 per cent). Fungicidal spray with copper oxychloride or Bordeaux mixture was reported to control leaf blotch infection in the field (Ali, 1982).

2.2   *Phytophthora* leaf blight

A widespread leaf blight incidence is observed in many cardamom plantations during the post-monsoon season. The infection starts on the young–middle aged leaves in the form of elongate or ovoid, large, brown coloured patches which soon become necrotic and dry off. These necrotic dry patches are seen mostly on leaf margins and in severe cases the entire leaf area on one side of the midrib is found affected. The disease appears during October–November months and may even extend up to January–February. Thick shade, low night temperature and fog prevailing in isolated pockets pre-dispose the plants to leaf blight infection. The causal organism is *Phytophthora meadii*, which can be easily isolated from infected leaf portions using water-floating technique. The infection is aerial and the infected plant debris serves as the source of primary inoculum. The pathogen grows internally and under moist or misty conditions produce abundant sporangia, which are disseminated by wind spreading the disease to other areas. Disease symptoms are seen only on the leaves. Although *Phytophthora* is a potential pathogen infecting all parts of cardamom, the leaf blight isolate is seen specific to only leaves under natural conditions. However cross inoculations of *P. meadii* leaf isolate on capsules and vice versa were found to be infective on plant parts tested under laboratory conditions.

2.2.1   Disease management

Leaf blight infection can rapidly spread to adjacent areas and can result in severe leaf necrosis and leaf drying unless the disease is controlled at the initial stage itself. One round of foliar spray with one per cent Bordeaux mixture, Aliette 0.3 per cent or Akomin 40 (Potassium phosphonate) at 0.3 per cent were found to be effective in limiting the spread of the disease.

2.3   Leaf rust

Thirumalachar (1943) reported the occurrence of a type of rust disease in cardamom plantations of Karnataka. This disease appears after monsoon during October–May. Disease symptoms appear on leaves in the form of numerous yellowish rusty coloured pustules distributed on leaf surface in several patches (Fig. 7.2d). These are mostly seen on the underside of leaves. As disease advances, pustules or uredosori become reddish-brown in colour and they protrude out from the leaf surface. The mature pustules break open and release uredospores. In severe cases infected leaves show several yellowish patches with numerous rusty coloured pinhead spots distributed in these yellowish areas on leaf surface. These areas later dry off as the disease advances.

The disease is caused by the rust fungus *Phakospora elettariae* (Racib.) Cummins (Syn: *Uredo elettariae* Racib.). Naidu (1978) reported a mycoparasite *Darluca filum* (Biv) Cast, hyperparasitising this rust fungus. The mycoparasite produces dark brown to black coloured pycnidia in large numbers, and they protrude out from the uredosori. The hyperparasitised uredospores shrivel off and they do not germinate. *Darluca filum* develops only in advanced stage of rust development. However it helps to prevent further secondary spread of the rust fungus. The spread of leaf rust infection can be minimized by spraying fungicides such as Mancozeb 0.2 per cent (Dithane M45) or Indofil M-45.

2.4   Leaf spot diseases

Cardamom foliage is affected by a number of leaf spot diseases caused by a variety of pathogenic fungi. They occur both at the seedling stage and in mature plants in the plantations. The types of leaf spots occurring in main plantation are dealt here and nursery leaf spots elsewhere in this chapter. Leaf spots generally occur as minor diseases and as such do not cause considerable damage to plants.

2.4.1   *Sphaceloma leaf spot*

Muthappa (1965) reported the occurrence of this leaf spot for the first time from Coorg area of Karnataka. Symptoms of the disease appear on leaves in the form of scattered spherical blotches measuring a few millimetre in diameter. The adjacent lesions coalesce to form large necrotic patches. In Coorg it was reported as a major disease problem. Although the disease is present throughout the year, its abundance and severity are more during the post-monsoon period. This disease is caused by *Sphaceloma cardamomi* Muthappa. Naidu (1978) reported that Ceylon and Alleppey Green cultivars in Coorg area showed resistance to *Sphaceloma* leaf spots. Cultivars having erect panicles are mostly resistant to leaf spot while cultivars with creeping or prostrate panicles are susceptible.

2.4.2   *Cercospora leaf spot*

Another leaf spot occurring in Coorg area was reported by Rangaswamy *et al.* (1968). Symptoms originate on the leaf blade as water-soaked linear lesions, which are rectangular and parallelly arranged alongside the veins. On upper leaf surface, lesions turn dark brown with dirty white long patches in the centre (Fig. 7.2c). In advanced stages, lesions become grayish-brown in colour and later these areas dry off. The disease is caused by *Cercospora zingiberi* Togashi Katsaki. The fungus produces conidiophores in clusters from many-celled dark brown stroma. Conidiophores are simple or branched rarely, septate, straight or curved, geniculate and often undulate at the tip and light brown coloured. The conidiophores measure 17.5–56 μm × 5.23–3.5 μm. Conidia are formed singly, linear, indistinctly septate with 3–6 septa, mostly curved with obtuse base 37–195 μm × 1.75–2.5 μm. Naidu (1978) observed that cultivars having erect (var. *Mysore*) panicle are relatively resistant to *Cercospora* leaf spot compared to var. *Malabar* having prostrate panicles.

2.4.3   *Glomerella leaf spot*

Nair (1979) reported the occurrence of a leaf spot disease characterized by the presence of circular–ovoid dark brown, concentric spots on the middle leaves. This disease appears during the post-monsoon period in isolated pockets. The disease is generally seen only in var. *Malabar*.

Infection starts as small pale yellow water-soaked lesions on leaves, which may be irregular in shape measuring 1–2 mm in size, later enlarge in size and form a depressed central area surrounded by a dark band of tissue. Later, alternate concentric dark and pale brown bands develop with yellow halo around the entire spot (Fig 7.2b). Large mature spots may coalesce and the lesion areas start drying. Sometimes the lesion areas measure as large as about 4 cm in diameter. The fruiting bodies of the fungus are seen as dark brown dot-like structures in the lesion areas.

The causal organism is identified as *Glomerella cingulata* Stoneum Spanding and Shronk. The fungus forms grayish white mycelial growth in potato dextrose agar medium which becomes dark gray with zonations. Acervuli are produced in cultures. Conidiophores short, hyaline, conidia cylindrical, hyaline and aseptate 12–25 μm × 3–5 μm in size. Perithecia are globose dark brown coloured, ostiolate and measure 85–135 μm in diameter.

2.4.4   *Phaeotrichoconis leaf spot*

Another type of leaf spot caused by *Phaeotrichoconis* was reported by Dhanalakshmy and Leelavathy (1976). Symptoms formed on young and old leaves are characterized by irregular papery white spots with brown margins on leaf blade. Under moist conditions the lesions enlarge and coalesce. During dry weather the central portion of lesions dries off. Causal organism is identified as *Phaeotrichoconis crotalariae* (Salam and Rao) Subram. The pathogen grows profusely in culture and produces yellowish-brown mycelium with numerous dark brown sclerotia. Conidiophores are indistinguishable from the hyphae, conidia solitary, elongate, fusoid, straight or slightly curved and 5–8 septate.

2.4.5   *Ceriospora leaf spot*

Yet another type of leaf spot seen rarely on cardamom leaves was reported by Ponnappa and Shaw (1978) from Coorg and is caused by *Ceriospora elettariae* Ponnappa and Shaw. The symptoms are appearance of numerous spots on the foliage, which are circular or oval, up to 8 mm diameter and they coalesce to form larger patches. The lesion centre is dirty white surrounded by light brown, circular necrotic areas.

2.4.6   *Management of leaf spot diseases*

Many of the leaf spot diseases described above occur sporadically in minor proportions and as such do not have much deleterious effect on crop yield. The spread of these can be prevented by one or two rounds of spraying with common fungicides such as Bordeaux mixture (1 per cent) or Mancozeb (0.25 per cent).

2.5   Sooty mould

A sooty mould infection on leaves of cardamom growing under the shade tree *Cedrella toona* Roxb. was reported by Nair (1979). The disease appears during January–February months when the shade trees are in blossom. Infection starts as minute scattered dark mycelial growth on the upper leaf surface. This spreads rapidly and covers the entire lamina and in severe cases extends to the petioles and leafy stems, which are later, covered with black mycelial growth. In advanced stages, leaves tear off at margin along the veins and dry prematurely. The sooty mould fungus is identified as *Trichosporiopsis* sp.

2.6   Stem lodging

A relatively new disease affecting the leafy stem (tillers) of cardamom has been found to occur in several plantations in Idukki district of Kerala and in lower Pulney area in Tamil Nadu (Dhanapal and Thomas, Unpublished). The disease attacks the middle portion of tillers in the form of pale discoloured patches, which lead to a sort of dry rotting. The leafy stem is weakened at this portion and leads to partial breakage. The partially broken tillers bend downwards and hang from the point of infection. Where infection occurs at lower region of tillers, they fall off giving a lodged appearance. In such tillers leaves and leaf sheaths dry up soon. The disease is caused by *Fusarium oxysporum*, and appears usually during post-monsoon period.

2.7   Anthracnose

Anthracnose occurring on cardamom capsules was reported as a minor disease in certain localities of cardamom cultivation (Suseela Bhai and Thomas, 1988). Symptoms appear on capsules as reddish brown round or oval spots of 1–2 mm diameter, often with a soft depressed centre (Fig. 7.2f). The lesions vary in number and size and in rare cases coalesce to form large lesions. Often less than 2 per cent disease incidence is noticed. But in Anamalai areas of Tamil Nadu as high as 10–28 per cent incidence also was noticed.

2.7.1   *Causal organism*

Colletotrichum gloeosporioides (Penz) Penz and Sacc has been shown to be the causative organism of anthracnose disease (Suseela Bhai *et al.*, 1988). The fungus grows profusely in potato dextrose agar medium producing dark, gray coloured, dense mycelium. Setae are dark brown, conidia abundant, cylindrical, straight 12–24 μm × 2.5–5 μm.

A similar infection of *C. gloeosporioides* on capsules resulting in the formation of much large lesions often extending upto three-fourth area of the capsules occur in plantations of Karnataka state. This severe form of anthracnose leads to decay and loss of infected capsules. Fungicides such as Cuman-L, Foltaf or Bavistin when sprayed 3 times at 0.3 per cent concentration was found to control the disease.

2.8   Capsule tip rot

A characteristic type of rotting of the tip of capsules is of common occurrence in plantations of Karnataka. The disease makes its appearance as small water-soaked lesions at the distal end of the capsule, which later spreads downwards. The tip portions of the infected capsules and often up to the middle from the tip, exhibit decaying symptoms. In advanced stages, the rotting extends to the entire capsule. *Rhizoctonia solani* is the causal agent of capsule tip rot. Spraying capsules with Foltaf (0.2 per cent) or Bavistin (0.2 per cent) or Copper oxychloride (0.2 per cent) reduces the disease spread.

2.9   *Fusarium* capsule disease

Wilson *et al.* (1979a) reported a type of capsule disease caused by *Fusarium moniliforme* Sheld. Infection appears as small lesions on capsule rind which later decays and the lesion's periphery turns reddish-brown in colour. In severe infection, entire capsules decay during rainy period. The disease symptoms described by Wilson *et al.* (1979a) closely resemble those of anthracnose, but *Fusarium* infection, often leads to decay of capsules.

2.10   Capsule canker

Agnihothrudu (1974) reported a type of capsule spot suspected to be caused by *Xanthomonas* sp. This is locally known as *Vythiri* spot and was initially found in Wynad areas. Later, occurrence of the disease was observed in several cardamom plantations. Symptoms develop on capsule rind as raised shining blisters or eruptions, which are pale to silvery white in colour (Fig. 7.2e), sometimes extending to cover almost half the area of the capsules. The nature of the causal organism is not established beyond doubt, as no pathogenic fungi or bacterium was found associated with these spots. The disease occurs only in minor proportions and is not alarming since no crop loss is observed due to infection. However, infected capsules fetch lesser price in cardamom auctions, as these blisters are clearly visible on cured capsules.

2.11   Capsule ring spot

A rare infection of capsules is noticed in certain plantations in Karnataka. The symptoms are characteristic reddish-brown concentric rings or zonations, which develop on capsule rind. These areas turn necrotic and later dry off. This infection is suspected to be caused by *Marasmius* sp., however it requires further detailed investigations.

2.12   Erwinia rot

Tomlinson and Cox (1987) reported a serious rot disease of cardamom in Papua New Guinea. Symptoms of the disease are manifested on the foliage as yellowing of leaves of mature plants. Rotting and collapse of leafy stems at ground level often accompany this. A pale brown colour develops on rhizomes, which later leads to decay of rhizomes. Roots become blackened and necrotic in advanced stages of infection. Infection is observed in the var. *Malabar* and often leads to collapse of the entire plant.

2.12.1   *Causal organism*

The disease has been reported to be caused by a gram-negative bacterium, which has been identified as a strain of *Erwinia chrysanthemi* Burkholder. Tomlinson and Cox (1987) have isolated this bacterium from infected cardamom rhizomes and roots and were found to be pathogenic. The bacterium has been biochemically characterized and identified as a pectolytic bacterium readily grows on crystal violet peptone agar (cvp agar). Colonies slightly raised with fried-egg appearance having distinct orange coloured centre. Pathogenic isolates are KOH soluble gram-negative rod-shaped bacteria. The bacterium survives in infected rhizomes and roots and also in the rhizosphere.

3   DISEASES IN NURSERIES

Cardamom is propagated mainly through seeds, which are raised in nurseries. The seedlings are retained for about 10–18 months in the nursery before they attain the age of field planting. Normally the nurseries consist of two stages, the primary nursery and the secondary nursery. Major diseases occurring in seedlings are given in Table 7.4.

Table 7.4 Diseases of cardamom in nurseries

Diseases	Parts affected	Causal agents
Primary nursery		
Damping off	Young seedlings	*Rhizoctonia solani*
		Pythium vexans
Seed/seedling rot	Seeds, young seedlings	*Fusarium oxysporum*
Seedling rot	Leaf/leaf sheath, Pseudostem	*Sclerotium rolfsii*
Leaf spot	Young leaves	*Phyllosticta elettariae*
Secondary nursery		
Seedling rot (clump rot)	Rhizomes, tillers roots and leaves	*Pythium vexans* *Rhizoctonia solani*
Leaf spot	Leaves	*Colletotrichum gloeosporioides*

3.1   Damping off

Wilson *et al.* (1979b) observed the incidence of damping off in young seedlings at the age of 1–6 months. Affected seedlings become pale green and wilt suddenly in masses, as their collar portion rots. Overcrowding of seedlings and excess soil moisture are the predisposing factors of this disease. The causal organism of damping off was identified as Rhizoctonia solani (Wilson *et al.*, 1979b) and *Pythium vexans* (Nambiar *et al.*, 1975).

3.2   Seedling rot or clump rot

This disease is similar to the rhizome rot disease of plantations. Usually, the disease is observed in nurseries where the seedlings attain an age of 6–12 months and is often seen during rainy season in overcrowded nurseries. The disease symptoms are characterized by wilting and drooping of leaves. Leaves turn pale yellow, followed by rotting of the collar portion of seedlings. As infection advances the young tillers fall off and the entire seedling collapses. The causal organisms reported are *R. solani* and *P. vexans*. In some nurseries, seedlings are affected by root rot alone. In such cases only *Fusarium* sp. was found to be pathogenic. Ali and Venugopal (1993) have reported the association of root knot nematode, *Meloidogyne incognita*, along with *R. solani* and *P. vexans*.

Siddaramaiah (1988a) reported the occurrence of seed rot disease resulting in the wilting of seedlings. The disease is caused by *Fusarium oxysporum*. Another seedling disease caused by *Sclerotium rolfsii*, which results in the rotting of leaves, leaf sheath and leafy stem, was also reported by Siddaramaiah (1988b).

3.2.1   Disease management

Pattanshetty *et al.* (1973) reported that pre-sowing treatment of nursery beds with 2 per cent formaldhyde improved seed germination and reduced damping-off incidence. Thomas *et al.* (1988) reported fungicidal control of seedling rot and damping off by soil drenching with Emisan 0.2 per cent or Mancozeb 0.4 per cent or Brassicol 0.2 per cent. Seed dressing with *Trichoderma harzianum* followed by one or two rounds of *T. harzianum* in nursery beds at 30 days intervals has been found to reduce the incidence of seedling rot disease.

3.3   Nursery leaf spot

Incidence of leaf spot is a serious problem in nurseries amounting to severe loss of seedlings. The disease was reported by Subha Rao (1939) and later Mayne (1942) identified the causal organism as *Phyllosticta* sp. The pathogen was isolated and studied in detail by Chowdhary (1948) who identified it as *Phyllosticta elettariae* Chowdhary. The disease occurs mainly in the primary nursery on tender leaves as minute water-soaked lesions almost circular in shape with light coloured periphery and a depressed necrotic centre. This central portion later dries off and becomes papery white. In later stages, shot holes are formed at the lesion centre. As disease advances numerous such lesions of varying sizes develop and the entire leaf dries off. Several minute dark pinheads like pycnidia of the fungus can be seen in the lesion areas. The older leaves of the seedlings are less susceptible to the disease. As the seedlings grow old they develop resistance to infection. The disease can be easily controlled by spraying with fungicides such as Difolatan (0.2 per cent) or Bordeaux mixture (1 per cent) or Dithane (0.2 per cent) when sprayed at 15-day intervals (Rao and Naidu, 1974).

3.4   Leaf spot in secondary nursery

Another type of leaf spot is observed in 6–12 months old seedlings in the secondary nursery. The disease is characterized by the development of many rectangular water-soaked lesions on the foliage. These lesions enlarge longitudinally and are parallely arranged along the side of the veins. As they mature, they exhibit a muddy red colour and become necrotic. The leaves dry off as too many lesions occur side by side. The disease is caused by *Colletotricum gloeosporioides*. Spraying the foliage with Mancozeb (0.25 per cent) is effective in reducing the disease spread.

4   CONCLUSION

Fungal diseases of cardamom are relatively easier to control than the more devastating systemic infections caused by viruses. However the use of fungicides and insecticides are being discouraged due to the strong antipathy of consumers for the use of phyto-chemicals. In view of the increased importance and interest in organically grown spices, it is essential to evolve effective biocontrol strategies against the more serious fungal diseases. A protocol for the production of organic cardamom needs to be developed and popularized in order to cater to the demand in the international market.

Extensive and intensive search for natural resistance to the pathogens needs to be initiated. The Western Ghats, being the centre of diversity for cardamom, the possibility of locating resistance is fairly high. Once such resistance genes are located, they can be incorporated in to the elite cultivars by conventional breeding. Where resistant genes are absent, biotechnological approaches may have to be resorted to for developing resistant genotypes.

REFERENCES

Agnihothrudu, V. (1968) Description of the fungus *Phaeodactylium venketesanum* on cardamom. *Proc. Indian Acad Sci. Sect. B.*, **68**, 206–209.

Agnihothrudu, V. (1969) A leaf disease of cardamom from Kerala with a note on fungi found on cardamom and allied genera all over the world. *Cardamom News Annual*, 1969, 35–40.

Agnihothrudu, V. (1974) Is there a bacterial disease in cardamom? *Cardamom News*, 6, 5.

Alagianagalingam, M.N. and Kandaswamy, T.K. (1981) Control of capsule rot and rhizome rot of cardamom (*Elettaria cardamomum* Maton). *Madras Agric. J.*, **68**, 564–567.

Ali, M.I.M. (1982) Field evaluation of fungicides against leaf blotch disease of cardamom. *Pesticides*, 11, 38–39.

Ali, S.S. (1983) Nematode problems in cardamom and their control measures. *Sixth Workshop of all India Coordinated Spices and Cashew nut Improvement project*, Calicut, November 10–13, 1983.

Ali, S.S. (1984) Effect of three systemic nematicides against root knot nematodes in a primary nursery of cardamom. *First Int. Cong. Nematol.. Ontario*, August, 1984 (Abs.).

Ali, S.S. and Venugopal, M.N. (1993) Prevalence of damping off and rhizome rot disease in nematode infested cardamom nurseries in Karnataka. *Current Nematology*, 4(1), 19–24.

Anonymous (1955) *Final Report of the ICAR.* Scheme for scientific aid to cardamom industry in South India. Madras state (From October 1944 to March 1954).

Anonymous (1972) *Eighth Annual Report.* University of Agricultural Sciences, Bangalore, p. 191.

Anonymous (1986) *Annual Report 1986.* ICRI, Myladumpara, pp. 51–53.

Anonymous (1989a) *Bi-annual Report 1987–89.* Indian Cardamom Research Institute, Myladumpara, pp. 41–47.

Anonymous (1989b) *Annual Report for 1988–89.* National Research Centre for Spices, Calicut, Kerala, India, pp. 37–38.

Chowdhary, S. (1948) Notes on Fungi from Assam. *Lloydia*, **21**, 152–156.

Dantanarayana, D.M., Peries, O.S. and Liyange, A.D.E. (1984) Taxonomy of *Phytophthora* species isolated from rubber in Sri Lanka. *Trans. Brit. Mycol. Soc.*, **82**, 113–126.

Dhanalakshmy, C. and Leelavathy, K.M. (1976) Leaf spot of cardamom caused by *Phaeotrichoconis crotalariae. Plant Dis. Reporter*, **60**, 188.

Dhanapal, K. and Thomas, J. (1996) Evaluation of *Trichoderma* isolates against rot pathogens of cardamom. In K. Manibhushan Rao and A. Mahadevan (eds) *Recent Trends in Biocontrol of Plant Pathogens*, Today and Tommorrow Publishers, New Delhi, pp. 67–65.

Eapen, S.J. and Venugopal, M.N. (1995) Field evaluation of *Trichoderma* sp. and *Paecilomyces lilacinus* for control of root knot nematodes and fungal disease in cardamom nurseries. *Indian J. Nematol.*, **25**, 115–116 (Abs.).

Ellis, M.B. (1971) *Dematiaceous hyphomycetes.* Commonwealth Mycological Institute. Kew, Surrey, England, p. 608.

George, M., Joseph T., Potty, V.P. and Jayasankar, N.P. (1976) A bacterial blight disease of cardamom. *J. Plantation Crops*, 4, 23–24.

George, M. and Jayasankar, N.P. (1977) Control of Chenthal (Bacterial blight) disease of cardamom with penicillin. *Curr. Sci.*, **46**, 237.

George, M. and Jayasankar, N.P. (1979) Distribution and factors influencing chenthal disease of cardamom. In *Proceedings of PLACROSYM-II*, CPCRI, Kasaragod, 343–347.

Govindaraju, C., Thomas, J. and Sudharsan, M.R. (1996) 'Chenthal' disease of cardamom caused by *Colletotrichum gloeosporioides* Penz and its management. In N.M. Methew and C.K. Jacob (eds) *Developments in Plantation Crop Research*, Allied pub., New Delhi, 255–259.

Manomohanan, T.P. and Abi Cheeran (1984) *Elettaria cardamomum*, A new host for *Phytophthora palmivora* (Butler). In *Proc. of PLACROSYM-VI*, CPCRI, Kasaragod, pp. 133–137.

Mayne, W.W. (1942) Report on cardamom cultivation in South India. *Misc. Bull.* No. 50, Ind. Counc. of Agric. Res, India, p. 67.

Menon, M.R., Sajoo, B.V., Ramakrishnan, C.K. and Remadevi, L. (1972) A new *Phytophthora* disease of cardamom (*Elettaria cardamomum (L.)* Maton). *Curr. Sci.*, 41, 231.

Menon, M.R., Sajoo, B.V., Ramakrishnan, C.K. and Rema Devi, L. (1973) Control of *Phytophthora* diseases of cardamom. *Agric. Res. J. Kerala*, 11, 93–94.

Muthappa, B.N. (1965) A new species of *Sphaceloma* on cardamom from India. *Sydowia*, 19, 143–145.

Naidu, R. (1978) Screening of cardamom varieties against *Sphaceloma* and *Cercospora* leaf spot diseases. *J. Plantation Crops*, 6, 48.

Nair, C., Zachariah, P.K. and George, K.V. (1982) Control of panicle rot disease of cardamom. In *Proc. of PLACROSYM-V*, CPCRI, Kasaragod, pp. 133–137.

Nair, R.R. (1979) *Investigations of fungal diseases of cardamom*. Ph.D Thesis, Kerala Agricultural University. Vellanikara, Trissur, p. 161.

Nair, R.R. and Menon, M.R. (1980) *Azhukal* disease of cardamom. In K.K.N. Nambiar (ed.) *Proceedings of the workshop on Phytophthora Diseases of Tropical Cultivated Plants*, CPCRI, Kasaragod, pp. 24–33.

Nambiar, K.K.N. and Sarma, Y.R. (1974) Chemical control of capsule rot of cardamom. *J. Plantation Crops*, 2, 30–31.

Nambiar, K.K.N. and Sarma, Y.R. (1976) Capsule rot of cardamom. *Pythium vexans* de Bary as a causal agent. *J. Plantation Crops*, 4, 21–22.

Nambiar, M.C., Pillai, G.B. and Nambiar, K.K.N. (1975) Diseases and pests of cardamom – a resume of research in India. *Pesticides Annual*, 1975.

Park, M. (1937) Report on the work of the Mycological Division. *Admn. Rep. Div. Agric.*, Ceylon, 1936, pp. 1728–1735.

Pattanshetty, H.V., Deshpande, R.S. and Sivappa, T.G. (1973) Cardamom seedlings can be protected against damping off disease by the treatment with formaldehyde. *Curr. Res.*, 2, 20–21.

Ponnappa, K.M. and Shaw, G.G. (1978) Notes on the genus *Ceriospora* in India. *Mycologia*, 70, 859–862.

Rao, D.G. and Naidu, R. (1974) Chemical control of nursery leaf spot disease of cardamom caused by *phyllosticta elletariae*. *J. Plantation Crops*, 2, 14–16.

Radha, K. and Joseph, T. (1974) Investigations on the bud rot disease (*Phytophthora palmivora* Butl.) of coconut. *Final Report* of PL.480 1968–1973, p. 30, CPCRI, Kayamkulam.

Rangaswami, G., Seshadri, V.S. and Lucy Channamma, K.M. (1968) A new *Cercospora* leaf spot of cardamom. *Curr. Sci.*, 37, 594–595.

Ramakrishnan, T.S. (1949) The occurrence of *Pythium vexans* de Bary in South India. *Indian Phytopath.*, 2, 27–30.

Ramana, K.V. and Eapen, S.J. (1992) Plant parasitic nematodes of black pepper and cardamom and their management. In *Proc. of National Seminar on Black pepper and Cardamom*, 17–18th May, 1992, Calicut, Kerala, pp. 43–47.

Sastry, M.N.L. and Hegde, R.K. (1987) Pathogenic variation in *Phytophthora* species affecting plantation crops. *Indian Phytopath.*, 40(3), 365–369.

Sastry, M.N.L. and Hegde, R.K. (1989) Variability of *Phytophthora* species obtained from plantation crops of Karnataka. *Indian phytopath.*, 42(3), 421–425.

Siddaramaiah, A.L. (1988a) Seed rot and seedling wilt – a new disease of cardamom. *Curr. Res.*, 17(3), 34–35.

Siddaramaiah, A.L. (1988b) Stem, leaf sheath and leaf rot diseases of cardamom caused by *Sclerotium rolfsii* from India. *Curr. Res.*, 17, 51.

Subba Rao, M.K. (1938) Report of the Mycologist 1937–38. *Admn. Rept. Tea Sci. Dept. United. Plant. Assoc. S. India*, 1937–1938, pp. 28–42.

Subha Rao, M.K. (1939) Report of the Mycologist 1937–1939, 1937–38. *Admn. Repts. Tea. Res. Dept. United. Plant. Assoc. S. India*, pp. 28–37.

Suseela Bhai, R., Thomas, J. and Naidu, R. (1988) Anthracnose – A new disease of cardamom. *Curr. Sci.*, 57, 1346–1347.

Suseela Bhai, R., Thomas, J. and Naidu, R. (1993) Biological control of 'Azhukal' disease of small cardamom caused by *P. meadii* Mc. Rae. *J. Plantation Crops*, 21(suppl.), 134–139.

Suseela Bhai, R., Thomas, J. and Naidu, R. (1994) Evaluation of carrier media for field application of *Trichoderma* sp. in cardamom growing soils. *J. Plantation Crops*, 22(1), 50–52.

Suseela Bhai, R., Thomas, J. and Sarma, Y.R. (1997) Biocontrol of Capsule rot of cardamom. Paper presented in '*International Conference on Integrated Plant Disease Management for Sustainable Agriculture*', November 10–15, New Delhi.

Thankamma, L. and Pillai, P.N.R. (1973) Fruit rot and leaf rot disease of cardamom in India. *F.A.O. Plant Prot. Bull.*, 21, 83–84.

Thirumalachar, M.J. (1943) A new rust disease of cardamom. *Curr. Sci.*, 12, 231–232.

Thomas, J., Naidu, R. and Suseela Bhai, R. (1988) Rhizome and root rot diseases of cardamom. A review. In *Proc. of the Workshop on Strategies of the Management of Root Disease Incidence in Plantation Crops*, 14th–18th January, 1988, pp. 38–45.

Thomas, J., Suseela Bhai, R. and Naidu, R. (1989) Comparative efficacy of fungicides against *Phytophthora* rot of small cardamom. *Pesticides*, 40–42.

Thomas, J., Suseela Bhai, R. and Naidu, R. (1991a) Capsule rot disease of cardamom *Elettaria cardamomum* (Maton) and its control. *J. Plantation Crops*, 18(suppl.), pp. 264–268.

Thomas, J., Suseela Bhai, R., Vijayan, A.K. and Naidu, R. (1991b) Management of rot diseases of cardamom through bio-agents. In *National Seminar on Biological Control in Plantation Crops*. (Abs.), June 27–28, RRII, Kottayam, Kerala, p. 21.

Thomas, J. and Vijayan, A.K. (1994) Occurrence, severity, etiology and control of rhizome disease of small cardamom.

Thomas, J., Suseela Bhai, R., Dhanapal, K. and Vijayan, A.K. (1997) Integrated management of rot diseases of cardamom. Paper presented in *International Conference on Integrated Plant Disease Management for Sustainable Agriculture*, November 10–15, New Delhi.

Tomlinson, D.L. and Cox, P.G. (1987) A New disease of cardamom (*Elettaria cardamomum*) caused by *Erwinia chrysanthemi* in Papua New Guinea. *Plant Pathol.*, 36, 79–83.

Wilson, K.I., Rahim, M.A. and Luke, P.L. (1974) *In vitro* evaluation of fungicides against azhukal disease of cardamom. *Agric. Res. J. Kerala*, 12, 94–95.

Wilson, K.I., Sasi, P.S. and James Mathew, J. (1979a) *Fusarium* capsule disease of cardamom. *Curr. Sci.*, 48, 1005.

Wilson, K.I., Sasi, P.S. and Rajagopalan, B. (1979b) Damping off of Cardamom caused by *Rhizoctonia solani* Kuhn. *Curr. Sci.*, 48, 364.

8 Insect pests of cardamom

B. Gopakumar and S.S. Chandrasekar

1 INTRODUCTION

Cardamom is infested by many pests, right from the seedling stage to the cured cardamom in storage. The pest problem has become a factor limiting the productivity of the crop (Anonymous, 1985c). Among its enemies, which range from viruses to mammals, insects undoubtedly are the most destructive. Nearly 60 species of insect pests infest cardamom (Kumaresan and Varadarasan, 1987) at different stages of its growth (Table 8.1). Based on severity of infestation, these pests are categorized as major and minor pests (Kumaresan *et al.*, 1988, 1989b; Premkumar *et al.*, 1994); the former include thrips, shoot borer, root grub, whitefly and hairy caterpillars and the latter capsule borers, root borer, rhizome weevil, shoot fly, lace wing bug, cut worm, aphid, scale insects, leaf folder, spotted grasshopper, leaf grub, mid-rib caterpillar, storage pests and red-spider mites. Kumaresan *et al.* (1989b) tried to classify them as foliar, subterranean and pests on reproductive parts based on the plant parts they infest. At different stage of the crop the pest complex varies.

2 MAJOR PESTS

2.1 Cardamom thrips {*Sciothrips cardamomi* (Ramk.)}

The cardamom thrips (Fig. 8.1) is the most destructive and persistent pest of cardamom. Ayyar (1935) reported its infestation on cardamom at Anamalai Hills of Tamil Nadu, and described the pest as *Taeniothrips cardamomi*, which was later renamed by Bhatti (1969) as *Sciothrips cardamomi*. Nature of damage and biology of the insect were described by Cheriyan and Kylasm (1941), Nair (1975), Anonymous (1985a), Kumaresan *et al.* (1988) and Krishnamurthy *et al.* (1989).

The larvae and adults lacerate tissues from leaf sheaths, unopened leaf spindle, panicles, flowers, tender capsules and suck the exuding sap resulting in qualitative and quantitative loss. Infestation on panicles results in its stunted growth (Fig. 8.2a), while infestation on flowers leads to shedding of flowers. Laceration of tissues from tender capsules leads to the formation of small scabs, which develop as prominent ugly growths when the capsules mature. These scabs generally appear as longitudinal lines over the ridges of the capsules or as patches over it (Fig. 8.2b). Such capsules appear malformed, shriveled with slits on the outer skin. They lack the usual aroma and fetch

Table 8.1 Insect pests of cardamom

Common name	Scientific name	Family	Order
Insect pests			
Wingless grasshopper	*Orthacris* sp.	Acrididae	Orthoptera
Spotted grasshopper	*Aularches miliaris*	Acrididae	Orthoptera
Thrips	*Sciothrips cardamomi* (Ramk.)	Thripidae	Thysanoptera
Thrips	*Panchaetothrips indicus* Bagnall	Heliothripidae	Thysanoptera
Leaf bug	*Riptortus pedestris* Fabr.	Alydidae	Heteroptera
Banana lace wing bug	*Stephanitis typicus*	Tingidae	Heteroptera
Plant lice (aphid)	*Pentalonia nigronervosa* f *typica*	Homoptera	Aphididae
Spittle bug	*Aphrophora nuwarans* Dist.	Cercopidae	Homoptera
Brown spittle bug	*Eosocarta nilgiriensis*	Cercopidae	Homoptera
Banded spittle bug	*Cosmoscarta thoracica* Distant	Cercopidae	Homoptera
Leaf hopper	*Tettigoniella feruginea*	Cicadellidae	Homoptera
Pink hopper	*Bothrogonia* sp.	Cicadellidae	Homoptera
Scale insect	*Diaspis* sp.	Diaspididae	Homoptera
Scale insect	*Aulacaspis* sp.	Diaspididae	Homoptera
Soft brown scale	*Saissetia coffeae* Walk.	Coccidae	Homoptera
Whitefly	*Kanakarajiella cardamomi*	Aleurodidae	Homoptera
Whitefly	*Aleurotuberculatus cardamomi*	Aleurodidae	Homoptera
Whitefly	*Dialeurodes citri*	Aleurodidae	Homoptera
Sooty mealy bug	*Planococous citri*	Pseudococcidae	
Leaf beetle	*Lema admiralis* Jac.	Chrysomelidae	Coleoptera
Leaf beetle	*Lema coromandeliana* (F.)	Chrysomelidae	Coleoptera
Leaf beetle	*Lema fulvimana* Jac.	Chrysomelidae	Coleoptera
Scolytid beetle	*Thammurgides cardamomi*	Chrysomelidae	Coleoptera
Beetle borer	*Onthophagus* sp.	Scarabaeidae	Coleoptera
Beetle borer	*O. coorgensis* Arrow	Scarabaeidae	Coleoptera
White grub	*Holotrichia* sp.	Melonthidae	Coleoptera
Spice beetle (storage pest)	*Stegobium panicum*	Anobiidae	Coleoptera
Rhizome weevil	*Prodioctes haematicus* Chev.	Curculionidae	Coleoptera
Bag worm	*Acanthopsyche bipar* Wlk.	Psychidae	Lepidoptera
Shoot, panicle and capsule borer	*Conogethes punctiferalis* Guen.	Pyralidae	Lepidoptera
Leaf webber	*Polythiipta macralis* Led.	Pyralidae	Lepidoptera
Early capsule borer	*Jamides* sp.	Lycaenidae	Lepidoptera
Early capsule borer	*Jamides alecto* Feld	Lycaenidae	Lepidoptera
Root borer	*Hilarographa caminodes* Meyr.	Plutellidae	Lepidoptera
Skipper	*Plasioneura alyosa* M.	Hesperiidae	Lepidoptera
Leaf folder	*Homona* sp.	Tortricidae	Lepidoptera
Cardamom skipper	*Notocrypta feisthaelii* BDV	Hesperiidae	Lepidoptera

Table 8.1 Continued

Common name	Scientific name	Family	Order
Cutworm	*Arcilasisa plagiata*	Noctuidae	Lepidoptera
Pale brown caterpillar	*Euproctis lutifolia* Hampson	Lymantridae	Lepidoptera
Hairy caterpillar	*Lenodera vittata* Wlk.	Lasiocampidae	Lepidoptera
Hairy caterpillar	*Alphaea biguttata*	Arctiidae	Lepidoptera
Black hairy caterpillar	*Pericalia ricini* Fb.	Arctiidae	Lepidoptera
Wild silkworm moth	*Attacus atlas* Linn.	Saturnidae	Lepidoptera
Hairy caterpillar	*Eupterote cardamomi*	Eupterotidae	Lepidoptera
Hairy caterpillar	*E. canarica* M.	Eupterotidae	Lepidoptera
Hairy caterpillar	*E. fabia* Cram.	Eupterotidae	Lepidoptera
Hairy caterpillar	*E. testacea*	Eupterotidae	Lepidoptera
Hairy caterpillar	*E. undata*	Eupterotidae	Lepidoptera
Flower and pod borer	*Lampides elpis* G.	Lycaenidae	Lepidoptera
Looper	*Eumelia rosalia* Cram.	Geometridae	Lepidoptera
Seedling dead heart	*Chloropisca* sp.	Chloropidae	Lepidoptera
Pale brown looper	*Anisodes denticulatus* Hamps.	Geometridae	Lepidoptera
Mid rib feeder	*Metapodistis polychrysa* Meyrick	Glyphipterigidae	Lepidoptera
Root gall midge	*Hallomyia cardamomi* Nayar	Cecidomyiidae	Diptera
Shoot fly	*Formosina flavipes*	Chloropidae	Diptera
Non-insect pests			
Red mite	*Dolichotetranychus floridanus*	Tetranychidae	Acarina
Mite	*Tetranychus cucurbitae*	Tetranychidae	Acarina
Nematode	*Meloidogyne incognita*	Meloidogynidae	Tylanchidae
	M. javanica	Meloidogynidae	Tylanchidae
	Radopholus similis	Pratylenchidae	Tylanchidae
	Pratylenchus sp.	Pratylenchidae	Tylanchidae
	P. coffea	Pratylenchidae	Tylanchidae
	Rotylenchus sp.	Pratylenchidae	Tylanchidae
	Trichodorus minor	Trichodoridae	Tylanchidae
	Tylenchor hynchuus sp.	Tylenchoida	Tylanchidae

only a very low price. Seeds of infested capsules are poorly developed and do not germinate. Unprotected crops may suffer upto 80 per cent damage causing about 45–48 per cent crop loss.

Unopened leaf spindles, space between leaf sheaths, flower bracts and flower tubes are the usual breeding sites of the pest. They reproduce by sexual and parthenogenetic methods. Adults are 1.25–1.50 mm long, grayish-brown with fringed wings. After a pre-oviposition period of 3–8 days, it lays minute kidney-shaped eggs singly on racemes and leaf sheaths. During its oviposition period of 4–25 days a female may lay 5–31 eggs. Eggs are 0.25–0.27 mm long and dirty white when freshly laid. After an incubation period of 8–12 days, pale white nymphs emerge. They are 0.5–0.6 mm long and moult into second

Figure 8.1 Major pests of cardamom – thrips (*Sciothrips cardamomi*) – adult.

instar nymphs within 3–5 days. These nymphs are about 1.2 mm long and after a period of 7–9 days moult as pre-pupa. Pre-pupa bears two pairs of wing pads and after about 2 days it moults to form a pupa. Pupal period lasts for 4–6 days. The entire life cycle is completed within 27–33 days. Warm weather of summer months is ideal for rapid multiplication of the pest while a cold humid weather is unsuitable (Chandrasekar and Balu, 1993). Susceptibility to the pest varies with the varieties; var. *Mysore* is the most susceptible followed by var. *Vazhukka*; var. *Malabar* is relatively tolerant (Chandrasekar and Balu, 1993). Population of thrips, which is more in the axils of lower leaf sheaths, establishes a negative correlation with rainfall and humidity, whereas a positive correlation associates with ambient temperature (Chandrasekar and Balu, 1993). Closer spacing between plants is also found to increase thrips infestation (Sridharan *et al.*, 1990).

Besides cardamom, *S. cardamomi* infest *Panicum longipes*, *Hedychium flavescens*, *H. coronarium*, *Amomum cannaecarpum*, *A. involucratum*, *Remusatia vivipara*, *Colocasia antiquorum*, *Zingiber* sp., *Curcuma pseudomontana*, *Costus speciosus*, *Crinum* sp., *Globba ophioglossa* and *Alpinia galanga* (Anonymous, 1985b; Kumaresan *et al.*, 1988).

Several field trials were conducted to evaluate the efficacy of various insecticides against cardamom thrips. Certain organochlorine insecticides recommended earlier for its management were later eliminated and effective low doses for organophosphorus and carbamate insecticides were determined. The earlier recommendations included Nicotinsulphate, benzene hexa chloride (BHC) and dieldrin (Anonymous, 1944–1952; Jones and Aiyyar, 1948; Nair, 1977). BHC dusting, a practice followed since 1945, was stopped completely in 1955 consequent to its failure in many instances. Subsequent experimentations with organophosphorus insecticides and other chemicals showed that quinalphos and dimethoate (each at 0.1 per cent) are effective against thrips (Pillai and Abraham, 1974; Nambiar *et al.*, 1975). Wilson *et al.* (1977) reported quinalphos, phenthoate, leptophose, dimethoate and formothion (each at 0.03 per cent) to be effective

Figure 8.2 Cardamom thrips: (a) affected panicles showing scab; (b) affected capsules. The scab on the capsules reduces value and quality.

against this pest when sprayed 8 times a year at monthly intervals from April. Sprays of quinalphos 0.03 per cent at 40 days interval from April to January was found equally effective (Wilson *et al.*, 1978). Pillai and Abraham (1978) were of the view that quinalphos, phenthoate, dimethoate and phosphamidon (each at 0.1 per cent) were better than other insecticides in controlling thrips. Another field trial for evaluation of certain newer insecticides including 12 formulations as sprays and five as dusts showed that quinalphos 0.05 per cent was most effective among sprays and quinalphos D 1.5 per cent among dusts (Nair *et al.*, 1979). Results of a field trial by Kumaresan (1982) revealed that methidathion, carbosulfan (each at 0.05 per cent) and bendiocarb 0.16 per cent were very effective in controlling thrips. Kumaresan (1983) tested six insecticides including two synthetic pyrethroids, and reported that permethrin, fenvalerate (each at 0.01 per cent) and quinalphos 0.05 per cent when sprayed 8 times a year controlled thrips effectively. Joseph (1983) tested two pyrethroids viz., permethrin (100 ppm) and cypermethrin (60 ppm) in comparison with monocrotophos (300 ppm) and quinalphos (500 ppm); permethrin was found to be better. Seven–eight times spray with quinalphos 0.025 per cent and monocrotophos 0.03 per cent were found effective against thrips (Chandrasekharan, 1984). Gopakumar and Kumaresan (1984) reported that spraying fenthion 0.05 per cent during February, March, April, May, August, September and October was effective against thrips. They found that frequency of insecticide applications can be reduced to five per year by giving 50 days interval between successive applications of quinalphos 0.025 per cent or 0.05 per cent or fenthion 0.05 per cent (Gopakumar and Kumaresan, 1988a). Varadarasan and Kumaresan (1987) reported the spraying schedule for lower Pulneys:

January   – Cypermethrin (0.01 per cent) or Permethrin 0.01 per cent
February – Monocrotophos (0.05 per cent) or Cypermethrin 0.01 per cent
June   – Monocrotophos (0.05 per cent) or Permethrin 0.01 per cent
August   – Monocrotophos (0.05 per cent) or Cypermethrin 0.01 per cent
October   – Cypermethrin (0.01 per cent) or Permethrin 0.01 per cent

An insecticide spray with monocrotophos/quinalphos at panicle initiation stage followed by another spray 30–40 days later with phosalone were found effective against thrips at Mudigere (Krishnamurthy *et al.*, 1989). A spray with fluvalinate 0.01 per cent or quinalphos 0.025 per cent during March, April, May, August and November reduced thrips damage to less than 15 per cent when the first spray in March was imposed after removal of dry drooping leaves (Kumaresan and Gopakumar, 1993). Eapen (1994) reported phorate as an effective granular insecticide against thrips in Karnataka region, where the severity of infestation was not high. Two new formulations of quinalphos viz., Ekalux AF and SAN 6538 1240 CS 045 both at 0.025 per cent reduced thrips damage on capsules to less than 10 per cent, when sprayed 7 times a year (Gopakumar and Chandrasekar, 1996). Detailed studies on efficacy of neem formulations against cardamom thrips were conducted (Gopakumar and Singh, 1994; Gopakumar *et al.*, 1996) and the results indicated that neem oil and other commercial neem formulations were not effective in controlling thrips. Insecticide spray schedule formulated for Kerala, Karnataka and Tamil Nadu regions of India is given below:

Kerala: (i)   February   ⎤
              (ii) March   ⎥
              (iii) April   ⎬ once in 30 days
              (iv) May   ⎦
              (v)   August   ⎤
              (vi) September   ⎬ once in 40 days
              (vii) October to November ⎦

Tamil Nadu: (i)   December/January
              (ii) March/April
              (iii) May/June

	(iv)	August
	(v)	October
Karnataka:	(i)	January
	(ii)	March
	(iii)	May
	(iv)	September / October

To combat thrips, insecticides need to be sprayed only up to 1/3rd portion of the tillers from the base giving adequate coverage on panicles. Approximately 350–450 ml of spray fluid with a high volume sprayer may be required for a spray per bush of 50–60 tillers. In times of acute water scarcity quinalphos 1.5 per cent D or methyl parathion 2 per cent D or phosalone 1.5 per cent D may be dusted @ 25 kg/ha using dusters. Caution may be taken to take harvest either prior to insecticide application or at least 2 weeks after it. The insecticides recommended for application at the specific time are given in the Table 8.2.

2.2   Panicle/capsule/shoot borer (*Conogethes punctiferalis* Guen.)

The panicle, capsule and shoot borer is a serious pest of cardamom in nurseries and main fields in Kerala, Karnataka and Tamil Nadu. However, infestation is severe in Tamil Nadu and Karnataka regions.

Larvae of this pest bore into panicles (Fig. 8.3a) or capsules (Fig. 8.3b) or shoots (Fig. 8.3c) and grow by feeding on the internal tissues. Infested tillers and panicles dry off and capsules become empty. The adult is a medium-sized moth with orange-yellow wings having several black dots over it. Eclosion generally happens towards the close of photophase and the emerging moths rest on under surface of cardamom leaves (Varadarasan *et al.*, 1989). Moths feed on nectar and cause no direct damage to cardamom. A female moth lays about 20–35 eggs singly or in groups of two or three on leaf margins or dry leaf sheath or along leaf veins. Eggs are flat, creamy white and turn pink before hatching. After an incubation period of 5–7 days minute, pale white, black-headed larvae emerge out of eggs. Larvae, up to their second or early third instar stages prefer to bore and feed on unopened leaf or tender panicle or immature capsule. The late third instar larvae bore into shoots, feed on its core and extrude frass and other wastes through the site of entry while tunnelling up. Appearance of extruded fresh frass material at the bore hole confirms the presence of larva within the shoot. Passing through the fourth and fifth instar stages, the larvae become full grown. Full-grown larvae are pale pink, 30–35 mm long and crawl to a place near to the bore hole within

Table 8.2 Insecticides recommended for the management of thrips

Insecticides	Strength (%)
Quinalphos	0.025
Fenthion	0.05
Phosalone	0.07
Chlorpyriphos	0.05
Dimethoate	0.05
Acephate	0.075
Triazophos	0.04
Monocrotophos	0.025
Methylparathion	0.05

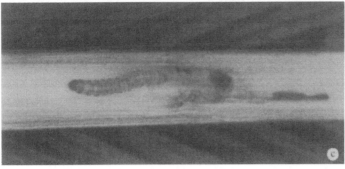

Figure 8.3 Major pests of cardamom – Panicle, capsule and shoot borer: (a) borer damage on panicle; (b) borer damage on capsule; (c) larve of *Conogethes punctiferalis* inside cardamom shoot.

the shoot. During the ensuing pre-pupal period of a day or two, they remain quiescent within a self-made cocoon and soon become pupae (Fig. 8.4b). Pupa is 10–15 mm long, pale green soon after its formation and gradually changes to deep-brown. Pre-pupa and pupa are non-feeding, harmless stages. After 10–12 days of pupal period, moth emerges out through the bore hole. Its life cycle during March–October is completed within 41–68 days, while during the winter months of November–February, it prolongs to 123 days (Varadarasan *et al.*, 1991). The pest population attains peaks during December–January, March–April, May–June and September–October under Lower Pulney conditions, whereas it is highest during January–February, May and September–October under Kerala conditions (Varadarasan *et al.*, 1989).

The pest is polyphagous; the alternate hosts recorded include guava (*Psidium guajava* L.), mango (*Mangifera indica* L.), peaches (*Prunus persica* Benth. and Hook.), pomegranate (*Punica granatum* L.), jack (*Artocarpus integrifolia* L.), ginger (*Zingiber officinale* Rosc.), turmeric (*Curcuma longa* L.), avocado (*Persia gratissima* Gaertn.), mulberry (*Morus alba* L.), loquat (*Eriobotrya japonica* Lindl.), pear (*Pyrus communis* L.), sorghum (*Sorghum bicolor* (L.) Moench.), cocoa (*Theobroma cacao* L.), castor (*Ricinus communis* L.), tamarind (*Tamarindus indica* L.) soapnut (*Sapindus emarginatus* Vahl.), *Caesalpinia bonducella* and *Annona cherimola* (Jacob, 1980). Nambiar *et al.* (1975) observed *Curcuma aromatica*, *C. amada*, *Alpinia* sp., *Amomum* sp. and *Aframomum melegueta* as its alternate hosts. Other recorded alternate hosts include *Amomum subulatum*, *A.microstephanum*, *Hedychium coronarium*, *H. flavescens* and *Alpinia galanga* (Anonymous, 1985b).

Under natural conditions *C. punctiferalis* is host for a number of parasites; *Angitia trochanterata* (Ichneumonidae), *Threonia inareolata*, *Bracon brevicornis* and *Apanteles* sp.

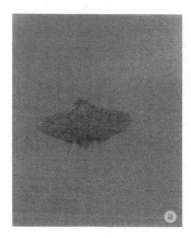 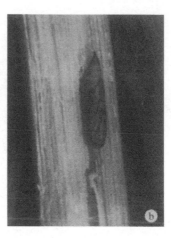

Figure 8.4 Panicle, capsule borer: (a) adult of *C. punctiferalis*; (b) pupa of *C. punctiferalis*.

parasitize its larvae, while *Brachymeria emploeae* parasitise pupae (David *et al.*, 1964). Patel and Gangrade (1971) noticed *Microbracon hebetor* as its larval parasite. Joseph *et al.* (1973) reported two hymenopterans *Brachymeria nosatoi* and *B.lasus* parasitizing on *C. punctiferalis*. Jacob (1981) reported *Myosoma* sp., *Xanthopimpla australis* and a nematode as parasites on *C. punctiferalis*. Varadarasan *et al.* (1990) reported *Temelucha* sp., *Agrypon* sp. and *Friona* sp. as parasites of *C. punctiferalis*. Natural parasitisation by *Agrypon* sp. on larvae of *C. punctiferalis* was maximum (19.8 per cent) during November in Udumbanchola area (Balu *et al.*, 1991).

Several insecticides were evaluated against this pest. David *et al.* (1964), Sulochana *et al.* (1968) and Saroja *et al.* (1973) recommended fenthion, BHC, DDT, endrin, malathion, trichlorfan, methyldemeton, imidan and carbaryl against this pest. In a field trial conducted at Lower Pulneys, monocrotophos 0.1 per cent was found most effective (Kumaresan *et al.*, 1978). From the field trial conducted by Regupathy (1979) carbofuran (2 kg a.i./ha) was found effective. Joseph and Kumaresan (1982) reported that collection and destruction of infested parts and alternate hosts followed by spraying any one of the insecticides viz., lebayacid 375 ml/300 l water/acre or nuvacron 750 ml/300 l water/acre or thiodan 900 ml/300 l water/acre or Ekalux 600 ml/300 l water/acre or zolon 900 ml/300 l water/acre or Rogor 1 l/300 l water/acre is effective control measure against this pest. In a trial with 13 insecticides, Naganathan *et al.* (1983) found garvox, acephate and sumicidin (each at 0.1 per cent) effective against this borer. A strategy of collection and destruction of moths and infested tillers followed by spraying monocrotophos 0.075 per cent or fenthion 0.075 per cent is recommended against this pest (Anonymous, 1985b). Krishnamurthy *et al.* (1989) recommended the removal of affected tillers during September–October if the infestation is less than 10 per cent and spraying quinalphos 0.03 per cent, if it is above 10 per cent.

Even the higher doses of insecticides imposed are not effective against the late stage larvae housed within shoots. Therefore sprays have to be targetted on first or second instar stages of the larvae, which feed on tender panicles or immature capsules. This can be achieved by monitoring adult emergence in field and spraying either monocrotophos 0.075 per cent or fenthion 0.075 per cent within 12–15 days after adult emergence (Varadarasan *et al.*, 1989).

2.3   Root grubs (*Basilepta fulvicorne* Jacoby)

Root grub is a serious subterranean pest of cardamom (Fig. 8.5a,b). It damages roots and thereby obstructs uptake of nutrients, leading to yellowing of leaves and gradual death of plants in case of severe infestation (Gopakumar *et al.*, 1987). The pest is noticed both in nurseries and plantations of Kerala, Karnataka and Tamil Nadu (Varadarasan *et al.*, 1988). However, in Karnataka Thyagaraj *et al.* (1991) observed it as a serious pest only in the primary and secondary nurseries. Biology and management of the pest was studied by Gopakumar *et al.* (1987, 1987a, 1988c, 1991), Kumaresan *et al.* (1989a), Varadarasan *et al.* (1990a, 1991a,b, 1992) and Thyagaraj *et al.* (1991).

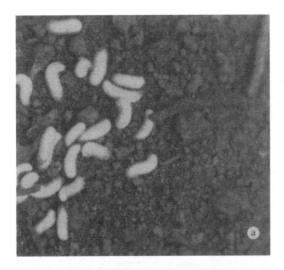

Figure 8.5 Pests of cardamom: (a) root grubs, *Basilepta fulvicerne*; (b) roots damaged by the grubs.

Adult of the pest is a small beetle 4–6 mm in length and of shiny metallic blue, green, or greenish-brown colour. Females are bigger than males. Beetles are polyphagous; jack (*Artocarpus heterophylla*), Indian almond (*Terminalia catapa*), mango (*Mangifera indica*), guava (*Psidium guajava*), ficus (*Ficus indica, F. bengalensis*), cocoa (*Theobroma cacao*), dadaps (*Erythrina lithosperma*), etc. are certain recorded alternate hosts of the beetle (Anonymous, 1993).

In plantations, infestation of this pest is experienced twice a year. Beetles occur during March–April and August–October, assuming peaks at April and September respectively. Beetles fly about in short distances, alighting on leaves of shade trees and cardamom. Copulation occurs during daytime and the mated females after a pre-oviposition period of 4–6 days extrude eggs in groups to a transparent fluid secreted on dry leaf sheaths or leaves, to which it remains glued. Females lay 124–393 eggs in batches of 12–63 during its oviposition period of 8–71 days. Freshly laid eggs are transparent and gradually turn yellow during the incubation period of 8–10 days (at a temperature of 28–31 °C) or 13–19 days (at a temperature of (19–24 °C) and hatch generally during morning hours liberating small creamy white grubs which fall on the ground, penetrate soil, reach root zone of cardamom and start feeding on the roots. Population of grubs and their infestation is more on cardamom plants under thin shade than under thick shade. As with beetles, grubs too have two periods of occurrence, the first during April–July and the second at August–September to December–January, assuming peaks during May–June and November–December respectively. Larvae feed on roots, become fully grown in 45–60 days and appear pale white, 1 cm long, stout and 'c'-shaped. Pupation takes place in an earthen shell where it remains for 10–17 days. It is seen that at a temperature of 19–24 °C, the pest completes its life cycle within 73–111 days. The development is slightly faster at 28–30 °C taking only 65–102 days for completion.

The pest problem can be effectively managed by a judicious integration of mechanical and chemical methods of control (Gopakumar *et al.*, 1987, 1987a, 1988c; Varadarasan *et al.*, 1990a, 1991a, 1993). Collection and destruction of beetles at the periods of their massive emergence and subsequent insecticidal control of grubs are the two methods incorporated in the strategy.

During periods of adult emergence (March–April and August–October) beetles alighting on cardamom plants can be easily collected using an insect net and destroyed. During peak periods of beetle emergence, 2500–3000 beetles could be collected in a day by a labourer. Such massive destruction of beetles drastically reduces the grub population in soil to a low level, which would otherwise have been enormous and caused heavy root damage. As it becomes impossible to trap and destroy the entire beetles emerging in a field, suitable measures for control of grubs in soil become inevitable in endemic areas. Earlier studies have shown that raking up the soil and subsequent application of granular insecticides, (phorate @ 30–40 g/clump or carbofuran @ 50 g/plant, during May–June and September–October) is an effective method (Gopakumar *et al.*, 1987). Subsequently HCHO 0.2 per cent and aldrin 0.1 per cent were found effective (Gopakumar *et al.*, 1987a, 1991). Varadarasan *et al.* (1990a) reported the application of 20–40 g of phorate or chlorpyriphos 0.06 per cent effective against grubs. Later, chlorpyriphos at lower doses of 0.04 per cent was found to give adequate control of grubs (Varadarasan *et al.*, 1991b).

The pest has been found susceptible to infection of entomopathogenic fungi both at grub and adult stages. *Beauveria bassiana* and *Metarrhizium anisopliae* were isolated

Figure 8.6 Root grubs of cardamom: (a) beetles infected by *Beauveria bassiana*; (b) grubs infected by *Metartizium anisopliae.*

from naturally infected beetles and grubs respectively (Varadarasan, 1995). The grubs were also infected by the nematode, *Heterorhabditis* sp. (Varadarasan, 1995). Laboratory studies as well as preliminary field trials with these bioagents have shown convincingly the efficacy of these bioagents, and it is expected that a suitable biocontrol strategy could be developed for management of root grubs (Fig. 8.6a,b).

2.4   Whitefly [*Kanakarajiella cardamomi* (David and Subr.) David and Sundararaj]

Infestation of whitefly was a rare phenomenon in cardamom till 1980s, its sporadic occurrence was limited to Nelliampathy and Vandiperiyar areas (Anonymous, 1980;

Butani, 1984). Its infestation is now noticed at many places in Udumpanchola and Peermedu taluks of Idukki District of Kerala and lower Pulneys of Tamil Nadu. The species of whiteflies reported to infest cardamom are *Dialeurodes cardamomi* David and Subr. (presently known as *Kanakarajiella cardamomi* (David and Subr.) David and Sundarraj (1993), *Aleuroclava cardamomi* (David and Subr.), *Aleurocanthus* sp., *Bemesia tabaci* (Genn.) and *Cockerella diascoreae* Sundararajan and David (Selvakumaran and Kumaresan, 1993); however, only *K. cardamomi* is destructive to cardamom (Fig. 8.7).

Nymphs of the insect are found in large numbers only on the adaxial surface of leaves and adults on different parts of the plant. Both suck the plant sap resulting in yellowing and gradual drying up of plants leading to drastic decline of yields and in certain cases complete destruction of the plants. Nymphs produce a sugary secretion, which drop on lower leaves, where sooty moulds develop, obstructing normal light interception and photosynthesis (Gopakumar *et al.*, 1988b). Adult is a small soft-bodied insect about 2 mm in length having two pairs of white wings. Male is smaller than female. Though adults are not active fliers, they fly about from plant to plant or even small distances, and are often swept off by wind, which is a major mode of migration of the pest. Gopakumar *et al.* (1988b) and Selvakumaran and Kumaresan (1993) studied the biology of the pest. Adults reproduce by parthenogenetic and sexual methods. Adults live for 7–8 days, females lay about 115 eggs singly which are inserted into the stomata on the adaxial leaf surface by means of short sub-terminal stalk. Eggs are pale yellow when freshly laid and turn brown before hatching. Parthenogenetic eggs develop exclusively into females. The first instar, called crawler, is ovate-elongate, greenish-yellow and wanders on the leaf surface in search of suitable feeding sites. Later it anchors itself with its piercing stylet at a suitable place on the leaf, from

Figure 8.7 White fly of cardamom: a nymphs and pupal exuvia of cardamom whitefly, *Kanakarajiella Cardamomi.*

where it sucks up plant sap. It becomes sedentary and grows to a dorso-ventrally-flattened structure possessing 14 pairs of marginal setae. The second instar is deep yellow, elliptical in shape and possesses paired anterior and posterior setae. Third instar is yellow, sub-elliptical, with better-developed legs and anterior and posterior abdominal spiracles. Fourth instar is distinct from other nymphal stages due to the presence of a pair of small red eyes and deep yellow-tinged slightly convex body. Its margin is irregularly crenulate. Puparia, which adhere to the leaves after adult emergence, appear as scabby patches on lower surface of affected leaves. Its life-cycle is completed within 43–60 days. Adults of the pest are very much attracted towards yellow colour and this behaviour is well exploited for trapping them on a yellow surface coated with sticky material. Yellow sticky traps made of metal sheets painted yellow and coated with castor oil has been recommended to trap and kill adult white-flies (Kumaresan *et al.*, 1993). Spraying neem oil 0.5 per cent + triton or sandovit 0.5 per cent on under surface of leaves, 2–3 times at fortnightly intervals during periods of pest infestation, is effective against nymphs (Gopakumar and Kumaresan, 1991). Acephate 0.075 per cent, Ethion 0.1 per cent and Triazophos 0.4 per cent were found equally effective against the nymphs (Gopakumar *et al.*, 1988a, 1988c; Kumaresan *et al.*, 1993; Selvakumaran and Kumaresan, 1993).

Under natural conditions the pest has been found susceptible to a number of natural enemies such as predators like *Mallada bonninensis*, unidentified neuropteran, dipteran, coleopteran and mite; parasitoids such as *Encarsia septentrionalis* and *E. dialeurodes*, and a pathogen *Aschersonia placenta* (Selvakumaran and Kumaresan, 1993; Selvakumaran *et al.*, 1996). Potential of these natural enemies as biocontrol agents of whiteflies is unexplored.

2.5   Hairy caterpillars

Hairy caterpillars of cardamom are a group of defoliators that appear sporadically and cause severe damage to the crop. Incidence of these pests was reported by Puttarudriah (1955), Rajan (1958), Ayyar (1961), John (1967), Joseph *et al.* (1983) and Selvan and Singh (1993). Nine species of hairy caterpillars are known to infest cardamom. They are *Eupterote canaraica* Moore, *E. cardamomi* Renga, *E. fabia* Cram., *E. testaceae* Walk., *E. undata*, *Linodera vittata* Walk., *Euproctis lutifacia* Hamp., *Alphaea biguttata* Walk. and *Pericallia ricini* Fabr. Biology of different species of these hairy caterpillars were described by Nair (1975), Kumaresan *et al.* (1988), Singh *et al.* (1992) and Selvan and Singh (1993). The different *Eupterote* sp. has striking similarities. They are polyphagous larvae, voraciously feeding on shade-tree leaves at early stages and on cardamom leaves at late stages, and have extensive larval period consisting of seven–ten larval instars. They congregate on tree trunks or on cardamom leaves. *Eupterote cardamomi*, *E. undata* and *E. canaraica* are the most destructive among them.

2.5.1   *Eupterote cardamomi*

Adults of *E. cardamomi* emerge from their pupae in June with the commencement of monsoon rains. The moth is brown in colour with double post-medial black lines and several light wavy lines on wings. Female is bigger (wing span of 90 ± 3.3 mm across and 36 ± 1.27 mm along body axis) than the male (wing span of 85.50 ± 1.45 mm across and 36 ± 1.27 mm along body axis). Antennae of the male are plumose while that of female is filiform. Males live longer than the females by 1 or 2 days. Females after pairing lay as many as 563 eggs in batches as clusters during their oviposition period of 5–8 days. Eggs are spherical with flat base, yellow and remain glued to the substratum. The chorion is hard and sculptured. Incubation period lasts for 20–25 days. Nearly 75 per cent of the eggs hatch, the neonate larvae cut a circular hole mid-dorsally and emerge out and they feed on empty egg shells. The first instar is yellowish with pale-brown head, 5.41 ± 0.15 mm long, gregarious and becomes a fully grown larva through 10 instar stages in 125–148 days. A fully grown larva measures nearly 77 mm in length, bears on its body tufts of long white-tipped black hairs on dorsal verrucae armed with needle-like hollow setae, which causes considerable irritation when pricked. The caterpillars, up to their sixth or seventh instar stage remain feeding on shade-tree leaves, from where they descend by silken thread to cardamom plants. The larvae are nocturnal, defoliating cardamom leaves from its margins, leaving only midrib. Prior to pupation, larvae stop feeding, enter 2–5 cm deep in soil, congregate in groups of 5–10, and become sluggish. The larvae after 3–6 days of pre-pupal period pupate inside cocoons spun with silken thread, debris and larval body hairs. The pupae are adecticous, obtect with spindle-shaped body and are dark brown in colour. They measure roughly 30 mm in length and 12 mm in width and have a long duration of 212–230 days. The entire life cycle is completed within 365–410 days.

2.5.2   *Eupterote undata*

Moths of *E. undata* are brownish-yellow or yellow, with black double post-medial lines and many black wavy lines on wings. Wingspan of female is 7 cm and that of male is 6.5 cm. Female moths have a longevity of 3–4 days, one day more than that of males. Female moths lay 140–350 eggs in clusters of 80–100, firmly glued to undersurface of leaves of shade trees. Eggs are pale yellow, hemispherical with a flat base, measuring 1.34 ± 0.03 mm in diameter and 1.14 ± 0.08 mm in height. Eggs become dark green towards the close of its incubation period of 20–24 days and yellowish; first instar larvae emerge out from 95 per cent of eggs. First instar larvae are black headed, bearing dorsally four rows of black dots from where tufts of hairs originate. They possess three pairs of thoracic and five pairs of abdominal prolegs. Two days after emergence, they start feeding at night voraciously on leaves of shade trees. During daytime they remain inactive in compact congregations on barks or leaves of shade trees (Fig. 8.8). They remain on shade trees till they become fourth or fifth instar larvae. Then they descend to cardamom plants by silken threads and defoliate the plants leaving only mid ribs. The larvae become black by this time, their thoracic legs become black, abdominal legs pale red, and the whole body gets covered with gray hairs, which upon contact cause irritation. They undergo eight moults within 182–202 days of larval life and fully developed ninth instar larvae of about 11 cm lengths are formed. They stop further feeding and move about during their pre-pupal period of 2–3 days in search of suitable sites for pupation on tree trunks or in soil. At pupation they enclose themselves in a self-made flimsy cocoon of larval hairs and soil particles. They shed the larval skin and become pupae, which appear dark brown, roughly oval and obtect. The pupal period lasts for about 65 days. The entire life cycle is completed within 265–294 days. Selvan and Singh (1993) noticed *Eugenia hemispherica, Coffea arabica, Maesa indica, Macaranga indica, Veronia arborea* and *Persia macrantha* as alternate hosts of *E. undata*. Ingestion of cardamom leaves sprayed with 1 per cent aqueous suspension of the extract of *Lantana camara* produced larval and pupal mortality in *E. undata* under laboratory condition (Gopakumar et al., 1996).

Figure 8.8 Hairy caterpillar: aggregation of the hairy caterpillar of cardamom, *Eupterota undata* on a shade tree.

2.5.3 Eupterote canarica

Moths of *E. canarica* emerge in June–July. The moths have dark wings of an expanse of 6.7 cm with post-medial lines on each. Female moths lay about 480 eggs in groups of 40–120 during its 9–12 days oviposition period. Eggs are light yellow, hemispherical, having a diameter of 1.5 mm and height 1 mm. After an incubation period of 20 days, larvae emerge out, which grow up to the second or third instar stages by feeding on leaves of shade trees. The third or fourth instar larvae descend to cardamom plants, feed on cardamom leaves and grow through successive stages to the fully-developed larvae, taking altogether 4–5 months to complete the larval period. A full-grown caterpillar is about 62 mm long with dark brown body, red head and prothoracic shield. They pupate in soil inside oval cocoons made by the last instar larvae with sand, debris and body hairs. After 7–8 months pupation, adults emerge. The caterpillars may get parasitised by *Sturmia sericariae* and infected by an entomogenous fungus (Nair, 1978; Kumaresan, 1988).

2.5.4 Eupterote fabia

Moths of *E. fabia* are brownish-yellow in colour with distinct post-medial black lines and wavy lines on wings. It has a wingspan of 12 cm. A fully-grown caterpillar is 7.5 cm long with black head, gray-tipped hairs and red prolegs. Pupation takes place in a self-made cocoon consisting of silk, sand particles and body hairs. Pupae measure 31 × 12 mm and the pupal period lasts for 7–8 months. Infestation on cardamom occurs usually during August–October (Nair, 1978; Kumaresan, 1988).

2.5.5   Eupterote testaceae

Caterpillars of *E. testaceae* cause only mild damage to cardamom. Its moths have yellowish wings with faint wavy black lines on it. Moths emerge usually in June–July and during its oviposition period of 8–9 days lay nearly 350 pale yellow eggs in clusters of 60–90 on adaxial surface of leaves of shade trees. Eggs are hemispherical with a diameter of 1 mm and height of 0.5 mm. Larvae hatch out of the eggs on 13th–15th day of incubation. The larva undergoes six moults and becomes fully grown in 100–106 days, measuring 7.2 cm in length. The grown-up caterpillars have pale yellow head, streaked with black lines and each segment bears dorsally conical tufts of short gray hairs and long white tipped hairs. A silvery white line on the mid-dorsal side is bordered on either side by red lines. On lateral side of the body is a longitudinal broad red band. They pupate inside loose cocoons made of silk, sand and body hairs. After a pupal period of nearly 8 months, adults emerge (Nair, 1978; Kumaresan, 1988).

2.5.6   Lenodera vittata

Adults of *L. vittata* are thickset moths with under-developed wings having a span of 5.6 cm. The adults emerge in June, lay eggs in single rows on cardamom leaves. A female may lay a maximum of 130 eggs during its oviposition period of 6–9 days. The eggs are off-white in colour, dome-shaped, with a diameter of 2.9 mm and a height of 1.6 mm. Chorion is leathery and translucent. After an incubation period of 10–13 days, caterpillars emerge out of the eggs which by feeding on cardamom leaves become fully grown in about 120 days, attaining a length of 106–110 mm. Dense capitate hairs cover its body. A pair of transverse rows of black bristles is present dorsally on each segment. On its meso- and meta-thoracic segments are long yellow hairs and from the dorsal, lateral and ventro-lateral spots arise black bristles. Fully-grown caterpillars, during their pre-pupal period, burrow in the soil around the plant base to a depth of about 6 cm and then pupate. Pupae are 32 mm long and 14 mm broad. Pupation is completed in 5–7 months. The parasite *Carcelia kockiana* is seen to parasitise the larvae (Nair, 1978; Kumaresan, 1988).

2.5.7   Euproctis lutifacia

Its hairy caterpillars infest tender foliage of cardamom. Adult is a vinous brown moth. Its wings have a span of 4 cm with an antimedial orange-red line over it. Fully-grown caterpillars are about 3 cm long, pale-brownish with a mid-dorsal black line. First and second abdominal segments bear tufts of brown hairs mid-dorsally. An eversible glandular organ is located on the dorsal surface of the seventh abdominal segment. Fully-grown larvae metamorphose into pupae inside a silken cocoon in soil. After a pupal period of 16–18 days, adults emerge generally during December (Nair, 1978; Kumaresan, 1988).

2.5.8   Alphaea biguttata

This is an arctiid black hairy caterpillar infesting cardamom. Its moths are comparatively small, having a wingspan of about 5 cm with a curved white band along its forewings. Fully-grown caterpillar is 6–7 cm long. Pupation takes place in a self-made black silken cocoon made of soil particles and body hairs. Adults emerge after about 22–23 days of pupation (Nair, 1978; Kumaresan, 1988).

2.5.9 *Pericalia ricini*

Moths are medium-sized. Forewings have pale red rings, wings and abdomen are crimson coloured. Black bands are present on its crimson coloured abdomen. Females lay about 170 eggs in clusters on under surface of leaves. After an incubation period of 4–5 days, dark brown larvae, bearing fine long reddish-brown hairs on bluish warts emerge (Nair, 1978, Kumaresan, 1988).

Beeson (1941) reported *Bombax malabaricum*, *Careya arborea*, *Cedrella toona*, *Dalbergia volubilis*, *Erythrina indica*, *Shorea robusta*, *Tectona grandifolia*, *T. grandis*, *Terminalia* sp. and *Vitex negundo* as alternate hosts of hairy caterpillars. These caterpillars congregate on tree trunks or cardamom plants during daytime and can be collected in large numbers and destroyed. Sekhar (1959) recommended fish oil rosin soap sprays against the pest. Nambiar *et al.* (1975) found BHC 0.2 per cent or malathion 0.1 per cent or carbaryl 0.1 per cent effective for its management. Collection and destruction of moths using light traps and spraying methyl parathion 0.1 per cent are also recommended (Anonymous, 1985a). The natural enemies of hairy caterpillars reported include *Apanteles tabrobanae* Cram., *Sturmia sericariae*, *Aphamites eupterotes*, and *Beauveria* sp. (Rajan, 1965; Nair, 1975; Varadarasan, 1986).

3 MINOR PESTS

3.1 Capsule borers

3.1.1 *Jamides alecto*

At times damage due to this lycaenid borer, had been severe in Karnataka region (Siddappaji and Reddy, 1972; Kumaresan *et al.*, 1988; Krishnamurthy *et al.*, 1989). Caterpillars of this pest bore and feed on inflorescence, flower buds and capsules. Affected capsule becomes empty, decays and drop in rainy season. Biology of the pest has been studied by Singh *et al.* (1993). Adult is a swift flying butterfly having metallic blue colour bordered with a white line and black shade on the dorsal surface of wings and with the same colour on the ventral surface. Hind wings have small delicate tail-like prolongations. Wings of male butterfly have a span of 36.5 ± 0.25 mm and that of female 39 ± 0.78 mm. Antennae are clubbed. A rim of white scales surrounds eyes. Female butterflies live (8–12 days) longer than males (6–8 days). A female lays about 65 eggs during its oviposition period of 5–7 days. Eggs are laid singly, scattered over flower buds, panicles, and tender capsules on which they remain glued. Eggs are light greenish, flat based and elliptical with a slightly depressed dorsal surface. Its chorion is beautifully sculptured. Eggs turn dark towards the close of its 5–6 days incubation period. Nearly 72 per cent of the eggs hatch liberating greenish, active caterpillars through an opening formed on the dorsal surface of the eggs. Larvae bear prominent hairs on its dorsal and lateral surfaces. The larvae grow feeding on flower buds and flowers to the third instar stage, which then bore and feed on immature capsules. Second and third stage larvae appear pale brown. They are devoid of dorsal hairs and remain inactive. Fourth and fifth instar larvae are pale yellow, sluggish with a comparatively small head capsule, which is almost concealed by the prothorax and bear five pairs of prolegs ventrally and fine setae dorso-laterally. The last three abdominal segments are flat and trowel shaped. These larvae feed on seeds of immature capsules; seeds of ripened capsules are rejected after sensing the seeds. A fifth instar larva requires seeds of 25–27 capsules as food to attain maturity (Siddappaji and Reddy, 1972). One larva has been found to feed on seeds of 2–3 capsules in a day. A fully-grown larva measures $11.4 \pm 0.358 \times 3.90 \pm 0.089$ mm. The entire larval period is completed in 20–25 days. Mature larva pupates inside the capsule or outside within the debris. After a pre-pupal period of 2 days it becomes a short yellowish-brown pupa measuring $9.63 \pm 0.48 \times 4.13 \pm 0.21$ mm. With a pupal period of 12–13 days, it completes the life cycle in 41–48 days. Spraying quinalphos 0.05 per cent or methylparathion 0.05 per cent during early blooming period has been found effective against the pest (Kumaresan et al., 1988). Application of fish oil rosin soap @ 1 kg in 45 l water is also found effective against the borer (Kumaresan, 1988).

3.1.2   Thammurgides cardamomi

Adults and larvae of T. cardamomi bore and feed on flowers and immature capsules. In Karnataka, pest infestation is noticeable during July–August especially on plants under thick shade. Adult is a dark brown beetle covered over with short thick hairs. Adults lay barrel-shaped eggs in clusters within capsules. The soft, white, wrinkled-bodied larvae complete their life inside cardamom capsules. Proper regulation of shade and spray with quinalphos or methyl parathion 0.05 per cent are recommended for its management (Anonymous, 1985a).

3.1.3   Onthophagus coorgensis

This pest bore and feed on flowers and young capsules during monsoon months. They are small, dark brown in colour with short thick erect hairs over the body. They lay clusters of barrel-shaped eggs, normally 6–12 in a capsule. Larvae are white in colour, soft and wrinkle-bodied. Pupae are ivory white in colour. The entire life cycle is completed within cardamom capsules. The pest can be controlled by proper shade regulation and spraying of quinalphos or methyl parathion 0.05 per cent (Kumaresan et al., 1988; Krishnamurthy et al., 1989).

3.2   Root borer (Hilarographa caminodes Meyr.)

Caterpillars of this insect bore into roots and feed on it resulting in yellowing of leaves and gradual drying of plants. Moths emerge during April–May and lay eggs on exposed parts of roots. Emerging caterpillars tunnel into the roots. A fully-grown larva is 8–10 mm long, slender and creamy yellow in colour. They diapause during January–April (Jones, 1944) and then become pupae within the tunnelled root. Pupa is creamy yellow and measures 8–12 mm. Pupal period lasts for about 15 days.

3.3   Rhizome weevil (Prodioctes haematicus Chev. F.)

Grubs of this insect tunnel into rhizomes and rarely into aerial stems. Infestation is noticed more on seedlings than on grown-up plants. Adult is a brown weevil 12 mm long having three black lines on pronotum, one mid-dorsally and other two on its either side. On each elytron there are three black dots, two anteriorly and one posteriorly. Adults emerge immediately after receipt of summer rains in April. Mated females lay creamy white eggs singly on feeding punctures made by it. Eggs are reniform, 2 mm long and one mm wide and hatch in 8–10 days. Emerging grubs are 2.5 mm long, glossy white in colour and with a light brown head. Tunnelling through and feeding on rhizomes, grubs grow to a length of 12.5 mm in about 50 days. Later they become stouter with reduced length. The entire grub phase prolongs up to three months (Jones, 1941, 1944a). At a wide part of the tunnel in the rhizome the grub positions itself, closes the tunnel above and below with frass and remains quiescent with head downwards. It pupates to a milky white pupa, which later changes to yellowish brown. Pupal period lasts for 20 days. Adults live for 7–8 months. Destruction of infested plants will help reduce the pest intensity. Drenching BHC 0.2 per cent or soil application of Phorate 10 G @ 20–40 gm/clump are also recommended against the pest (Anonymous, 1985a; Kumaresan, 1988).

3.4   Shoot fly (*Formosina flavipes* Mall.)

This is a pest prominent during warm summer months infesting seedlings and young tillers under thin shade. Mature flies lay cigar-shaped white eggs singly in rows of 4–6 between the terminal leaf sheaths. However only one maggot is seen to penetrate down the pseudostem feeding on core tissues, which results in drying of the terminal unfurled leaf (dead heart symptom). Pupation takes place within the pseudostem at its base. Its life cycle is completed within 20–25 days. The pest can be checked by destruction of affected plants and application of carbofuran @ 8–10 kg/acre or by spraying dimethoate or quinalphos or methyl parathion 0.05 per cent (Kumaresan *et al.*, 1988). Sufficient shade also should be provided to the plants.

3.5   Lace wing bug (*Stephanitis typicus* Dist.)

This is a bug with shiny, transparent lace-like reticulate wings. Females lay about 30 eggs singly, which are inserted into adaxial surface of leaves. Eggs hatch in 12 days and the emerging nymphs congregate on a suitable feeding site on leaves and suck plant sap resulting in development of necrotic spots on leaves. Banana, *Colacacia*, coconut and turmeric are its alternate hosts (Nair, 1978; Kumaresan, 1988).

3.6   Cutworm (*Acrilasisa plagiata* M.)

Cutworms are commonly seen to feed on tender cardamom leaves in nurseries. Infestation is noticed during January–March. The fully-grown caterpillar is dark brown with an orange-red head and a hump-like projection dorsally on the eighth segment. It pupates in soil for a period of 17–18 days (Nair, 1975).

3.7   Cardamom aphid (*Pentalonia nigronervosa* f. *caladii* van der Goot)

The cardamom aphid is of concern not as a pest of the crop, but as a vector of the virus, which causes the serious disease '*Katte*' in cardamom. Siddappaji and Reddy (1972a) confirmed the form of the vector occurring on cardamom as *P. nigronervosa* f. *caladii*. The virus–vector relationship was studied by Uppal *et al.* (1945), Varma (1962), Rao and

Naidu (1973, 1974), Rajan (1981) and Naidu *et al.* (1982). Biology of the insect was studied and reported by Rajan (1981).

Of the winged and wingless forms of the aphid, the former is longer and slimmer than the latter. Adults are dark brown in colour. They reproduce by viviparous and parthenogenetic means. A female may give birth to 8–28 offsprings. The nymphs moult thrice and become adults in about 15 days. The aphid colonies are scavengered by ants. Population of the insect is high during January–February. They are found also on *Colocasia* sp., *Alocasia* sp. and *Caladium* sp.

The insect population gets reduced drasticaly during rainy season due to the infection of *Verticillium intertextum* (Deshpande *et al.*, 1972). *Peragum indica, Cocinella transversalis*, and *Ischiodon scutellaris* were found to predate over the aphids.

Spraying phosphamidon or dimethoate (0.05 per cent each) during April and November is recommended to control aphids (Anonymous, 1985a).

3.8   Scale insect (*Aulacaspis* sp.)

Infestation of this pest is noticed during summer periods. Capsules, panicles and pseudostems are the usual sites of infestation. As a result of infestation, capsules shrivel.

3.9   Leaf folder (*Homona* sp.)

Caterpillars of this pest fold tip of cardamom leaves and feed on the leaf by remaining within the leaf fold. Fully-grown caterpillar is almost 3 cm long, pale green and with a black head.

3.10   Spotted grass hopper (*Aularches* sp.)

Adults and nymphs of this polyphagous pest scrape and feed on leaves of cardamom voraciously. Infestation is usually noticed during March and continues till November. Adult hoppers have pretty green wings with yellow spots over it. Wings extend beyond the abdominal tip. Pronotum is warty with short thick peg-like outgrowths. Males are smaller than females.

Adults congregate on shade tree tops presumably for basking; they descend to cardamom in swarms and defoliate the plants. Mated females thrust their abdomen to a depth of 5–8 cm in the soil and lay about 60 eggs in cluster into an egg case formed of secretions of the female and close it by a frothy secretion. Eggs are elongated, spindle-shaped, about 7 mm long and 2 mm wide. After about 5 months small wingless nymphs emerge. Its pinkish body is intermingled with red dots and white longitudinal stripes. Nymphal stage lasts for 86–221 days. Females live for 80–85 days while males have longevity of only 5–30 days. Coconut, coffee, areca palm, cashew, banana, pepper, teak, jack, dadaps, etc. are its alternate hosts.

Exposure of the eggs to desiccation by sun and application of contact insecticides against nymphs are recommended for management of the pest.

3.11   Leaf grub (*Lema fulvimana* Jacoby)

Beetles and grubs of *L. fulvimana* feed on tender foliage of cardamom seedlings. Biology of the pest was reported by Singh (1994). Beetles emerge during May. Females lay about

95 eggs on tender foliage. Eggs are yellow, cylindrical and hatch in 5–6 days. The larvae undergo three moults and develop into fully-grown fourth instar grubs of 8–9 mm length in about 20 days. Fully-grown grubs have a dull-white body with brown streaks, black head and disproportionately swollen abdomen. Usually they carry on their back faecal matter. They pupate in soil inside a papery cocoon. Pupal period lasts for 10–24 days. Adults live for 45–95 days. *Zingiber cernuum* and *Curcuma neilgherrensis* are alternate hosts of the pest. Removal of alternate hosts and spraying of quinalphos 0.025 per cent or monocrotophos 0.025 per cent are effective against the pest (Singh, 1994).

3.12   *Metapodistis polychrysa* Meyrick

Caterpillars of this insect feed on unopened leaves of cardamom (Gopakumar *et al.*, 1989). First instar larvae makes a hole almost at the middle of an unopened leaf, enter through the hole and feed on one half of the leaf lamina as well as part of the mid rib from its point of entry. The caterpillar is pale green and 1 cm long when fully grown. Large black dots are present dorsally on head and the last abdominal segment. Larval period lasts for 12–15 days. It pupates in a silken cocoon and after 15–18 days of pupal life black-brown moths having two golden stripes on their wings emerge. The insect completes its life cycle in 30–35 days.

3.13   Storage pests

Tribolium castaneum and *Lasioderma serricorne* are the major storage pests of cardamom. *L. serricorne* completes its life cycle in about 115 days on stored cardamom (Balu, 1991). Fumigation with methyl bromide or storing capsules in alkathene-lined jute bags sprayed with malathion 0.5 per cent is effective against the pest (Abraham, 1975).

3.14   Red spider mites

Sporadic infestation of red spider mites on cardamom leaves has been noticed during summer months (February–May). The mites remain on the undersurfaces of leaves within a self-made web. They suck plant sap from leaves resulting in the formation of white bloches on leaves. Undersurface of the infested leaves, with its eggs, excreta and nymphs embedded in a web of fine delicate silken threads appear ashy white. Affected leaves gradually dry up. Infestation on panicles and tillers is rarely noticed. This mite completes its life cycle in about 20 days. Dicofol 2 ml/l or sulphur 80 WP 2.5 g/l or ethion 2 ml/l or dimethoate 1.67 ml/l or phosalone 2 ml/l is effective against the pest (Selvan *et al.*, 1996).

CONCLUSION

A lot of information on biology of most of the major and minor pests of cardamom as well as methods for their management is known. These management methods are mostly confined to the use of chemical insecticides. Some parasitoids and predators of certain pests have been identified. There is an urgent need to bring down the use of chemical insecticides in the cardamom sector, mainly through substitution with plant derivatives as well as by biological control agents. Efforts are to be focused on developing suitable biocontrol methods against the major pests such as thrips, borers and root grubs. The potential of bioagents that have already been identified in suppressing pest population and the impact of agro-climatic conditions prevailing in the cardamom ecosystem on them should be thoroughly studied. An integrated pest management (IPM) system with due importance to non-insecticidal means of pest control has to be evolved. Also, awareness has to be created among planters on the importance of an IPM for sustainable cardamom cultivation.

REFERENCES

Abraham, C.C. (1975) Insect pests of stored spices and their control. *Arecanut and Spices Bull.*, 7, 4–6.

Anonymous (1944–1952) *Annual Reports of the Scheme of Scientific Aid to Cardamom Industry in South India*, Govt. of Madras, India.

Anonymous (1980) *Pests and Diseases in the Main Plantation in Cardamom*, Package of practices, Cardamom Board Cochin, India.

Anonymous (1985a) *Cardamom Package of Practices*, Pamphlet No. 9, Central Plantation Crops Research Institute, Kasaragod, India, p. 30.

Anonymous (1985b) *Annual Report for 1983*, Central Plantation Research Institute, Kasaragod, India, p. 249.

Anonymous (1985c) *Cardamom Package of Practices*, Pamphlet No. 28, Central Plantation Crops Research Institute, Kasaragod, India, p. 20.

Anonymous (1993) *Consolidated Report on Rootgrubs*. Indian Cardamom Research Institute, Spices Board, India, p. 116.

Ayyar, T.V.R. (1935) A new species of Thysanoptera from S. India *(Taeniothrips cardamomi* sp. nov.). *Bull. Ent. Res.*, 26, 357–358.

Ayyar, R.G. (1961) Studies on the hairy caterpillar pests of cardamom in Kerala. *Agric. Res. J. Kerala*, 1, 18–30.

Balu, A. (1991) Studies on Storage pests of cardamom. *Third Annual Research Council Meeting agenda*, Indian Cardamom Research Institute, Spices Board, India, p. 27.

Balu, A., Gopakumar, B. and Chandrasekar, S.S. (1991) *Third Annual Research Council Meeting agenda*, Indian Cardamom Research Institute, Spices Board, India, 25–26.

Beeson, C.F.C. (1941) *The Ecology and Control of Forest Insects of India and Neighbouring Countries*, F.R.I., Dehradun, p. 1007.

Bhatti, J.S. (1969) The taxonomic status of *Megalurothrips* Bagnall (Thysanoptera: Thripidae). *Oriental Insects*, 3, 239–244.

Butani, P.K. (1984) Spices and pest problems – Cardamom. *Pestology*, 8, 28–39.

Chandrasekar, S.S. and Balu, A. (1993) Vertical distribution of thrips *Sciothrips cardamomi* (Ramk.) in cardamom. *J. Plantation Crops*, 21(suppl.), 227–230.

Chandrasekharan, R. (1984) Field control of cardamom thrips. *Proc. PLACROSYM-II*, Indian Society for Plantation Crops, Kasaragod, Kerala, 644–652.

Cheriyan, M.C. and Kylasam, M.S. (1941) Preliminary studies on the cardamom thrips *(Taeniothrips cardamomi* Ramk.) and its control. *Madras Agri. J.*, 29, 355–359.

David, B.V., Narayanaswami, P.S. and Murugesan, M. (1964) Bionomics and control of the castor shoot and capsule borer *Dichocrocis punctiferalis* Guen. in Madras State. *Indian Oil Seeds J.*, 8, 146–158.

David, B.V. and Sundarraj (1993) Studies on *Dialeurodes* of India, *Kanakarajiella* gen. nov. *J. Ent. Res.*, 17, 233.

Deshpande, R.S., Vishwanath, S. and Rahman, M.V. (1972) A new entomogenous fungus on banana aphid *(Pentalonia nigronervosa* Coq.) vector of katte disease of cardamom *(Elettaria cardamomum* Maton.). *Mysore J. Agri. Sci.*, 6, 54.

Eapen, S.J. (1994) Effect of three granular pesticides on damage by thrips *(Sciothrips cardamomi* (Ramk.) in small cardamom *(Elettaria cardamomum* Maton.). *J. Ent. Res.*, 18, 181–183.

Gopakumar, B. and Kumaresan, D. (1984) A strategy for controlling cardamom thrips *(Sciothrips cardamomi* (Ramk.) (Thysanoptera: Thripidae), *Paper presented in III Oriental Entomology Symposium*, Trivandrum, India, Abstract, p. 103.

Gopakumar, B., Kumaresan, D. and Varadarasan, S. (1987) Management strategy of root grub in *cardamom. Cardamom*, 20, 15–17.

Gopakumar, B., Kumaresan, D. and Varadarasan, S. (1987a) Occurrence of flea beetle, *Basilepta (Nodostoma) fulvicorne* (Jacoby) (Eumolpidae: Coleoptera) in cardamom and preliminary studies on its management. *J. Coffee Res.*, 17(suppl.), 154–155.

Gopakumar, B. and Kumaresan, D. (1988a) Use of common insecticides at reduced dosage and frequency for the control of cardamom thrips. In *Proc. National Symposium on Integrated Pest Control – Progress and Perspectives*, pp. 342–344.

Gopakumar, B., Kumaresan, D. and Varadarasan, S. (1988b) Whitefly management in cardamom plantation. *Cardamom*, 20, 5–6.

Gopakumar, B., Kumaresan, D. and Naidu, R. (1988c) Survey and distribution of root grubs *(Basilepta fulvicorne* [Jacoby]) and white grub *(Holotrichia* sp.) and their management strategy on cardamom. *Paper presented at the Workshop on All India co-ordinated Research Project on White grub*, Jaipur, India, May, pp. 13–14.

Gopakumar, B., Kumaresan, D. and Varadarasan, S. (1989) Note on the incidence of *Metapodistis polychrysa* Meyrick (Lepidoptera: Gtyphipterigidae) on small cardamom *(Elettaria cardamomum* Maton.). *Entomon.*, 14, 170.

Gopakumar, B. and Kumaresan, D. (1991) Evaluation of certain insecticides against cardamom whitefly, *Dialeurodes cardamomi* (David and Subr.). *Pestology*, 15, 4–5.

Gopakumar, B., Kumaresan, D., Varadarasan, S. and Naidu, R. (1991) Bioecology, damage potential and management of root grub *Basilepta fulvicorne* (Coleoptera: Chrysomelidae). In G.K. Veeresh, D. Rajagopal and C.A. Virktamath (eds) *Advances in Management and Conservation of Soil Fauna*, Oxford & IBH Publ. Co. Ltd., New Delhi, India.

Gopakumar, B. and Singh, J. (1994) Evaluation of neem-based insecticides against cardamom thrips *Sciothrips cardamomi* (Ramk.). *Paper presented at the international symposium on allelopathy in sustainable agriculture. forestry and environment*, Delhi, India, (Ab.), p. 124.

Gopakumar, B. and Chandrasekar, S.S. (1996) Evaluation of two new formulations of quinalphos against cardamom thrips, *Sciothrips cardamomi* (Ramk.). *Paper presented at PLACROSYM XII*, RRII, Kottayam, India, (Ab.), p. 60.

Gopakumar, B., Kumaresan, D., Chadrasekar, S.S. and Varadarasan, S. (1996) Evaluation of neem derivatives and *Lantana camara* L. against cardamom pests. In R.P. Singh, M.S. Chari, A.K. Raheja and W. Krans (eds) *Neem and Environment*, Oxford & IBH Publ. Co. Ltd., India, pp. 589–594.

Jacob, S.A. (1980) Pests of ginger and turmeric and their control. *Pesticides*, 14, 36–40

Jacob, S.A. (1981) Biology of *Dichocrocis punctiferalis* Guen. on turmeric. *J. Plantation Crops*, 9, 119–123.

John, J.M. (1967) It is time to fight against cardamom hairy caterpillars. *Cardamom News*, 1, 4–5.

Joseph, D. and Kumaresan, D. (1982) Beware of panicle borer in cardamom. *Cardamom*, 14, 33–34.

Joseph, D. (1983) Effect of two synthetic pyrethroid insecticides on the control of cardamom thrips *Sciothrips cardamomi* (Ramk.) infesting cardamom *(Elettaria cardamomum). Agric. Res. J. Kerala*, 21, 77–78.

Joseph, K.J., Narendran, T.C. and Joy, P.J. (1973) Studies on *Oriental Brachymeria (Chalacidoidae) Report*, P.L. 480 Research Project, Taxonomic studies on the oriental species of Brachymeria (Hymenoptera: Chalcididae), University of Calicut, Calicut, India.

Joseph, K.J., Narendran, T.C. and Haq, M.A. (1983) Outbreak of hairy caterpillars *(Eupterote* sp.) as serious pests of cardamom in the Meghamalai area of south India and recommendations for their integrated management. *Tropical Pest Management*, 29, 166–172.

Jones, S. (1941) The cardamom weevil *Prodioctes haematicus* Chev. in South India. *Curr. Sci.*, 10, 172.

Jones, S. (1944) An interesting case of diapause in the caterpillars of the cardamom root borer *Hialographa caminodes* Meyr. *Curr. Sci.*, 13, 188.

Jones, S. (1944a) *Prodioctes haematicus* Chev., a new weevil pest of cardamom in South India. *Indian J. Ent.*, 6, 49–52.

Jones, S. and Aiyar, G.R. (1948) On pests of cardamom. *Rep. Dep. Res.* for the Septennium 1939–1946, University of Travancore, Trivandrum, India, pp. 70–76.

Krishnamurthy, K., Khan, M.M., Avadhani, K.K., Venkatesh, J., Siddaramaiah, A.L., Chakravarthy, A.K. and Gurumurthy, S.R. (1989) *Three Decades of Cardamom Research* at *Regional Research Station, Mudigere* (1958–1988), *Technical Bulletin No. 2*, Regional Researsch Station, Mudigere, Karnataka, India, p. 94.

Kumaresan, D., Regupathy, A. and George, K.V. (1978) Control of cardamom stem borer *Dichocrocis punctiferalis* Guen. *J. Plantation Crops.*, 6, 85–86.

Kumaresan, D. (1982) Efficacy of modern synthetic insecticides against cardamom thrips. *Pesticides*, 16, 26–27.

Kumaresan, D. (1983) Field evaluation of insecticides for the control of cardamom thrips. *South Indian Hort.* 31, 151–152.

Kumaresan, D. and Varadarasan, S. (1987) Review and current status of research on insect pest control in cardamom cropping systems. *Paper presented in the Workshop on Insect Pest Management of Coffee, Tea, Cardamom cropping systems*, Central Coffee Research Station, Chikmagalur, India, January 23–24, 1987.

Kumaresan, D. (1988) Cardamom Pests. In *Pest and Disease Management (A guide to planters)*, Indian Cardamom Research Institute, Spices Board, India, p. 40.

Kumaresan, D., Regupathy, A. and Bhaskaran, P. (1988) *Pests of Spices*, Rajalakshmi Publications, Nagercoil, Tamil Nadu, India, p. 241.

Kumaresan, D., Gopakumar, B. and Varadarasan, S. (1989a) A novel method for root grub management in cardamom. *Spice India*, 2, 9–11.

Kumaresan, D., Varadarasan, S. and Gopakumar, B. (1989b) General accomplishments towards better pest management in cardamom. *Spice India*, 2, 5–8.

Kumaresan, D. and Gopakumar, B. (1993) Evaluation of fluvalinate and methamidophos against cardamom thrips *Sciothrips cardamomi* (Ramk.). *Pestology*, 17, 34–35.

Kumaresan, D., Selvakumaran, S. and Gopakumar, B. (1993) Cardamom whitefly and its management. *Extension folder No. 12*, Indian Cardamom Research Institute, Spices Board, Myladumpara, India.

Naganathan, T.G., Regupathy, A. and Kumaresan, D. (1983) Efficacy of certain insecticides in controlling the cardamom stem borer *Dichocrocis punctiferalis*. *Pesticides*, 17, 22–23.

Naidu, R., Venugopal, M.N. and Rajan, P. (1982) Influence of different aphid biotypes and virus strains on mode of transmission of 'katte' disease of small cardamom. *35th Annual Meeting of Indian Phytopathological Society*, Mysore.

Nair, M.R.G.K. (1975) *Insects and Mites of Crops in India*. ICAR, New Delhi, p. 408.

Nair, M.R.G.K. (1977) Pest Problems of cardamom. In *Towards Higher Yields in Cardamom*, Central Plantation Crops Research Institute (ICAR), Kasaragod, Kerala, India, pp. 30–31.

Nair, M.R.G.K. (1978) Cardamom. In *A Monograph on Crop Pests of Kerala and Their Control*, Kerala Agricultural University, Vellanikkara, Kerala, India, pp. 65–74.

Nair, C.K., Zachariah, P.K., George, K.V. and Nair, M.R.G.K. (1979) Field evaluation of newer insecticides for control of cardamom thrips, *Sciothrips cardamomi* (Ramk.). *Pesticides*, 13, 49–50.

Nambiar, M.C., Pillai, G.B. and Nambiar, K.K.N. (1975) Diseases and pests of cardamom, a resume of research in India. *Pesticides Annual*, pp. 122–127.

Patel, R.K. and Gangrade, G.A. (1971) Note on the biology of castor capsule borer, *Dichocrocis punctiferalis*. *Indian J. Agric. Sci.*, 41, 443–444.

Pillai, G.B. and Abraham, V.A. (1974) *CPCRI Annual Report, 1974*, p. 146.

Pillai, G.B. and Abraham, V.A. (1978) Field evaluation of some insecticides in the control of cardamom thrips *Sciothrips cardamomi* (Ramk.) (Thysanoptera: Thripidae). *Pesticides*, 12, 32–33.

Premkumar, T., Devasahayam, S. and Abdulla Koya, K.M. (1994) Pest of Spice Crops. In K.L. Chadha and P. Rethinam (eds) *Advances in Horticulture, Vol. 10: Plantation and Spice Crops* Part 2, Malhotra Publishing House, New Delhi, India.

Puttarudriah, M. (1955) An epidemic outbreak of cardamom hairy caterpillar. *Indian Coffee*, 19, 151–156.

Rajan, S.V. (1958) The hairy caterpillar pest of cardamom. *Mysore Agric. J.*, 33, 71–75.

Rajan, S.V. (1965) Protect your cardamom against insect pests. *Indian Farming*, 14, 21–24.

Rajan, P. (1981) Biology of *Pentalonia nigronervosa f. caladii* van der Goot vector of 'katte' disease of cardamom. *J. Plantation Crops*, 9, 34–41.

Rao, D.G. and Naidu, R. (1973) Studies on 'katte' or mosaic disease of small cardamom. *J. Plantation Crops*, 1, 129–136.

Rao, D.G. and Naidu, R. (1974) Additional vectors of 'katte' disease of small cardamom. *Indian J. Hort.*, 31, 380–381.

Regupathy, A. (1979) Effect of soil application of insecticides on cardamom stem borer *Dichocrocis punctiferalis* Guen. In C.S. Venkataram (ed.) *Proc. PLACROSYM I*, Indian Society for Plantation Crops, Kasaragod, Kerala, India.

Saroja, R., Lewin, H.D. and Padmanabhan, M.D. (1973) Control of the pests of castor with the insecticides. *Madras Agric. J.*, 60, 484–486.

Sekhar, P.S. (1959) Control of cardamom hairy caterpillar. *Indian Coffee*, 23, 256–268.

Selvakumaran, S. and Kumaresan, D. (1993) Final report – ICAR research scheme on *Studies on bio-ecology. damage potential and control of the whitefly Dialeurodes cardamomi David and Subr. (Aleyrodidae: Homoptera) on small cardamom.* Indian Cardamom Research Institute, Spices Board, Myladumpara, India, p. 44.

Selvakumaran, S., David, B.V. and Kumaresan, D. (1996) Observations on the natural enemies of the whitefly, *Kanakarajiella cardamomi* (David and Subr.) a pest on cardamom. *Indian J. Environ. Toxicol.*, 6, 26–27.

Selvan, M.T. and Singh, J. (1993) Studies on the biology of *Eupterote undata* Blanchard, a defoliator on cardamom. *J. Plantation Crops*, 21 (suppl.), 231–233.

Selvan, M.T., Varadarasan, S. and Chandrasekar, S.S. (1996) Beware of mite infestation in cardamom plantation. *Spice India*, 1, 18–19.

Siddappaji, C. and Reddy, N.R.N. (1972) Some new insect and mite pests of cardamom and their control. In *Third International Symposium on Subtropical and Tropical Horticulture*, Today and Tomorrows' Printers and Publishers, New Delhi, India, p. 174.

Siddappaji, C. and Reddy, D.N.R.N. (1972a) A note on the occurrence of the aphid *Pentalonia nigronervosa* form *caladii* van der Goot (Aphididae: Homoptera) on cardamom *(Elettaria cardamomum* Maton). *Mysore J. Agric. Sci.*, 6, 194–195.

Singh, J., Selvan, M.T. and Kumaresan, D. (1992) On the developmental stages of a cardamom defoliator *Eupterote cardamomi* Renga (Lepidoptera: Eupterotidae). *Biocontrol & Control Ins. Pests*, Vol. 79–84.

Singh, J., Gopakumar, B. and Kumaresan, D. (1993) Biological studies on blue butterfly, *Jamides alecto* (Felder) (Lycaenidae: Lepidoptera), a major cardamom capsule borer in Karnataka. *Entomon*, 18, 85–89.

Singh, J. (1994) Studies on *Lema* sp. *Sixth Annual Research Council Meeting on small cardamom and other spices* – agenda, Indian Cardamom Research Institue, Spices Board, Myladumpara, Spices Board, India, p. 23.

Sridharan, S., Nagarajun, N., Thamburaj, S. and Moideen, M.K. (1990) Effect of planting density on the capsule damage by *Sciothrips cardamomi*. *South Indian Hort.*, 38, 120–121.

Sulochana Bai, B., Radha, N.V. and David, B.V. (1968) Control of castor shoot and capsule borer *Dichocrocis punctiferalis* Guen. *Madras Agric. J.*, 55, 470–473.

Thyagaraj, N.E., Chakravarthy, A.K., Rajagopal, D., Sudharsan, M.R. (1991) Bioecology of cardamom root grub *Basilepta fulvicorne* Jacoby (Eumolpinae: Chrysomelidae: Coleoptera). *J. Plantation Crops*, 18, 316–319.

Uppal, B.N., Varma, P.M. and Capoor, S.P. (1945) A mosaic disease of cardamom. *Curr. Sci.*, 14, 208–209.

Varadarasan, S. (1986) New record of a braconid parasite on cardamom hairy caterpillar, *Eupterote cardamomi* Renga and cardamom looper *Eumelia rosalia* Cram. *Entomon*, 11, 260–261.

Varadarasan, S. and Kumaresan, D. (1987) Field evaluation of certain insecticides for the control of cardamom stem/capsule borer, *Dichocrocis punctiferalis* Guen. and thrips, *Sciothrips cardamomi* (Ramk.). In M.R. Sethuraj (ed.) *Proc. of PLACROSYM VI.*, Indian Society for Plantation Crops, Kasaragod, India, pp. 189–198.

Varadarasan, S., Kumaresan, D. and Gopakumar, B. (1988) Occurence of root grubs as a pest of cardamom *(Elettaria cardamomum* Maton.). *Curr. Sci.*, 57, 36.

Varadarasan, S., Kumaresan, D. and Gopakumar, B. (1989) Bioecology and management of cardamom shoot/panicle/capsule borer, *Conogethes punctiferalis* Guen. *Spice India*, 2, 15–17.

Varadarasan, S., Kumaresan, D. and Gopakumar, B. (1990) Bi-annual report 1987–1988 and 1988–1989, Indian Cardamom Research Institute, Spices Board, Myladumpara, India, pp. 65–66.

Varadarasan, S., Manimegalai, R., Sivasubramonian, T. and Kumaresan, D. (1990a) Integrated management of cardamom root grubs. *Spice India*, 3, 9–11.

Varadarasan, S., Gopakumar, B. and Kumaresan, D. (1991a) Cardamom root grub. *Spice India*, 4, 6–10.

Varadarasan, S., Sivasubramonian, T. and Manimegalai, R. (1991b) Control of root grub – Trial cum demonstration. *Spice India*, 4, 7–8.

Varadarasan, S., Kumaresan, D. and Gopakumar, B. (1991) Dynamics of the life cycle of cardamom shoot borer, *Conogethes punctiferalis* Guen. *J. Plantation Crops*, 18 (suppl.), 302–304.

Varadarasan, S., Manimegalai, R., Sivasubramonian, T., Naidu, R. (1992) Ethology of beetle, *Basilepta fulvicorne* Jacoby (Eumolpinae: Chrysomelidae: Coleoptera) – a pest of small cardamom. *J. Plantation Crops*, 20 (suppl.), 103–105.

Varadarasan, S., Sivasubramonian, T., Manimegalai, R. and Naidu, R. (1993) Integrated management of cardamom root grub *Basilepta fulvicorne* Jacoby (Eumolpinae: Chrysomelidae: Coleoptera). *J. Plantation Crops*, 21 (suppl.), 191–194.

Varadarasan, S. (1995) Biological control of insect pests of cardamom. In T.N. Ananthakrishnan (ed.) *Biological Control of Social Forest and Plantation Crops Insects*, Oxford & IBH Publ. Co. Ltd., New Delhi, India, pp. 109–119.

Varma, P.M. (1962) The banana aphid (*Pentalonia nigronervosa* Coq.) and transmission of *katte* disease of cardamom. *Indian Phytopath.*, 15, 1–10.

Wilson, K.I., Josoeph, D., Rahim, M.A. and Nair, M.R.G.K. (1977) Use of some newer insecticides for the control of cardamom thrips, *Sciothrips cardamomi* (Ramk.). *Agri. Res. J. Kerala*, 15, 192–194.

Wilson, K.I., Joseph, D. and Rajagopalan, B. (1978) On the frequency of insecticidal application against cardamom thrips. *Pesticides*, 12, 27.

9 Harvesting and processing of cardamom

T. John Zachariah and V.S. Korikanthimath

1 HARVESTING

Cardamom plants start bearing 2–3 years after planting seedlings or suckers. Panicles appear from the bases of plants from January onwards and flowering continues from April–August or even later. Generally flowering is highest during May–June. Fruits mature in about 120 days after flowering. Fruits are small trilocular capsules, containing 15–20 seeds. On maturity seeds turn dark brown to black in colour. A healthy cardamom plant on an average produces annually about 2000 fruits weighing about 900 g, which on drying and curing gives about 200 g marketable produce.

1.1 Time and stage of harvesting

Due to the prolonged flowering period, cardamom capsules ripen successively at intervals over an extended period, necessitating several pickings. Harvesting commences in August–September and extends till February–March. Generally in the peak season, harvesting is carried out at an interval of 15 days and completed in 8–10 rounds (Korikanthimath, 1983). In Kerala and Tamil Nadu harvesting starts from August–September and continues till February–March, whereas in Karnataka areas picking starts in August and continues till December–January. Fruits that are just ripened or physiologically ripened (*Karikai*) are picked by experienced workers. Korikanthimath and Naidu (1986) in their study on stage of harvesting on recovery per cent in Malabar cardamom observed highest dry weight of 285 g per kg of wet (green) capsules in the sample picked at ripened stage, followed by 240 g and 140 g in case of samples picked at physiologically mature and immature stages respectively. Thus there is 100 per cent weight gain from immature to mature harvesting stage. When harvesting is done at immature stage the splitting of capsules during curing will be much less compared to ripe fruits, and also retention of green colour will be better. It has been shown in a study that splitting of capsules is less when picked at immature stage; 13.5 per cent when harvested at physiologically mature stage, and 41.5 per cent when harvested after full maturity (Korikanthimath, 1993). Essential oil content was 20–30 per cent more in the physiologically mature or immature stages compared to ripe stage.

About 2860 ripe capsules weigh 1 kg while ca. 3330 physiologically mature and ca. 5000 immature capsules weigh 1 kg. Percentage recovery was 29 per cent in ripened stage, 24 per cent in physiologically mature and 14 per cent in immature stage. Hence it is ideal to pick cardamom at the just ripened stage or physiologically mature stage. At this stage the seeds inside the capsules will be black in colour.

Two types of pickings are adopted – light picking and hard picking. In the first one only mature capsules are harvested. In hard picking semi-mature crop is also removed. While this may reduce the curing percentage it could increase the picking average, secure green coloured capsules and also reduce the chance of fruit drop and splitting in the field. The choice depends on the availability of labour. Cardamom harvesting is done usually by employing women labourers (Fig. 9.1).

2   POST-HARVEST HANDLING

2.1   Retention of green colour

Colour of processed produce is an important factor in consumer market. Most markets, especially in the Middle East, prefer green coloured cardamom. The highest quality cardamom traditionally was the Alleppey Green, which is still regarded as the best. The traditional Alleppey green came from the high ranges of Idukki district of Kerala, where the predominant variety grown is *Mysore*. Studies have been carried out to understand the mechanism of colour retention during the processing of harvested cardamom capsules.

Synthesis and degradation studies indicate that total chlorophyll content declines after about 100 days from flowering. A comparative evaluation of chlorophyll contents in dark green, medium and light green capsules have shown that the depth of green colour is directly proportional to concentration of chlorophyll content of capsules (Table 9.1). Chlorophyll a is more than chlorophyll b in fresh as well as in cured capsules. In husk, 60 per cent of the total chlorophyll is present in the surface layer. In the three clones tested (viz. *Thachangal*, *Mudigiri* and PV1) total chlorophyll content was more

Figure 9.1 Harvesting cardamom; usually harvesting is done by women labourers.

Table 9.1 Chlorophyll profile of different capsules

Capsule colour	Chlorophyll content (ppm)			a/b ratio
	a	b	Total	
Light green	509 (186)*	561 (167)*	1070 (352)*	0.9 (1.1)
Medium green	727 (384)	700 (349)	1424 (731)	1.0 (1.1)
Dark green	1677 (446)	1890 (382)	3567 (828)	1.12 (1.2)

Source: Anonymous (1991).

Note
* Chlorophyll content of cured capsules.

in 100 days old capsules of *Thachangal* (2186 ppm) followed by *Mudigiri* (1756 ppm) and PV1 (1488 ppm) (Anonymous, 1991). The above studies suggested that dry matter continued to increase till capsules reached maturity while chlorophyll content started declining after 100 days from flowering (Anonymous, 1991). The fall in chlorophyll content during post-ripening period is more in *Mudigiri* (var. *Malabar*) compared to var. *Vazhukka* and *Mysore*, indicating delay in picking of this clone could affect the final greenness of capsules. (Anonymous, 1991).

Different chemical treatments have been tried to retain green colour of harvested capsules, because such a product fetches premium price in the market. Among such treatments, soaking green (wet) capsules immediately after harvest in 2 per cent sodium carbonate solution for 10 min fixes green colour during subsequent drying and storage (Natarajan *et al.*, 1968). Immature capsules retained greater intensity of green colour. Except sun drying all other means of drying are found to protect the green colour. Meisheri (1993) has developed a dehydration unit, which can retain the green colour and dry the produce fast at ambient temperature (27–40 °C).

2.2 Pre-drying operations

Capsules after harvest are washed thoroughly in water to remove adhering soil before taking for drying in kilns. In various trials conducted earlier, it was found that pre-soaking (quick dip) of capsules in hot water at 40 °C and dipping capsules for 10 min in 2 per cent sodium carbonate had helped in better retention of green colour of cured capsules. Dipping capsules in hot/warm water at lower temperatures viz., 30 and 35 °C also was tried. Additives like sodium carbonate (2 per cent) in warm/hot water (particularly at 35 °C) helped to increase green colour of capsules. Dipping in hot water may arrest the activity of certain hydrolytic enzymes. Volatiles extracted from capsules pre-soaked in hot water and sodium carbonate solution were subjected to gas-liquid chromatography (GLC) analysis. Results indicated that there were no significant changes in oil profile due to hot water or sodium carbonate treatments (Anonymous, 1991).

Pre-soaking of capsules in copper formulations and chemicals like NAA, IAA, GA and magnesium sulphate helped to retain more chlorophyll compared to other treatments tried. However, when pre-soaking time was extended to 60 min significant depletion of chlorophyll was observed in all, except in ascorbic acid treatment. Other treatments viz., Urea, 2,4-D, cycocel at 100 ppm each, kinetin 10 ppm, glycerol 5 per cent and polyethylene glycol 5 per cent recorded either no effect or marginal negative effect on the stability of chlorophyll (Anonymous, 1991).

Exposure to sun will lead to degradation of chlorophyll or bleaching of green colour. Post harvest delay prior to curing is known to cause chlorophyll breakdown, a better storage system could help to minimize such loss of chlorophyll.

Various trials conducted to study the impact of pre-curing storage indicated that: (1) capsule cured immediately after picking retained more green colour; (2) loss of greenness was more significant if capsules were stored for more than 12 h from picking and (3) bagging capsules helped to minimize the rate of loss of green colour. Jute bag was found to be better for storing fresh capsules compared to polypropylene woven bag. Low temperature storage of fresh capsules was found to reduce post-harvest precuring loss of greenness. Capsules stored in low energy or zero energy cool chambers were found to be distinctly greener than the capsules stored under open condition.

In large estates two RCC tanks constructed side by side are used, one for initial washing of capsules to get rid of dirt/soil etc. and the other for washing with washing soda. After washing, capsules are spread in a single layer in portable drying trays for draining water. Later the trays are arranged in kilns for drying.

3   CURING

Cardamom capsules at harvest, depending on the degree of maturity, carry moisture levels of 70–80 per cent. For proper storage, the initial moisture level has to be brought down to 8–10 per cent (wet-weight basis) by curing. Curing also plays an important role in preserving greenness of capsules since as much as 60–80 per cent of the initial colour is lost while processing.

Most widely adopted system is a slow or passive process stretching from 18 to 30 h. with an initial temperature around 50 °C. The entire curing time can be divided into four stages as follows

Stages	Time lag
I	0–3 h
II	3–6 h
III	6–9 h
IV	9 to final curing

Both the degree of maturity and curing temperature influence the percentage of splits in cured capsules, however temperature has greater influence. (Anonymous, 1991). During the process of curing, if temperature exceeds the threshold levels or the inflow of air is insufficient, capsules develop brownish streaks as a result of heat injury. In case of fairly high temperature, oil from seeds oozes out. Maintaining the temperature at 40 °C in all the four stages of curing process helps in greater retention of colour. Percentage of split and discoloured capsules increase with increase in temperature. Curing at temperature of 60 and 55 °C significantly increase the percentage of yellow capsules.

The husk of raw capsules contains about 80 per cent water, which has to be removed completely in the process of drying. Maximum loss of chlorophyll occurs in the initial 6 h of curing. Higher air flow rates, increase the loss of chlorophyll. Lesser energy is required for extraction of moisture in the initial stages when the evaporation is from the surface layers of capsules. Comparatively greater energy is required to remove the same amount of moisture when moisture content of capsules falls in the later stages of curing. Cardamom oil extracted from samples dried at 45 and 60 °C did not show much difference in the GLC profile (Anonymous, 1991). Cardamom capsules are moderately hygroscopic and absorb and desorb moisture depending upon the changes in the ambient relative humidity and temperature.

Two types of drying are generally adopted, viz., natural sun drying and artificial drying by using fire wood, fuel or electric current. Drying operation demands heavy input of energy (the energy required to dry 1 kg of green cardamom under 100 per cent efficiency can be used to light 250 numbers of 100 watt bulbs for 1 h).

3.1   Sun drying

Sun drying is generally undesirable for cardamom. Main reason is the bleaching effect due to the action of UV light present in sunlight. Sun drying requires 5–6 days or more depending upon the availability of sunlight. As the capsules are turned frequently during sun drying splitting of capsule is more. Cloudy atmosphere and frequent rains hinder proper sun drying. The method is prevalent among the small holdings of Sirsi and surrounding places in Karnataka.

3.2   Artificial drying

3.2.1   *Electrical dryer*

A dryer having dimensions of 90 × 84 cm is more common. Inside the dryer 24 numbers of aluminum trays of size 81 cm length and 40 cm breadth can be piled one over the other with a gap of 2 cm between trays. Green capsules after harvest are to be uniformly spread in trays and arranged in the dryer. Uniform heat distribution is ensured by means of fans. This way 50 kg of capsules can be dried in 10–12 h. It is possible to obtain medium green coloured cardamom by this method by maintaining the temperature between 45 and 50 °C.

3.2.2   *Pipe curing (Kiln drying)*

This is one of the best methods of drying to obtain high quality green cardamom. The structure usually consists of walls made of bricks or stones and tiled roof with ceiling. A furnace is situated on one side of the chamber; heat is generated by burning firewood or farm waste. A pipe made of iron or zinc sheet starting from the furnace passes through the chamber and opens outside the roof. The heated air currents generated in the furnace passes through the pipe and increases the temperature of the room. The fans located on either sides of the wall uniformly spread the temperature. Inside the room wooden/aluminum trays are to be piled one over the other with spacing of 20–22.5 cm between the trays. The fire in the furnace is regulated to maintain a temperature of 45–50 °C. Using this facility high quality green cardamom can be prepared in 18–22 h. A drying chamber of dimension 4.5 m length and breadth is sufficient for a plantation producing 1800–2000 kg of raw cardamom. Some of the kilns make use of brick constructed heat conveyer lines. (Fig. 9.2; Kachru and Gupta, 1993).

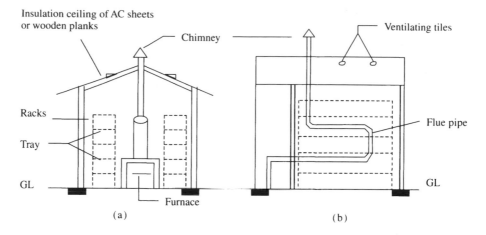

Figure 9.2 Kiln dryer for cardamom (Flue pipe curing). Diagrammatic views of the Kiln dryer (see text for details).

3.2.3   *Bin dryer*

This is a dryer designed by University of Agricultural Sciences, Banglore (Karnataka) (Fig. 9.3). Drying unit consists mainly of a blower with motor, electrical heating unit and drying chamber. The blower is of backward curve vane type coupled to 373 KW motor, 2820 revolutions per min. Volume of air driven through the dryer can be adjusted from 1.5–8 cm^3/sec. The dryer is made of mild steel, asbestos sheet and wood. Aluminum or steel trays of the size 0.4 × 0.6 m can be arranged one over the other. Required amount of air passes below the trays by means of centrally located flue pipe. Cardamom capsules are to be uniformly spread on these trays. Hot air passing through the pipes increases the temperature ranging from 30 °C to 80 °C. Good quality cardamom can be produced by drying capsules at 55 °C by maintaining the volume of air at 3.7 m^3/sec. The relative humidity ranges from 65 to 92 per cent

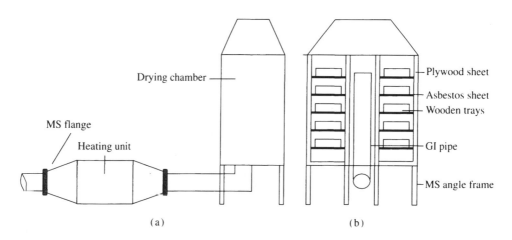

Figure 9.3 Bin dryer: (a) general diagram of the drier; (b) inside arrangement in the drying chamber.

during drying. Cost of drying by this method is around Rs. 0.66/kg compared to Rs. 1/kg in the conventional method (Gurumurthy *et al.*, 1985).

3.2.4   Melccard dryer

This is a fire wood operated dryer being used at Bodinayakanur region (Tamil Nadu). It consists of a fully insulated (fire bricks with mud coating) oven kept 3 m below the dryer. The hot flue gas from the oven is passed to an iron tank through the insulated pipes. Four iron tubes fitted at the four corners of the smoke tank carry the flue gases inside the drier and is finally exhausted through two chimneys. The surface of the smoke tank and flue pipes transfer heat to the dryer. A central opening at the ceiling (with an exhaust fan) of the dryer ensures removal of moist air. A double wall structure with a gap filled by insulation materials prevents heat loss. Roofing is also insulated with thick glass wool. The dryer can be charged easily from outside by opening four doors at the front. All the trays (with wheels) move smoothly on rails fixed inside the dryer. Trap doors attached can be opened periodically to clean off the soot formed in the interior of the flue pipes (Palaniappan, 1986). Dried capsules are rubbed by hand or with coir mat or wire mesh and winnowed to remove other plant residues and foreign matter. They are then sorted out according to size and colour.

3.2.5   Cross-flow electric dryer

This is a tray-type cross-flow dryer having capacities ranging from 25 to 400 kg (Fig. 9.4). The air is heated by 15 KW electric heaters and circulated over the material by 0.5 hp electric fan. The drying time requirement at full loading condition is about 18–20 h (Kachru and Gupta, 1993).

3.2.6   Solar cardamom dryer

Direct type solar drier developed by CPCRI, Kasargod (Kerala) for copra drying can also be used for cardamom. The dryer has an area of 1 m^2 drying surface made of black painted wire mesh tray over black painted corrugated GI sheet inclined at 12.5°. The aluminium foil reflectors of 1.5 m^2 are provided from three sides of the drier. Material load density can be three times than that used in open drying system. Complete drying of cardamom could be achieved with in 3 days using this dryer in comparison to 5 days in the open sun (Fig. 9.5). Bleaching of cardamom capsules due to the action of UV rays in sunlight is a disadvantage of this dryer.

3.2.7   Mechanical cardamom dryer

Developed by RRL, Trivandrum (Kerala) this dryer consists of a centrifugal blower, electrical furnace, conducting arrangement for uniform hot air flow and a drying chamber (Fig. 9.6). It can be used for cardamom drying at a load of 120 kg fresh cardamom/batch. It takes about 22 h for complete drying at a temperature of 50 °C. The final product is claimed to possess superior green colour, flavour and appearance (Kachru and Gupta, 1993).

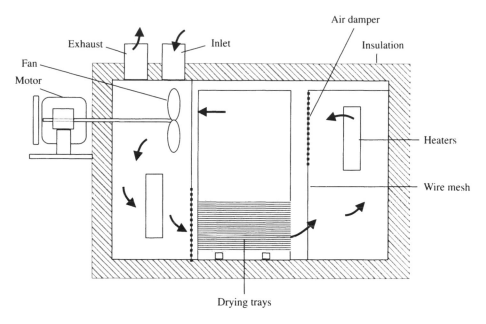

Figure 9.4 Cross flow electric dryer.

3.2.8   *Through flow dryer*

This is fabricated by CFTRI, Mysore. The drier consists of a centrifugal blower, electrical furnace ducting with arrangements to distribute the flow of hot air uniformly and a drying chamber where 120 kg fresh cardamom capsules can be loaded to a bed thickness of 20 cm. The air velocity at 60 cm/sec and the drying temperature was thermostatically controlled. The hot air carrying the humidity was not allowed to recycle. It was found to take about 22 h to complete drying of 120 kg fresh capsules at a temperature of 50 °C.

As the cardamom plantations are generally located in forest areas where electricity is not available, flue pipe dryers are more dependable and suitable to ensure continuous working. There is still good scope in developing a dryer ideal to produce green capsules without any volatile loss and with minimum expenditure for drying.

3.3   Bleached cardamom

Bleached cardamom is creamy white or golden yellow in colour. Bleaching can be done either with dried cardamom capsules or freshly harvested capsules as starting material.

3.3.1   *Bleaching of freshly harvested capsules*

Fresh capsules soaked for 1 h in 20 per cent potassium metabisulphite solution containing 1 per cent hydrogen peroxide solution degrade the chlorophyll. Drying of these capsules yield golden yellow colour.

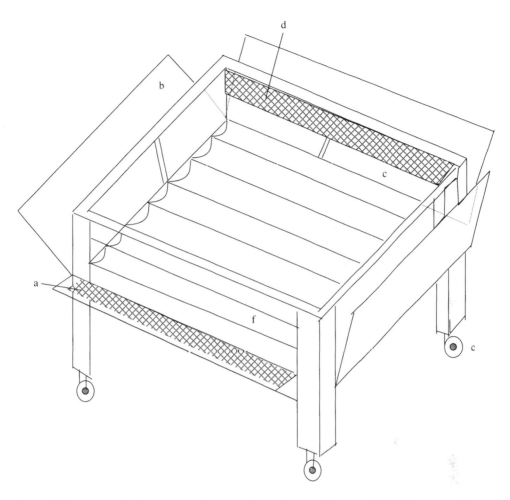

Figure 9.5 Low cost solar dryer for small scale drying: (a) sun tracking rod; (b) 24 G commercial grade aluminum foil reflector; (c) 3 mm plastic sheet; (d) wire mesh covered exhaust; (e) coster wheel; (f) 3 mm window glass; (g) 22 G corrugated GI sheet painted black.

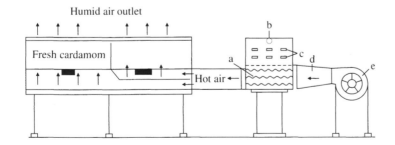

Figure 9.6 Mechanical dryer for drying pepper and cardamom: (a) electric heating system; (b) thermostat; (c) control switch; (d) cold air; (e) blower.

3.3.2　*Bleaching of dry capsules*

(a) Sulphur bleaching: It involves sulphur fumigation with alternate periods of soaking and drying. Capsules are soaked in 2 per cent bleaching powder (20 g/l of water) for 1 h and spread on wooden trays, which are arranged inside air-tight chambers. Sulphur dioxide is produced by burning sulphur (15 g/kg of capsules) and made to pass over the trays. The process of soaking and drying is repeated 3–4 times depending upon the intensity of white colour required.

(b) Potassium metabisulphite bleaching: In this method capsules are treated with 2 per cent potassium metabisulphite containing 1 per cent hydrochloric acid for 30 min. Further they are transferred to 4 per cent hydrogen peroxide solution for 6 h.

(c) Hydrogen peroxide bleaching: Hydrogen peroxide at low concentration (4–6 per cent, pH4) can bleach capsules in 6–8 h of soaking. These capsules are then dried to moisture content of 10–12 per cent. Bleached capsules contain sulphur which protect cardamom from pests. However it was found that bleaching lead to loss of volatile oil.

(d) Conventional bleaching: In Karnataka state of India, bleaching of cardamom is carried out by steeping the dried capsules in soap nut water. The fruits of soap nut (*Sapindus saponaria*) are mixed with water in a large vessel and stirred vigourously to produce plenty of lather. Dried cardamom capsules are then steeped in this water with occasional stirring. After an hour or so the fruits are collected in wicker baskets, water is allowed to be completely drained off and then spread out in mats for drying. Clean water is occasionally sprinkled over the cardamom capsules. The process of sprinkling water and drying is continued for a couple of days till a good quality bleached product is obtained.

In general, bleaching of dried capsules lead to loss of volatile oil probably because the bleaching process makes the husk brittle. However, bleached cardamom has white appearance and is resistant to weevil infestation due to sulphur dioxide content (Krishnamoorthy and Natarajan, 1976; Govindarajan *et al.*, 1982).

3.4　Moisture content

Moisture content of commercial samples from market ranges from 7 to 20 per cent depending on the regions and mode of curing (Varkey *et al.*, 1980). It is found that 10 per cent moisture is ideal for the retention of green colour, which also depends on the type of drying. Well-dried capsules produce a typical tinkling sound on shaking.

4　GRADING

The quality requirement of a produce varies with the primary raw material producer, intermediary collector, the trader, exporter, importer, processor, distributor and final consumer. Product quality is related to moisture level, cleanliness, content of substandard product, extraneous matter, appearance and colour. The processor values the extractives, volatile oil and specific ingredients. Specifications are restricted to attributes, which can be simply and rapidly analyzed. Many of them related to physical parameters such as colour, size, weight per specified volume, freedom from microbial, insect and filth contaminations (Govindarajan *et al.*, 1982). Specification for Indian cardamom is given in Table 9.2.

Alleppey green cardamom is the dried capsule of *Eletteria cardamomum* grown in South India, kiln dried, having a reasonably uniform shade of green colour, three cornered and having a ribbed appearance. Coorg clipped cardamom is the dried capsules of *E. cardamomum* var. *Malabar* grown in Coorg (Karnataka), colour ranging from pale yellow to brown, global shape, skin ribbed or smooth; the pedicels separated. Bleachable white cardamom is the fully developed dried capsule of *E. cardamomum* grown in Karnataka state to a reasonably uniform shade of white, light green or light gray colour and suitable for bleaching.

The Agmark grades of Coorg clipped cardamom and bleachable cardamom are given in Tables 9.3 and 9.4.

4.1 Bleached or half bleached cardamom

Cardamom should be fully developed, dried capsules, bleached or half bleached by sulphuring; colour ranging from pale cream to white, globose or three cornered with skin ribbed or smooth.

Table 9.2 Specifications for Indian cardamoms; physical characteristics

Grade	Description	Size (mm)	Weight (g/l) min	Colour	General characteristics
AG Alleppey green					
AGB	Extra bold	7	435 ⎫	Green	Kiln dried
AGS	Superior	5	385 ⎭		3 cornered and with ribbed appearance
AG S1	Shipment	4	320–350	Light	
AGL	Light	3.5	260	green	
CG Coorg green					
CGEB	Extra bold	8	450	Golden to light green	
CGB	Bold	7.5	435		Round,
CG-1	Superior	6.5	415	Light green	ribbed or smooth skin
CG-2	Mota, green	6	385	Green	
CG-3	Shipment	5	350	Cream	
CG-4	Light	3.5	280	Brown	
Bleached (half bleached)					
BL-1		8.5	340	Pale	Fully developed
BL-2		7	340	Creamy	round/3 cornered ribbed or smooth skin
BL-3		5	300	Dull white	

Source: Indian Standard specification for Cardamom.
IS: 1907–1966. Indian Standards Institution, New Delhi-1.

Table 9.3 Coorg clipped cardamoms: Agmark specifications

Grade designation	Trade name	Empty and malformed capsules by count (max) (%)	Unclipped capsules by count (max) (%)	Immature and shrivelled capsules (%) by weight	Size (mm)	Weight g/l (min)
CCS 1	Bold	5.0	0.0	0.0	8.5	435
CCS 2	Coorg green or Mota green	5.0	3.0	4.0	6.0	385
CC 3	Shipment	3.0	5.0	7.0	4.0	350
CC 4	Light				3.5	260

Table 9.4 Bleachable white cardamoms: Agmark specifications

Grade designation	Trade name	Empty and malformed capsules (%) by count (max)	Immature and shrivelled capsules (%) by weight	Size (mm)	Weight g/l (min)
BW 1	Mysore/Mangalore bleachable cardamom clipped	1.0	0.0	7.0	460
BW 2	Mysore/Mangalore bleachable cardamom unclipped	1.0	0.0	7.0	460
BW 3	Bleachable bulk cardamom clipped	2.0	0.0	4.3	435
BW 4	Bleachable bulk cardamom unclipped	2.0	0.0	4.3	435

Alleppey cardamom seeds are the decorticated and dry seeds of any variety of *E. cardamomum* grown in Kerala state and southern districts of Tamil Nadu, and the specifications are given in Table 9.5.

Mangalore cardamom seeds are the decorticated and dry seeds of *E. cardamomum* grown in Coorg and adjoining districts of Karnataka. The specifications are given below in Table 9.6.

Some of the general specifications for cardamom are:

(i) The capsules should be well formed, packed with sound seeds inside. The cardamom may be graded on the basis of place of origin, colour, size, mass per litre, bleach level, proportion of lower grades and extraneous matter.

(ii) The aroma and taste of cardamom in capsules and seeds should be characteristic, fresh and free from foreign aroma and taste, including rancidity and mustiness.

(iii) Cardamom capsules and seeds should be free from living insects, moulds and should be practically free from dead insects, its fragments and rodent contamination.

(iv) The mass of cardamom capsules or seeds contained in 1 l should be as specified for different grades.

Table 9.5 Alleppey cardamom seeds: Agmark specifications

Grade designation	Trade name	Extraneous matter (%) by wt.	Light seeds (%) by wt.	Weight g/l min.
AS 1	Prime	1.0	3.0	675
AS 2	Shipment	2.0	5.0	460
AS 3	Brokens	5.0		

Table 9.6 Mangalore cardamom seeds: Agmark specifications

Grade designation	Trade name	Extraneous matter (%) by wt.	Light seeds (%) by wt.	Weight g/l min.
MS 1	Prime	1.0	3.0	675
MS 2	Shipment	2.0	5.0	460
MS 3	Brokens	5.0		

Notes
Extraneous matter – Includes calyx pieces, stalk bits and other foreign matter.
Light seeds – Includes seeds brown or red in colour and broken, immature and shrivelled seeds.

(v) Cardamom should be free from visible dirt or dust. Extraneous matter such as bits of calyx, stalks and others shall not be more than 5 per cent by weight in cardamom capsules and 0.5–2 per cent by wt. in different grades of cardamom seeds.

(vi) The proportion of empty or malformed capsules, from opening and examining 100 capsules taken from the sample, should not be more than 1–7 per cent by count, varying with grade specifications.

(vii) The proportion of immature and shrivelled capsules separated according to specified methods should not be more than 2–7 per cent (m/m).

(viii) Capsules having black colour and those which are split open at corners for more than half the length should not be found in bold grades, and not be more than 10 and 15 per cent by count in the 'shipment' and 'light' grades.

(ix) The proportion of cardamom seeds which are light brown, broken or immature (shrivelled) should not be more than 3–5 per cent (m/m).

The chemical and physical specifications for whole and ground cardamom are given in Tables 9.7 and 9.8.

4.2   Commercial cardamom grades in Sri Lanka

Most of the cardamom produced in Sri Lanka is exported. The traders use various designations for cardamom, such as the one given below (Guenther, 1952).

1   Green cardamoms:
   • Kandy type – relatively large, dark greenish colour
   • Copernicus type – Slightly smaller than the kandy, colour generally green
   • General faq. type – Small cardamoms with grey–green colour

Table 9.7 Whole cardamom: chemical and physical specifications

Specification	Suggested limits
ASTA cleanliness specifications	
Whole dead insects, by count	4
Mammalian excreta, by mg/lb	3
Other excreta, by mg/lb	1.0
Mould, % by weight	1.0
Insect defiled, infested, % by weight	1.0
Extraneous, % by weight	0.5
FDA DALs	None
Volatile oil	3% min
Moisture	12% max
Ash	10% max
Acid insoluble ash	2% max
Average bulk index (mg/100 g)	
Bleached	320
Green	250

Table 9.8 Ground cardamom: chemical and physical specifications

Specifications	Suggested limits
FDA DALs	None
Volatile oil	3% min
Moisture	12% max
Total ash	10% max
Acid insoluble ash	2% max
Military specifications	
(EE-S-631J, 1981)	
(Decorticated Cardamom)	
Volatile oil (ml/100 g)	3% min
Moisture	12% max
Total ash	7% max
Acid insoluble ash	3% max
Granulation	95% min through a USS 40
Bulk index (2 ml/100 g)	190

2 Bleached cardamoms:
   - Malabar half-bleached – fair average quality, small capsules
   - Curtius – fair sized, rather long capsules
   - Cleophas – fair sized, roundish capsules

3 Seeds:
   - Crispus type – freshly removed seeds obtained by the husking of either green or bleached capsules

In general the trade distinguishes between decorticated cardamoms, green cardamoms and bleached cardamoms.

5 GRADING, PACKING

Cardamom, cured by sun drying or in a dryer, has to be protected from absorption of moisture, contamination with foreign odours, microorganisms or insect infestation. The specific requirement of packing cardamom is the protection of the product against sunlight in order to maintain the husk colour, green colour or golden colour of the bleached cardamom.

Cardamom is a high value crop and all care should be given for efficient processing and grading besides curing. Sieves of different mesh sizes viz., 8, 7.5, 7 and 6 mm are available which are manually operated. After sieving the capsules and grouping them into different grades, it is essential to sort out splits, thrips and borer-infested capsules separately. Presently sorting out is done by skilled women labourers. As cardamom harvest alone requires nearly 60 per cent of total labourers, there is an urgent need to fabricate mechanical sorting machines so as to get cardamom of different sizes with a provision to separate out capsules infested with insect and splits.

After grading, cardamom needs to be stored over a period of time, which is normally kept in double lined polythene bags. During storage some of the storage pests do impair the quality of produce. Hence, there is need to evolve storage systems to minimize such infestation.

Equilibrium relative humidity studies have shown that cardamom dried and maintained at or below 10 per cent moisture retains the original colour and avoids mould growth (Govindarajan, 1982). If black polyethylene is used, the effect of light is further minimized and safe storage is possible for 4 months, required for port storage and transshipment.

It is advisable to make use of the dried cardamom capsules preferably within 12–15 months of harvest, failing which the pleasant flavour and aroma are likely to be affected. The stored samples may be frequently tested for storage pests.

6 CONCLUSION

Processing of cardamom requires close monitoring right from harvest to the drying and final grading. The initial moisture that the capsules release while drying should be removed immediately by providing cross ventilation and exhaust fans. Any breakdown in the heat energy supply affects the appearance and quality of the produce. Over heat affects the quality of cardamom capsules. Even in the pipe dryers where heat is generated by firewood, a mechanism should be developed to regulate temperature suitably.

The small and marginal farmers, who constitute nearly 70 per cent of the cardamom growers, face difficulties in possessing their own drying kilns due to economic constraints. As cardamom estates are located in deep interior forests, transporting of wet capsules to far off places causes a lot of practical difficulties. Hence, there is an urgent need to design and fabricate cardamom dryer, which involves comparatively lesser cost and provide efficient drying facility.

Cardamom capsules need to be dried within 24 h, any further delay would result in deterioration of green colour and appearance. It is uneconomical to operate cardamom dryers with smaller quantities, hence research efforts need to be made to store wet capsules for 2–3 days without impairing quality.

Use of non-conventional energy sources for drying cardamom merits consideration in the fast changing agro-ecological conditions of cardamom tracts. Exploration of locally available farm wastes and their bio-recycling to generate heat energy merits immediate attention.

REFERENCES

Anonymous (1991) Post Harvest Technology of Cardamom – Project Report, Tea research substation, United Planters Association of South India (UPASI), Vandiperiyar, Kerala.

Govindarajan, V.S., Shanthi Narasimhan, Raghuveer, K.G. and Lewis, Y.S. (1982) Cardamom – Production, Technology, Chemistry and Quality. *CRC Critical Reviews in Food Science and Nutrition.* Vol. 16, CRC Press Inc., Florida, p. 326.

Guenther, E.(1975). The cardamom oils. In E. Robert (ed.) *The Essential Oils*, Vol. 5, Krieger Pub. Co., New York. pp. 85–106.

Gurumurthy, B.R., Nataraj, S.P. and Pattanshetti, S.P. (1985) Improved methods of drying. *Cardamom*, 18(9), 3–7.

Kachru, K.P. and Gupta, R.K. (1993) Drying of Spices – status and challenges. *Proc. National Seminar on Post Harvest Technology of Spices*, ISS, Calicut, pp. 15–27.

Korikanthimath, V.S. (1983) Seminar on production and prospects of cardamom in India – Univ. of Agricultural sciences, Dharwad, January 7, 1983.

Korikanthimath, V.S. (1993) Harvesting and on farm processing of cardamom. *Proc. National Seminar on Post Harvest Technology of Spices*, ISS, Calicut, pp. 62–68.

Korikanthimath, V.S. and Naidu, R. (1986) Influence of stage of harvest on the recovery percentage of cardamom. *Cardamom J.*, 19(11), 5–8.

Krishnamoorthy, M.N. and Natarajan C.P. (1976) Preliminary studies on bleaching cardamoms. *Cardamom J.*, 8(8), 17–19.

Meisheri, L.D. (1993) Dehydration of horticulture produce at room temperature (27–34 °C) National Sym. on food processing, New Delhi (February 16–17, 1993). Mimeographed copy.

Natarajan, C.P., Kuppuswamy, S. and Krishanmoorthy M.N. (1968) Maturity, regional variations and retention of green colour in cardamom. *J. Food Sci. and Tech.*, 5(2), 65–68.

Palaniappan, C. (1986) Analysis of cardamom curing in conventional chamber and Melccard Drier. *Cardamom*, 19(11), 5–8.

Tainter, D.R. and Grenis, A.T. (1993) *Spices and Seasoning.* VCH Pub., New York.

Varkey, G.A., Gopalakrishnan, M. and Mathew, A.G. (1980) Drying of cardamom, *Workshop on Agricultural Engineering and Technology in Kerala*, Agricultural University, Trichur.

10 Industrial processing and products of cardamom

N. Krishnamurthy and S.R. Sampathu

1 INTRODUCTION

Cardamom of commerce is the dried fruit (capsule) of the cardamom plant, *Elettaria cardamomum*. Cardamom is processed into various products like cardamom seeds, cardamom powder, cardamom oil, cardamom oleoresin, encapsulated cardamom flavour etc. Proper maturity with good characteristic aroma and high volatile oil content are the prime considerations for processing cardamom into various products.

Cardamom is sold in the form of dried capsules in trade. However, there is some demand for seeds in USA, UK and Scandinavian markets. The shelf life of cardamom seeds is poor because the aromatic volatile principle is present in a single layer just below the epidermis. The flavour of cardamom is entirely due to its volatile oil and flavour strength is directly related to the quantity of oil present in it. Hence a suitable packaging material has to be employed for storing the seeds. On the other hand cardamom capsules with the husk intact can be stored for a year without the loss of any volatiles (Guenther, 1952; Gerhardt, 1972). Whole cardamom does not deteriorate on storage because of the natural protection of the outer cover. Cardamom is stored in gunny bags lined with 300 gauge polyethylene and sometimes packed in wooden chests lined with moisture proof kraft paper or polyethylene.

Cardamom seeds, powder, volatile oil, oleoresin and encapsulated flavours obtained from cardamom are the important products in the trade. Industrial processing and related technological aspects of these products are covered here.

2 CARDAMOM SEEDS

Cardamom seeds are obtained by decorticating capsules. Decortication is achieved by using a flourmill or plate mill also called as disc mill. The distance or the gap between the discs plays a crucial role during the decortication process. The gap is adjusted in such a way that only husk is detached without damaging seeds. The entire mass comprising of seeds and broken husk passes through a vibratory sieve (Fig. 10.1), which separates them. Normally the ratio of seeds to husk is 70:30. With proper disc adjustment there should not be any loss of material during the dehusking operation. Good quality seeds will be black to dark brown in colour. The seed of cv. *Malabar* is sweet to taste due to the presence of a sweet mucilaginous matter on it (Purseglove, 1981). The quality specifications for cardamom seeds are the following.

Component (%)	Requirement	Country
Moisture (max)	9.0	US, Great Britain
Volatile oil (min)	4.0	ISO, Sri Lanka, Great Britain India IS: 1797–1961
Total ash (max)	5–6	Great Britain, US
Acid insoluble ash (max)	3–3.5	US and Great Britain

Source: Data from Govindarajan *et al.*, 1982.

Figure 10.1 Plate mill with vibratory sieve for dehusking and separation of cardamom seeds.

2.1   Packaging and storage of cardamom seeds

Storage of cardamom seeds requires greater attention because the natural protection given by the husk is removed. Bulk packaging of seeds is done in wooden chests lined with aluminium foil laminate. Loss of oil from seeds is reported to be as high as 30 per cent in 8 months under ambient conditions, while from the dried capsules the loss is negligible (Guenther, 1952). Clevenger (1934) observed that there was 30 per cent loss of volatiles from the seeds in the course of 8 months storage. The importance of distilling seeds immediately is well understood but the reported values (Table 10.1) are confusing and not clear (Mahindru, 1978; Wijesekera and Nethsingha, 1975). Freshly harvested and processed capsules gave an yield of 9.8 per cent oil while capsules after exposure to air for 1 month yielded 2.9 per cent oil indicating a loss of 70 per cent oil which appears to be unreasonable. The loss of oil from the ground seed is rapid when not protected and in 13 weeks' time gave only trace amounts of oil.

Griebel and Hess (1940), Gerhardt (1972) and Koller (1976) have reported the effect of different storage conditions on the rate of oil loss from seeds and ground cardamom and particularly Koller mentions that the temperature of storage has a greater influence on the rate of oil loss rather than the type of container or the period of storage.

Table 10.1 Yield of cardamom oil at different storage periods

Type of material	Yield of volatile oil (%)
Freshly gathered whole fruit	9.8
Seeds exposed to air for 1 month	2.9
Seeds exposed to air for 6 months	2.4
Seeds exposed to air for 14 months	2.0
Cardamom seeds exposed to air for 1 week	2.4
Cardamom seeds exposed to air for 13 weeks	Trace
Seeds freshly removed from capsules	4.8
Seeds exposed to air for 6 weeks	2.4
Seeds exposed to air for 14 months	1.0
Ground seed exposed to air for 1 week	2.4
Ground seed exposed to air for 13 weeks	Trace

Sources: Mahindru, 1978; Wijesekera and Nethsingha, 1975.

2.2   Cardamom powder

Cardamom in its powder form gives the maximum flavour to the food products. But the disadvantage with powder is that it loses aroma quality by rapid loss of volatiles. Hence the powder needs more protection than whole capsules or seeds. The industrial and institutional requirements of cardamom are met by grinding seeds just before use (ITC/SEPC, 1978; ITC, 1977).

2.2.1   *Grinding*

Grinding is an important step in the process of converting a spice into powder and one has to be very cautious with a spice like cardamom because of its delicate aroma.

The aroma principles of cardamom seed are present near the surface and hence more attention is needed during grinding because of the heat produced in attrition. The temperature during grinding can go as high as 95 °C in mass production (Wistreich and Schafer, 1962; Pruthi, 1980). For grinding, conventional mills like plate mill or hammer mill or pin mill are employed. The particle size of the ground spice may vary from 250 to 700 microns depending on the end use. For oil distillation the size will be 600–700 microns while as a flavourant for addition in food products the preferred size will be 250–300 microns. Finer particle size helps in easy release of aroma and better mixing with food products.

Gopalakrishnan et al. (1990) carried out studies on grinding of cardamom at ambient conditions (using plate mill) as well as at low temperatures (using centrifugal mill). At ambient conditions, by using 0.25 mm mesh sieve, loss of volatiles was 52.8 per cent while with 0.5 mm the loss was only 34 per cent. However, with a coarse powder obtained from 0.75 mm sieve, loss of volatiles was 26.2 per cent, but the trend got reversed when a coarser powder obtained from 1.0 mm sieve gave 39 per cent loss of volatiles. The higher loss or poor recovery in the latter case was attributed to the incomplete release of oil from the very coarse powder. Grinding of frozen cardamom seeds or grinding seeds with liquid nitrogen using 0.25 mm sieve resulted in 35.4 per cent and 37.8 per cent volatile oil loss respectively. The grinding time was also more when compared with ambient grinding. However, cryo-grinding seeds with dry ice gave the best result and the loss of volatile oil was only 8.74 per cent, but during grinding moisture absorption by the material was noticed. Other studies have also shown that the loss of volatiles was considerably minimized by prechilling the spice and grinding at low temperature (Anonymous, 1975, 1977).

Cryo-grinding or freeze grinding of spices is a novel approach to get better quality spice powders with increased retention of volatiles (Wistreich and Schafer, 1962). Advantages of cryo-grinding are minimum oxidation loss of volatiles, increased output of powder (product) and prevention of gumming up of screens or discs during milling (Russo, 1976). The product so obtained has good dispersibility in food preparations. It is also reported that low-temperature reduces the microbial load on spices. The cost of cryo-process gets reduced when the milling operations are carried out on a bigger scale and with efficient recycling of the refrigerant. Maximum yield of oil has been obtained when the cardamom seeds are precooled by using liquid nitrogen to a temperature of −180 to −190 °C and grinding the spice to a less finer powder (250 micron size).

2.2.2   Storage of powder

Ground cardamom loses its aroma quality rapidly by loss of volatiles and hence proper care should be taken during storage. Gerhardt (1972) found that lacquered cans, PVDC and high-density polyethylene (HDPE) were suitable for storage of powder. Koller (1976) found that vacuum-packaged ground cardamom stored at 5 °C retained flavour for longer periods. Polyester/aluminium foil/polyethylene laminate with its outstanding, moisture, oxygen and odour barrier properties can offer a long shelf life of over 180 days under normal conditions for cardamom powder. For shorter storage life of 90 days and below metalized polyester/polyethylene laminate can be considered.

3   CARDAMOM OIL

Cardamom oil is obtained by distillation of powdered seeds of cardamom. Steam distillation is the common method employed for the production of oil. Use of the cohabation technique for distillation has been discontinued due to the hydrolysis of esters during the process operation. The quality of oil depends on the variety, rate and time of distillation. The important trade varieties are Alleppey green, Coorg green and Sakleshpur bleached. Yield of volatile oil from the seeds of these three varieties was 10.8 per cent, 9.0 per cent and 8.0 per cent respectively (Lewis *et al.*, 1967). External appearance, size or bleached colour are not the parameters to be considered while selecting cardamom for distillation. The high-grade cardamom is not economical for distillation, since it fetches a better price as whole cardamom in the trade. Lower grades, which do not fetch higher value because of defective appearance, but still good from the flavour point of view, are ideally suited for distillation. The husk is almost devoid of any volatile oil (Anonymous, 1985). The flavour of cardamom is mainly due to 1,8-cineole, terpinyl acetate, linalyl acetate and linalool. The total flavour profile as given by Lawrence (1978) is given in Table 10.2 (Also see Chapter 3).

The relative proportions of these constituents have a direct bearing on the cardamom quality. It is well known that cineole is responsible for the camphoraceous odour while other esters and alcohols give the pleasant fruity odour characteristic of cardamom. Coorg cardamom is considered more camphoraceous than Alleppey green (Govindarajan *et al.*, 1982). During grinding, composition of the flavourants gets affected depending on the type of mill and the temperature attained during milling.

Table 10.2 Flavour profile of cardamom oil (main components)*

Components	(%)	Trace components	
α-pinene	1.5	Hydrocarbons	Alcohols and Phenols
β-pinene	0.2	α-thujene	3-methyl butanol
Sabinene	2.8	Camphen	*p*-menth-3-en-1-ol
Myrcene	1.6	α-terpinene	Perillyl alcohol
α-phellandrene	0.2	*cis*-ocimene	Cuminyl alcohol
Limonene	11.6	*trans*-ocimene	*p*-cresol
1,8-cineole	36.3	Toluene	Carvacerol
γ-terpinene	0.7	*p*-dimethylstyrene	Thymol
p-cymene	0.1	Cyclosativene	Carbonyls
Terpinolene	0.5	α-copaene	3-methyl butanal
Linalool	3.0	α-ylangene	2-methyl butanal
Linalyl acetate	2.5	γ-cadinene	Pentanal
Terpinen-4-ol	0.9	γ-cadinene	Furfural
α-terpineol	2.6		8-acetoxycarvotanacetone
α-terpinyl acetate	31.3	Acids	Cuminaldehyde
Citronellol	0.3	Acetic	Carvone
Nerol	0.5	Propionic	
Geraniol	0.5	Butyric	Others
Methyl eugenol	0.2	2-methyl butyric	Pinole
trans-nerolidol	2.7	3-methyl butyric	Terpinyl-4-yl acetate
			α-terpinene propionate
			Dihydro-α-terpinyl acetate

Source: Lawrence (1978).

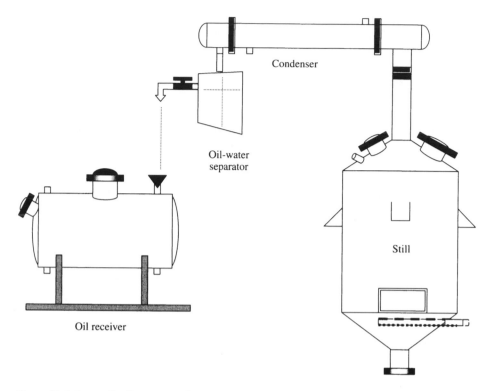

Figure 10.2 Steam distillation unit for cardamom oil.

The United Kingdom was earlier distilling oil from the cardamom obtained from India, Sri Lanka and Tanzania. The oil used to be called as 'English distilled cardamom oil' and priced higher when compared to the oils produced from these cardamom-producing countries. With the development of better technology, for the distillation of cardamom oil in the cardamom growing countries, the production of oil in UK has been considerably reduced and the oil is being imported now.

3.1   Industrial production of cardamom oil

Cardamom capsules of proper maturity having moisture content of 10–12 per cent are selected for oil distillation. The capsules are cleaned through a destoner and air classifier to remove the undesirable extraneous matter. The cleaned capsules are dehusked in a disc (plate) mill. The gap between the discs is critical in order to avoid damage to seeds. Seeds and broken husk are separated in a vibratory sieve. The average composition of capsules of different varieties of cardamom is given in Table 10.3.

Cardamom seeds free of husk are passed through the plate mill, wherein the gap between the discs is brought closer to get a coarse powder to pass through 2 mm sieve. The oil glands exist just below the epidermal layer and hence great care should be exercised while powdering. Fine milling should be avoided to prevent loss of volatiles.

Table 10.3 Composition of different varieties of cardamom

Variety	Husk (%)	Seeds (%)	Volatile oil in seeds (% v/w)
Kerala			
Alleppey	26.0–38.0	62.0–72.3	7.5–11.3
Karnataka			
Coorg	25.2–28.0	69.6–73.3	7.5–9.1
Tamil Nadu			
Yercaud	24.0–33.0	73.0–76.0	6.5–9.6

Sources: Data compiled from Nambudiri *et al.*, 1968; Shankaracharya and Natarajan, 1971.

Note

Moisture in the above raw materials ranged between 8–12%.

The powdered material is subjected to distillation as quickly as possible. If for any reason there is delay, the powder is packed in airtight containers till it is used. A commercial steam distillation unit consists of a stainless steel (SS) vessel (1 mt diameter × 2 mt height) with a volume of about 1600 l capacity (Fig. 10.2). This vessel can easily take 500 kg of powdered material whose bulk density is in the range of 500–550 g/l. The vessel is provided with a perforated false bottom on which the powdered material is placed uniformly and loosely packed without applying any pressure, which ultimately helps in the prevention of channelling of steam. The powder is charged from top of the vessel and leveled periodically using a wooden leveler. The vessel is not packed to full capacity and a headspace of 40–50 cm is maintained. The lid at the top is properly secured and connected to a water-cooled condenser of suitable capacity, which in turn is led to an oil-water separator. From the bottom, steam is let in slowly which passes through the bed of material. By entrainment, the steam carries along with it the volatile principles of cardamom and gets condensed when it passes through the condenser. The condensate enters the oil-water separator and the oil floats at the top. When sufficient volume of oil is collected, it is taken out through a tap provided for this purpose. The condensate water is continuously drained from the other side of the separator. It is not desirable to leave the oil in contact with water till the end of distillation, since this could result in loss of oil by saturation with water and may cause compositional variation due to differential dissolution of the components. The distillation for a 500 kg batch powder usually takes 5–6 h and some times is continued up to 10 h. The rate of distillation and the condensate temperature are carefully regulated and it has been observed that keeping the condensate warm helps in clear separation of oil from water (Nambudiri *et al.*, 1968).

After commencement of distillation, in the first 1 h about 60–70 per cent of the oil is collected. Distillation is continued for longer hours to recover the high boiling fractions, which are equally important from the flavour point of view. It has been observed that early fractions are rich in low boiling terpenes and 1,8-cineole and the subsequent fractions are rich in esters like terpinyl acetate (Krishan and Guha, 1950). Baruah *et al.* (1973) have also confirmed that the proportion of 1,8-neole and α-terpinyl acetate varies with the time of distillation and at 30 min period the composition was 47 and 30 per cent while at 180 min it was 17 and 58 per cent respectively. Oil of var. *Malabar* contains 41 per cent 1,8-cineole, 30 per cent α-terpinyl acetate, 0.4 per cent linalool, and 1.6 per cent linalyl acetate, while the corresponding values for var. *Mysore* are 26.5, 34.5, 3.7 and 7.7 per cent respectively. Because of the high cineole content Malabar oil is harsh and camphoraceous while the

Mysore oil has sweet and fruity-floral odour due to the lower amount of cineole and higher amounts of terpinyl acetate, linalool and linalyl acetate (Lewis, 1973). Incidentally var. *Mysore* is the largest selling Indian cardamom grade called as 'Alleppey green'.

3.2   Improvement in flavour quality of cardamom oil

Flavour quality of cardamom oil containing high amounts of 1,8-cineole has been improved by fractional distillation (Narayanan and Natarajan, 1977). The authors in their experiment subjected 200 g of cardamom powder for distillation and in the first 2.5 min collected 6.5 ml oil, which on analysis was found to contain 78.86 per cent of 1,8-cineole and traces of α-terpinyl acetate. In the subsequent period of distillation time of 2.5–120 min collected 10.5 ml oil which was found to contain 47.5 per cent, 1,8-cineole and 36.8 per cent α-terpinyl acetate. Hence it is possible to get good quality cardamom oil by using inferior grade cardamom by suitably collecting the oil fractions at different intervals of time. Careful blending of the fractions is carried out by keeping the aroma profile and specifications in view (Table 10.4). The oil yield

Table 10.4 Specification for cardamom volatile oil

Definition, source	Volatile oil distilled from the seeds of *Elettaria cardamomum* (Linn.) Maton; family; Zingiberaceae; cardamom grown in South India, Ceylon, Guatemala, Indonesia, Thailand, South China
Physical and chemical constants	Appearance: colourless to very pale yellow liquid Odour and taste: Aromatic, penetrating, somewhat camphoraceous odour of cardamom; persistently pungent; strongly aromatic taste Specific gravity: 0.917–0.947 at 25 °C (temperature correction factor, 0.00079 per °C). Optical rotation: + 22 ° to + 44 ° Refractive index: 1.463 to 1.466 at 20 °C
Descriptive characteristics	Solubility: 70% alcohol: in five volumes; occasional opalescence: Benzyl alcohol: in all proportions Diethyl phthalate: in all proportions Fixed oil: in all proportions Glycerine: insoluble Mineral oil: soluble with opalescence Propylene glycol: insoluble Stability: Unstable in presence of strong alkali and strong acids; relatively stable to weak organic acids; affected by light
Containers and storage	Glass, aluminium, or suitably lined containers, filled full; tightly closed and stored in cool place, protected from light

Source: Adopted from EOA, 1976.

will be less by about 25 per cent by this method, but will be economical since the subsequent fraction fetches a higher price.

Raghavan *et al.* (1991) have standardized a method for the separation of 1,8-cineole from cardamom oil by adduct-formation using ortho-phosphoric acid. In this method 100 ml of cardamom oil is first treated with 30 ml of ortho-phosphoric acid and then with 50 ml of petroleum ether with constant stirring. The adduct (ppt) formed is filtered. The precipitate is air dried and extracted with 500 ml of hot water. Cineole fraction is released as a separate layer and recovered. The aqueous layer is extracted with 200 ml of petroleum ether and desolventized to get terpinyl acetate rich fraction. The GC analysis of these fractions showed that cineole fraction (28 ml) contained 80 per cent cineole and 18 per cent terpinyl acetate while the terpinyl acetate fraction (58 ml) contained 76 per cent terpinyl acetate and 16 per cent cineole.

An attempt was made to produce terpeneless cardamom oil by column chromatography technique using activated silica gel (Raghavan *et al.*, 1991). Petroleum ether was used to elute out terpene fractions and acetone was used to elute oxygenated compounds. It was concluded that overall flavour value of the deterpenated oil was not better than the original non-deterpenated oil. The finer notes of the original oil were missing in the deterpenated oils.

The yield of oil from husk is reported to vary from 0.2 to 1 per cent and the oil possessed similar properties as that of seed oil (Rao *et al.*, 1925; Rosengarten, 1969). Nambudiri *et al.* (1968) have found that husk does not give more than 0.1 per cent volatile oil while the reported higher values may be due to the admixture of seeds along with the husk during sieving operation. Govindarajan *et al.* (1982) mention that the available data are not clear on this point. The chemical quality of the oil obtained from seeds and husk was evaluated by Verghese (1985) using GLC and IR. Though there was excellent correlation and the spectra were super imposable the organoleptic profile differed. The author concluded that distillation of oil from seeds along with husk is detrimental as it is likely to impair the flavour spectrum of the oil. Purseglove (1981) mentions that oil obtained from green and bleached cardamom will be of similar composition.

Hydro-distillation of cardamom is not practiced commercially because the distillation time is more and the release of oil is slow due to gelatinization of starch besides hydrolysis of the esters present in the oil (Wijesekera and Clodagh, 1975). Another disadvantage is, the resulting mass after hydro-distillation is not easily amenable for oleoresin extraction with solvents.

3.3   Storage of cardamom oil

Cardamom oil before storage should be free of trace amounts of moisture, and this is accomplished by addition of anhydrous sodium sulphate. Cardamom oil is stored in aluminium or SS containers. Polyethylene terepthalate (PET) bottles, which possess very good odour barrier properties, can also be considered. Food grade high molecular weight high-density polyethylene (HMHDPE) containers are also being used. The oil is filled to the full capacity of the container and stored at 8–10 °C and protected from light.

Flow sheet for the production of cardamom oil

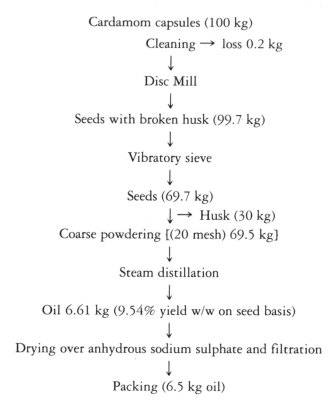

Cardamom capsules (100 kg)

Cleaning → loss 0.2 kg

↓

Disc Mill

↓

Seeds with broken husk (99.7 kg)

↓

Vibratory sieve

↓

Seeds (69.7 kg)

↓ → Husk (30 kg)

Coarse powdering [(20 mesh) 69.5 kg]

↓

Steam distillation

↓

Oil 6.61 kg (9.54% yield w/w on seed basis)

↓

Drying over anhydrous sodium sulphate and filtration

↓

Packing (6.5 kg oil)

4   CARDAMOM OLEORESIN

Oleoresin is made of two components, viz., the volatile oil and resin. Volatile oil represents the aroma while the resin is made up of non-volatile matter like, colour, fat, pungent constituents, waxes etc. The total flavour effect of a spice is obtained only after blending the oil and resin. Volatile oil is obtained by steam or hydro-distillation while the resin is obtained by solvent extraction. Of late supercritical fluid extraction (SCFE) is also being adopted.

The demand for cardamom oleoresin is small unlike other oleoresins. Cardamom oil itself represents almost all the aroma and flavour of the capsules. However, the consumption of cardamom oleoresin is slowly picking up probably due its mellower and less harsh flavour characteristics (Sankarikutty et al., 1982). Though the cardamom oil represents the flavour of cardamom it lacks the 'richness' which is attributed to the absence of non-volatile components (Lewis et al., 1974). Sensory differences have also been noticed between oils and oleoresins of cardamom (Govindarajan et al., 1982). Oleoresin, besides containing flavouring principles also contains non-aromatic fats, waxes, resinoids and colour which act as fixatives for the volatiles. This advantage has been particularly seen in the bakery and confectionery processing (Farell, 1985). Oleoresins are truer to the spice in flavour and are capable of withstanding high temperature processing to a better extent when compared to the volatile oils obtained by direct steam distillation (Eiserle and Rogers, 1972).

For oleoresin extraction, either fresh ground cardamom or essential oil-free cardamom powder (cardamom powder from which oil has been distilled off) is employed. The main considerations involved in the oleoresin preparations are selection of suitable raw material, grinding to the optimum particle size for extraction, choice of solvent, type of extraction, miscella distillation and blending.

4.1   Solvent extraction

Cardamom seed is ground to coarse powder of particle size of 500–700 microns, which helps in the rupture of flavour cells and is amenable for ready extraction by solvents. Fine grinding should be avoided, which not only results in the volatile loss but also creates problems during extraction, like slow percolation of the solvent, channelling and engagement of the extractor for longer periods. The powdered spice is loaded into the extractor also called as percolator and extracted with a suitable solvent. The choice of solvent can be from acetone, alcohol, methanol, ethylene dichloride, hexane, isopropyl alcohol, ethyl acetate, ethyl methyl ketone, methylene chloride, iso-propyl ether or a mixture of solvents. The selection of solvent for extraction is a crucial step and it should be standardized on a small scale at the laboratory level before venturing into commercial production. Points to be considered for selection of solvent are extrac-tion efficiency, boiling point, inflammability, miscibility with water, residual notes, toxicity, cost and availability. Available information on the physico-chemical properties of some solvents is given in Table 10.5.

The selected solvent is allowed to percolate through the bed of material by keeping the bottom drain valve open for the escape of air. When the entire material is soaked in solvent, the bottom drain is closed and sufficient contact time given for leaching of the solutes into the solvent. After the contact time, the extract called as 'miscella' is drained and collected.

For oleoresin production either soxhlet extraction method (Goldman, 1949) or batch counter current extraction (CCE) is industrially practised (Nambudiri *et al.*, 1970). In the soxhlet method the solvent is allowed to percolate through the bed of material several times. But for the initial addition of fresh solvent, the subsequent solvent for extraction is derived from the continuous heating of the extract, which may ultimately cause some heat damage to the finished product resulting in some-what poor sensory quality. In the batch CCE method that is widely adopted, the spice powder is packed in a series of extractors and extracted with a suitable selected solvent. Though it is called CCE method, here the bed material is stationery while the solvent or extract moves from extractor to extractor; only the weaker or dilute extract is allowed to percolate through the spice while the concentrated extract obtained first from each extractor (after the stage of equilibrium) is withdrawn for solvent stripping to make the product viz., oleoresin. The first withdrawal of the concentrated extract is usually fixed at 30–40 per cent of the weight of material charged into the extractor i.e. for 100 kg material, miscella withdrawal is 30–40 kg. The concentrated extract is not added to the next extractor because it is already rich in solids and viscous in nature. In the process of extraction the addition of dilute extracts to the extractor is carried out in such a way that the more concentrated extract is added first followed by the less concentrated extract. This is to ensure that the counter current technique is fully exploited. Ultimately each extractor is washed with a fresh lot of solvent.

Table 10.5 Physico-chemical properties of solvents

Solvent	Formula	Formula wt.	Density 20 °C	Boiling point °C	Flash point °C	Solubility in 100 parts	Availability
Acetone	CH_3COCH_3	58.08	0.7908	56.24	−20	misc., aq., alc., eth.	Commercially available
Ethylene dichloride (1,2-dichloro ethane)	CH_2Cl-CH_2Cl	98.96	1.2531	83.5	15	0.8 aq., misc., alc., chl., eth.	"
Isopropyl ether	$(CH_3)_2CH$-O-$CH_2(CH_3)_2$	102.18	0.7250	68.5	−12	0.2 aq., misc., alc., chl., eth.	"
Ethyl acetate	$CH_3COOC_2H_5$	88.11	0.9006	77.1	−3	9.7 aq., misc., alc., acet., chl., eth.	"
Ethyl methyl ketone (2-butanene)	$CH_3CH_2COCH_3$	72.11	0.8049	79.6	−3	24 aq., misc., alc., bz., eth.	"
Alcohol	CH_3CH_2OH	46.07	0.7894	78.3	8	misc., alc., ace., chl., eth., bz.	
Methanol	CH_3OH	32.04	0.7913	64.7	11	misc., aq., alc., chl., eth.	"
Isopropyl alcohol	$(CH_3)_2CHOH$	60.10	0.7855	82.4	22	misc., aq., alc., chl., eth.	"
Methylene chloride	CH_2Cl_2	84.93	1.3255	40.5	none	1.3 aq., misc., alc., eth.	"
Hexane	$CH_3(CH_2)_4$ CH_3CH_3	86.18	0.6594	68.7	−23	misc., alc., chl., eth.	"

Source: Dean (1987).

Notes
misc. – miscible, aq. – aqueous; alc. – alcohol; chl. – chloroform; eth. – diethyl ether; bz. – benzene; ace. – acetone.

The concentrated miscella obtained from each extractor is carefully distilled to get the finished product. Most of the solvent (about 90–95 per cent) present in the miscella is recovered by normal atmospheric distillation, while the remaining solvent is taken off by distillation under reduced pressure (Fig. 10.3). Great care should be exercised during distillation to minimize heat damage to the product. Constant stirring prevents over-heating near the walls of the still. The trace amounts of solvent are removed by azeotropic or extractive distillation method using an innocuous solvent like ethyl alcohol. Alternatively bubbling nitrogen into the thick viscous material is carried out to drive away the resid-ual solvent. The experience and skill of the operator plays a crucial role in bringing down the solvent levels to as low as 5–10 ppm in the product. The maximum permitted residual limits for some of the solvents are: 30 ppm for acetone and chlorinated solvents, 50 ppm for methyl alcohol and isopropyl alcohol and 25 ppm for hexane (CFR, 1995).

After completion of solvent stripping, the product, while hot, is discharged from the bottom of still and stored in suitable containers. A typical miscella distillation unit is shown in Fig. 10.3. Cardamom oleoresin is stored in aluminium or stainless steel containers. Epoxy-coated drums and HMHDPE containers are also being employed for storage and export. Desolventisation of the spent meal after extraction is carried out *in situ* by passing steam from the bottom of the extractor through the bed of material. The steaming operation is carried out slowly and carefully to avoid any channelling. Water immiscible solvents like ethylene dichloride, methylene chloride, ethyl acetate, hexane etc. are easily distilled off and collected for re-use while the water miscible solvents like, acetone, alcohol, methanol etc. tend to get diluted and need rectification before re-use. It has been observed that in a 100 kg batch extraction, the retention of solvent in the spent material is of the order of 60 to 70 kg and about 95 per cent of

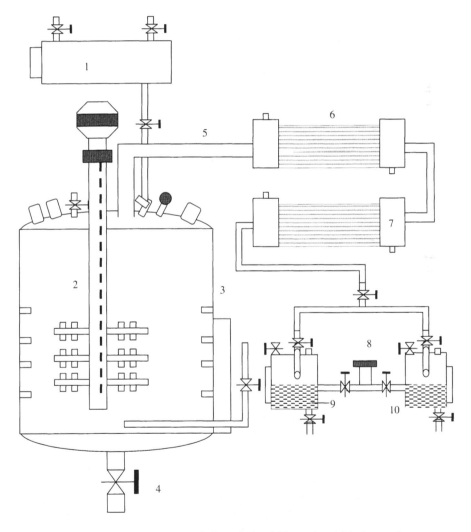

1. Miscella tank;   2. Stirrer;
5. Vapour line;   6. Condensers;

3. Steam jacketed SS vessel;   4. Discharge valve;
7 & 8. Vaccum line;   9 & 10. Condensate tanks.

Figure 10.3 Miscella distillation unit.

which is recovered during the desolventisation process. The spent meal after the recovery of solvent is discharged from the bottom side vent of the extractor and dried. Spent meal contains starch, fibre, carbohydrate, protein etc. and finds application in animal feed composition. It can also be used as a broiler feed and as a source for manure. Cardamom spent meal has been used in *Agarbathi* (scented sticks) formulation (Suresh, 1987). Quality, and yield of cardamom oleoresin depends upon the raw material variety, the solvent used and the method of extraction. By using hydrocarbon solvents, oleoresin having 10–20 per cent fixed oil has been obtained while with a polar solvent like alcohol, a fat-free product is obtained (Naves, 1974). Oleoresins containing 54–67 per cent volatile oil have been obtained by Salzer (1975), wherein the fixed oil content varied with the extracting solvent. The colour of the product varies from brown to greenish brown. Kasturi and Iyer (1955), extracted cardamom seeds from which volatile oil had already been distilled, using carbon tetrachloride as solvent and got 4 per cent yield of fixed oil. The fixed oil on analysis was found to contain oleic acid 62.6 per cent, stearic acid 18.3 per cent, palmitic acid 8.4 per cent, linoleic acid 10.5 per cent, caprylic and caproic acids 0.3 per cent. Miyazawa and Kameoka (1975) and Marsh *et al.* (1977) found palmitic (28–38 per cent), oleic (43–44 per cent) and linoleic (2–16 per cent) as the major fatty acids present in the fatty oil. The CFTRI, Mysore India has developed processes for the production of cardamom oil, spice oleoesins and encapsulated spice flavours, which have been commercially exploited.

Recently a company in UK claims to have produced a good quality cardamom oil by extracting the seeds with a hydrofluro solvent having bp about -26 °C. During extraction damages due to heat or oxygen or high pH is eliminated (Anonymous, 1996).

4.2   Supercritical carbon dioxide extraction of cardamom

Use of liquid and supercritical carbon dioxide as a solvent for flavour extraction from plant materials has been a subject of intense study (Schultz and Randall, 1970; Hubert and Vitzuthu, 1978; Caragay, 1981; Rizvi *et al.*, 1986; Udayasankar, 1988). Three decades back Shultz *et al.* (1967) and Aleksandrov (1969) used CO_2 for the extraction of spices like cardamom, clove, nutmeg, coriander and celery. The use of CO_2 for flavour extraction has several advantages over the traditional methods using other solvents.

Carbon dioxide is abundantly available, cheap, non-inflammable, non-toxic, non-corrosive and does not cause any environmental pollution. Carbon dioxide has been widely accepted as a permitted safe solvent for flavour extraction and does not leave any residue of its own. It behaves either as a polar or non-polar solvent depending on pressure and temperature employed. Carbon dioxide is liquid below its critical point (31.2 °C, 7.38 milli Pascal pressure) and above its critical point it exists as a SCF. Under normal conditions density of CO_2 is less than 100 g/l while under supercritical conditions density varies between 200–900 g/l. The diffusion coefficient and viscosity are higher in SCF and this helps in better and faster mass transfer rates from the matrix to the solvent. Hence SCF is preferred over liquid CO_2 for extraction. The technical advantages in using SCF extraction for spices like pepper, ginger and cumin were studied by Udayasankar (1989). Naik and Maheswari (1988) extracted cardamom using liquid CO_2 (20 °C, 55–58 bar pressure) and using a modified high pressure soxhlet apparatus. They got 9.4 per cent yield in 2.5 h extraction time while with steam distillation in 5 h time 9 per cent yield was obtained. The extracts on analysis by gas-liquid chromatography (GLC) and thin-layer chromatography (TLC) showed that liquid CO_2 extract contained slightly higher amounts of cineole, terpinyl acetate, geraniol and α-terpineol (35.72, 24.87, 4.53 and 11.06 per cent respectively) when compared to steam distilled oil where the corresponding values were 30.25, 22.05, 4.22 and 7.88 per cent. Gopalakrishnan and Narayanan (1988) extracted cardamom using SCF at 40 °C and 100 bar pressure and the product quality was evaluated in comparison with the hexane extract and distilled oil. The results are summarized in Table 10.6.

The yield with SCFE was more than in hexane extract. The SCF product had considerably lower amounts of non volatile matter and possessed fresh cardamom aroma, which was absent in hexane extract and distilled oil. The composition of the steam distilled oil and the oil distilled from liquid CO_2 extract were found to be very similar by the studies carried out by Meerov *et al.* (1971). Extraction of cardamom under different conditions of pressure, temperature, contact time and moisture content had not much influence with respect to yield or quality of the product. However, the extraction of non-volatiles and chlorophyll content increased with the increase in pressure and time (Gopalakrishnan and Narayanan, 1991).

It is rather disappointing to note that the CO_2 extracted cardamom oil, though has a better quality when freshly extracted, lost its fine aroma during 90 days storage. Quality deterioration of the commercial steam distilled oil was comparatively less under similar conditions (Gopalakrishnan, 1994).

A few SCFE plants are working on a commercial scale particularly for the extraction of hops and decaffeinating of coffee seeds in countries like Germany, Australia, Austria and Canada. In India a few imported SCFE plants are coming up for the extraction of spices. It is yet to be assessed as to how the products derived from SCF compete with the traditional solvent extracts, price-wise and quality-wise in the international market. A few well-established spice extractors in India are silent on this venture of putting SCFE plants. A 1 tonne capacity plant for SCFE will cost around Rs. 100 million while for a solvent extraction plant of similar capacity the cost is around Rs. 10 million. Though it is generally considered that SCF-derived products are superior in quality

Table 10.6 Quality of cardamom extracts obtained by different methods

	SCF	*Hexane extract*	*Clevenger distilled oil*
Parameters			
Yield (%)	7.7	6.2	8.3
Non-volatile matter (%)	4.6	22.5	–
Colour	Pale green	Pale green	Colourless
Aroma	Superior, close to fresh cardamom	Residual solvent note, fresh aroma absent	Varied due to artifact formation, and absence of non-volatiles
Major components			
1,8-cineole	29.7	16.6	31.2
α-terpenyl acetate	37.0	57.3	35.5
α-terpeneol	4.6	5.0	2.4
Linalool	2.6	2.3	3.8
Sabinene	4.1	2.2	3.4
β-pinene	2.8	1.7	2.8
D-limonene	2.4	2.2	3.3
Linalyl acetate	1.6	1.6	2.1

with no residual solvent, cost wise they will be much higher and their acceptability by the food processing units is a point yet to be answered.

5   ENCAPSULATED CARDAMOM FLAVOUR

Encapsulation is a technique in which the flavourant is covered by a suitable material thereby protecting the flavour from exposure to environment. In this method the liquid aroma concentrate is converted to a solid stable powder form having good shelf life. It is reported that some volatile liquid flavours are retained in microcapsules for periods up to 2 years (Bakan, 1973). The flavour protecting material is called as 'wall material' or 'encapsulating material' and is generally either gum acacia or a starch derivative like malto-dextrin. The actual flavourant, which is to be encapsulated, is called as 'core' material. The selected wall material should be of food grade and an effective film former and should stabilize the emulsified flavour in the process of encapsulation. The encapsulated product is spherical and miniature in size ranging from sub-micron to several millimeters (Bakan, 1978). When the particle size of the capsules is less than 5000 μm they are called as microcapsules (Anandaraman and Reineccius, 1980).

There are different methods of encapsulation of which spray drying is the one most widely used for cardamom oil encapsulation (Sankarikutty et al., 1988; Raghavan et al., 1990). Besides spray drying method, other encapsulation techniques such as phase separation, adsorption, molten extrusion, spray cooling or chilling, inclusion complex formation etc. are practiced. However, in the present context only spray drying is dealt with. The basic steps involved in this process are (i) preparation of the emulsion; (ii) homogenization of the emulsion with the flavourant and (iii) atomization of the mass into the drying chamber.

Raghavan et al. (1990) have carried out spray drying of cardamom oil using a small spray drier (Bower Engineering, New Jersey, USA) as well as a pilot scale spray drier (Anhydro). In a small-scale batch, 300 g of gum arabica was made into a paste using 300 ml of water and finally the volume made up to 3 l. After complete dissolution in about 2 h time, the solution was filtered through a strainer, then mixed with 30 ml of cardamom oil and 0.6 ml of Tween-80 and homogenized using a warring blender. Addition of tween helps in reducing the viscosity besides giving a stable emulsion. The emulsion thus obtained was spray dried by maintaining a feed rate of 70–80 ml/min, inlet air temperature 175–180 °C and outlet 105–115 °C. The yield of dry powder was 250 g having a moisture content of 4.5 per cent and volatile oil 7.5 per cent. In the pilot scale batch 7.5 kg gum arabica was dissolved with 25 l of water in about 10–12 h time. The mixture after good homogenization was mixed with 750 ml of cardamom oil and 15 ml of Tween-80 and again homogenized to a particle size of 1–4 microns. The emulsion was spray dried at an inlet air temperature of 160–170 °C and feed rate of 14–15 l/h. The emulsion is atomized into the drying chamber wherein it comes in contact with a stream of heated air. The atomized particles assume a spherical shape with the oil encased in the aqueous phase as they pass through the gaseous medium. The rapid evaporation of water from the emulsion during its solidification keeps the 'core' temperature below 100 °C in spite of the higher temperature used in the process (Brenner, 1983) and the particles exposure to heat in the range of a few seconds. The yield of dried product was 7.2 kg having 4 per cent moisture and 8.5 per cent volatile oil. After several trial batches, the optimum oil to encapsulant ratio was found to be 1:4, which very much agrees with the studies carried out by Sankarikutty *et al.* (1988). These authors in their experiments have used hot water (50–60 °C) to aid dispersion of the gum. After mixing cardamom oil and emulsification the globule size was 2 μm. The material was spray dried at an inlet air temperature of 155 ± 5 °C and exit air of 100 ± 5 °C. The water evaporation rate was maintained at 0.25 kg/hr. While carrying out these trials, it was observed that gum acacia, dextrin white, and maltodextrin showed promise as high solid carriers and the optimum concentrations were found to be 26, 21 and 36 per cent respectively. The viscosity of the emulsion was determined by using a viscometer and particle size by using a microscope. The scanning electron micrographs of capsules obtained with different wall materials showed that maltodextrin gives a more spherical shape with minimum dents having a tendency to develop crack, while in gum acacia with maltodextrin in equal proportions, cracks were not noticed. However, gum acacia alone because of its excellent emulsification and film forming properties and low solution viscosities even at high solids content is the preferred wall material as an encapsulant.

Under ideal conditions, the encapsulated powder should not have any flavour. However, in practice, it has been observed that the product will have mild odour due to some amount of flavour left unencapsulated and also due to rupture of a few capsules. Attempts have been made to remove the surface flavour by washing the particles with hexane (Omankutty and Mathew, 1985). About 1–2 per cent surface oil was found on the spray-dried flavours made by using different encapsulants.

Gas chromatographic examination of the cardamom oils obtained from the emulsion and their spray dried product showed a similar pattern when compared to the original oil. However, the oil derived from the spray dried encapsulated product showed a slight decrease in the cineole content (44 per cent) and increase in terpinyl acetate (38.9 per cent) and the corresponding values in the oil obtained from emulsion were 47.4 per cent and 34.8 per cent (Sankarikutty *et al.*, 1988; Raghavan *et al.*, 1990). Property-wise cineole is more volatile and has lower molecular weight than terpinyl acetate.

Stability of the encapsulated cardamom oil product was found to be satisfactory when stored in airtight glass containers at room temperature. The moisture pick up was negligible and there was only 5 per cent loss of volatile oil during the 2 years storage period (Raghavan *et al.*, 1990).

The advantages of using spray dried encapsulated cardamom flavours are; they are non-volatile, dry and free flowing; can be readily incorporated into food mixes to obtain uniform flavour effect; flavour stability is good over longer storage periods even at higher temperatures, and in aqueous system the capsules break and flavour is released.

6   LARGE CARDAMOM (NEPAL CARDAMOM)

Another type of cardamom of commercial importance is *Amomum subulatum* known by the name Sikkim or Nepal large cardamom. In India, Sikkim state alone produces the major quantity of large cardamom almost equal to small cardamom produced in the southern states of India. Thailand, Indonesia and Laos also produce large cardamom to a limited extent. Large cardamom is used as a flavouring agent in curry powders, sweet dishes, cakes and for masticatory and medicinal purposes.

Recently physico-chemical studies on five cultivars of large cardamom viz., *Ramsey, Golsey, Sawney, Ramla* and *Madhusey* have been carried out (Pura Naik, 1996). The studies revealed that percentage husk varied between 27 and 31.5 per cent and seeds

68.2–72.0 per cent. Volatile oil content in seeds ranged from 2.7–3.6 per cent. Large cardamom contains less volatile oil than small cardamom and is more camphoraceous and harsh in aroma with a flat cineole odour. The oil is rich in 1,8-cineole and devoid of α-terpinyl acetate (Govindarajan *et al.*, 1982). The chemical composition of the large cardamom oil is well documented (Lawrence, 1970; Patra, 1982; Balakrishnan *et al.*, 1984) (see Chapter 15).

Products like volatile oil, oleoresin, encapsulated flavour etc. can be produced from large cardamom also using the processing methods described for cardamom.

7   OTHER PRODUCTS

A number of products having cardamom as the major flavourant can be prepared. Some such products having commercial potential are indicated (Raghavan *et al.*, 1991).

7.1   Sugar cardamom mix

It is a blend of sugar powder with encapsulated cardamom flavour along with sunset yellow colourant and may contain tricalcium phosphate as anticaking agent. This finds application in culinary sweets, flavouring of milk and milk products. Incorporation of the above mix with malted *ragi* (finger millet) flour makes a good *ragi* malt beverage.

7.2   Cardamom flavoured cola beverage

This is an amber-coloured sparkling carbonated beverage containing sugar, caramel, acid and flavours. A market survey on this product was very encouraging and the product was quite acceptable.

7.3   Cardamom-flavoured 'Flan'

This product is made from milk, sugar, starch with added colour, flavour and gelling agents. It tastes like custard dairy dessert. The formula for the preparation of this product has been standardized (CFTRI, Mysore, India).

7.4   Cardamom chocolate

Cardamom-flavoured milk chocolate is prepared by using cocoa mass with butter, sugar powder, milk powder, encapsulated cardamom flavour and emulsifiers. The resultant product has good acceptability.

7.5   Cardamom plus

In West German market 'pepper plus' has made a good impact. This product is prepared by fortifying pepper powder with encapsulated pepper flavour. The advantage of this blend would be the rich naturalness with high flavour strength. On similar lines 'cardamom plus' can be prepared by mixing freshly ground cardamom with encapsulated cardamom flavour. In the place of cardamom powder one can use cardamom residue (after oil distillation), which contains fixed flavours, and resinous mass, which can be removed by solvent extraction only. Spent residue is a valuable source for making 'cardamom plus' preparation.

7.6 Cardamom tincture

This product is made by the extraction of crushed cardamom seeds alongwith other spices like caraway and cinnamon, and cochineal using 60 per cent alcohol as solvent. Glycerine is added at 5 per cent level to the extract and is used as a caraminative (British Pharmacopoeia, 1980).

7.7 Cardamom coffee and tea

The cardamom flavour is very compatible with beverages like coffee and tea. In Middle East the major use for cardamom is in coffee. This cardamom coffee is called as 'Gahwa' and is a traditional drink with Arabs of the Gulf area (Survey of India's Export Potential of Spices, 1968). Encapsulated cardamom flavour is very handy and useful while making 'Gahwa' coffee. The flavour is incorporated into roast and ground coffee which on brewing gives a predominantly cardamom-flavoured extract. This can be consumed as 'black' coffee or with milk. Similarly cardamom tea is available in the market in which cardamom powder is mixed with tea powder. Like cardamom coffee cardamom tea can be conveniently made using encapsulated cardamom flavour, which will have longer shelf life.

8 CONCLUSION

Cardamom flavour finds application in various items like processed foods, beverages, confectionery, health foods, medicines, perfumery and cosmetics. Cardamom products like essential oil, oleoresin and encapsulated flavour may find good potential in food and non-food industries because of advantages like standardized flavour strength, good shelf life, hygienic quality, ease of handling and stabilized flavour. However, the increase in demand for the processed cardamom products is related to the increase in population and income in the established markets. Research and marketing activities may prove crucial in creating new forms of value-added products and also new markets. It is also important for the cardamom producing countries to create and maintain internal demand for cardamom and its products.

REFERENCES

Aleksandrov, L.G., Serdyuk, V.I. and Anoshin, I.M. (1969) Determination of the co-efficients of molecular diffusion of some extracts of vegetable raw materials in liquid carbon dioxide *Fzv Vyssh. Ucheb. Zaved. Pishch. Tekhnol.*, 2, 143–144 (Russian).
Anandaraman, S. and Reineccius, G.A. (1980) Microencapsulation of flavours. *Food*, 1(9), 14.
Anonymous (1975) C.C. Spice open new mill. *Food Process. Ind.*, 44(529), 36.
Anonymous (1977) Pulverising System. *Food Process*, 38(3), 108.
Anonymous (1985) *Wealth of India*, vol. 1 A (Revised) Council of Scientific and Industrial Research, New Delhi, pp. 227–229.

Anonymous (1996) Extracting the benefits. *Speciality Chemicals*, 16(6), 228.

Bakan, J.A. (1973) Microencapsulation of foods and related products. *Food Technol.*, 27(11), 33.

Bakan, J.A. (1978) Microencapsulation. In M.S. Peterson and R. Johnson (eds) *Encyclopedia of Food Science*, Avi Pub. Co., Inc., Westport, Conn.

Balakrishnan, K.V., George, K.M., Mathulla, T., Narayana Pillai, O.G., Chandran, C.V. and Verghese, J. (1984) Studies in cardamom-1, Focus on oil of *Amomum subulatum* Roxb. *Indian Spices*, 21(1), 9–12.

Baruah, A.K.S., Bhagat, S.D. and Salkia, B.K. (1973) Chemical composition of Alleppey cardamom oil by gas chromatography. *Analyst*, 98(1164), 168.

Brenner, J. (1983) The essence of spray dried flavours. The state of the art. *Perf. Flav.*, 8(2), 40.

British Pharmacopoeia (1980) *2 HMSO*, Cambridge, UK, p. 834.

Caragay, A.B. (1981) Supercritical fluids for extraction of flavours and fragrances. *Perfumer and Flavourist*, 6, 43.

CFR (1995) Code of Federal Regulations, 21, CFR 173.2, Washington, USA.

Clevenger, J.F. (1934) Volatile oil from cardamom seed. *J. Assoc. Agric. Chem.*, 17, 283.

Dean, J.A. (1987) *Hand Book of Organic Chemistry*, McGraw-Hill Book Company, USA.

Eiserle, R.J. and Rogers, J.A. (1972) The composition of volatiles derived from oleoresins. *J. Am. Oil. Chem. Soc.*, 49, 573.

EOA (1976) Specification: oil cardamom, 289, Essential Oil Association of United States of America, New York.

Farell, K.T. (1985) *Spices, Condiments and Seasonings*. The AVI Publishing Company, USA.

Gerhardt, U. (1972) Changes in spice constituents due to the influence of various factors. *Fleischwirtschaft*, 52(1), 77–80.

Goldman, A. (1949) How spice oleoresins are made? *Am. Perfum. Esse. Oil Rec.*, 53, 320–323.

Gopalakrishnan, N. and Narayanan, C.S. (1988) Composition of volatile constituents of cardamom oleoresin obtained by carbondioxide extraction. *PAFAI J.*, 10, 21–24.

Gopalakrishnan, N., Laxmivarma, R., Padmakumari, K.P., Symon, B., Umma, H. and Narayanan, C.S. (1990) Studies on cryogenic grinding of cardamom. *Spice India*, 3(7), 7–13.

Gopalakrishnan, N. and Narayanan, C.S. (1991) Supercritical CO_2 extraction of cardamom. *J. Agri. Food Chem.*, 39, 1976–1978.

Gopalakrishnan, N. (1994) Studies on the storage quality of carbondioxide extracted cardamom and clove oils. *J. Agric. Food Chem.*, 42, 796–798.

Govindarajan, V.S., Narasimhan, S., Raghuveer, K.G. and Lewis, Y.S. (1982) Cardamom-production, technology, chemistry and quality. *CRC Critical reviews in Food Science and Technology*, 16(3), 229–326.

Griebel, C. and Hess, G. (1940) 'Die Haltbarkeit abgepackter gemahlener Gewurze'. *Zeits. Untersuch der Lebensmittel*, 79, 184–91.

Guenther, E. (1952) Cardamom. *The Essential Oils*, Vol. 5, Robert E. Krieger Publishing, New York, p. 85.

Hubert, P. and Vitzuthum, O.G. (1978) Fluid extraction of hops, spices and tobacco with super-critical gases. *Angew Chem. Intl. Engl.*, 17, 710.

ITC (1977) Spices – A survey of the world market. Vol. 1, 2, International Trade Centre, UNC-TAD/GATT, Geneva.

ITC/SEPC (1978) Market Survey of Consumer packed spices in selected countries, International Trade Centre, UNCTAD/GATT, Geneva.

Kasturi, T.R. and Iyer, B.H. (1955) Fixed oil from *Elettaria cardamomum* seeds. *J. Indian Inst. Sci.*, 37A, 106.

Koller, W.D. (1976) The importance of temperature on storage of ground natural spices. *Z. Lebensm. Unters. Forsch.*, 160, 143–147.

Krishnan, R.P. and Guha, P.C. (1950) Mysore Cardamom oil. *Curr. Sci.*, 19, 157.

Lawrence, B.M. (1970) Terpenoids in two *Amomum* Species. *Phytochem.*, 9, 665.

Lawrence, B.M. (1978) *Major tropical spices – cardamom (Elettaria cardamomum).* In Essential Oils, Allured Publ., Wheaton, pp. 105–155.

Lewis, Y.S., Nambudiri, E.S. and Natarajan, C.P. (1967) Studies on some essential oils. *Indian Fd. Packer,* 11(1), 5.

Lewis, Y.S. (1973) The importance of selecting the proper variety of a spice for oil and oleoresin extraction. Tropical Product Institute Conference papers, London.

Lewis, Y.S., Nambudiri, E.S. and Krishnamurthy, N. (1974) Flavour quality of cardamom oils. *Proc. VI Int. Congr. Essential Oils,* London.

Mahindru, S.N. (1978) Spice extraction – cardamom. *Indian Chemical Journal,* 13(1), 28–30.

Marsh, A.C., Moss, M.K. and Murphy, E.W. (1977) Composition of Foods, Spices and Herbs. Raw, processed, prepared. USDA, *Agric. Res. Serv. Hand Book,* No. 8, 2. Washington, D.C.

Meerov, Ya, S., Popova, S.A. and Ponomarenko, I. Ya. (1971) Quality of cardamom oil produced by different methods. *FSTA* 1971, 3, 11T, 552.

Miyazawa, M. and Kameoka, H. (1975) The constitution of the essential oil and non-volatile oil from cardamom seed. *J. Japanese Oil Chemists Soc.* (Yukaguku), 24, 22.

Naik, S.N. and Maheshwari, R.C. (1988) Extraction of essential oils with liquid carbon-dioxide. *PAFAI J.,* 10(3), 18–24.

Nambudiri, E.S., Lewis, Y.S., Rajagopalan, P. and Natarajan, C.P. (1968) production of cardamom oil by distillatiion, *Res. and Ind.,* 13, 140.

Nambudiri, E.S., Lewis, Y.S., Krishnamurthy, N. and Mathew, A.G. (1970) Oleoresin of pepper. *The Flav. Ind.,* 1, 97–99.

Narayanan, C.S. and Natarajan, C.P. (1977) Improvements in the flavour quality of cardamom oil from cultivated Malabar type cardamom grown in Karnataka State. *J. Fd. Sci. & Tech.,* 14, 233.

Naves, Y.R. (1974) *Technologie et chimie des parfums Natures Paris.* Masson & Cie.

Omanakutty, M. and Mathew, A.G. (1985) Microencapsulation. *PAFAI J.,* 7(2), 11.

Patra, N.K. (1982) G.C.examination of the oil from fruits of *A. subulatum* Roxb. growing wild in Darjeeling. *PAFAI J.,* 4(4), 29.

Pruthi, J.S. (1980) *Spices and Condiments – Chemistry, Microbiology and Technology.* Academic Press Inc., New York (USA).

Pura Naik, J. (1996) *Physico-chemical and technological studies on large cardamom (A. subulatum) for its use in foods.* Ph.D. thesis, University of Mysore, Mysore, India.

Purseglove, J.W., Brown, E.G., Green, C.L. and Robbins, S.R.J. (1981) *Spices,* 2, 606–608, Longman, New York.

Raghavan, B., Abraham, K.O. and Shankaranarayana, M.L. (1990) Encapsulation of spice and other flavour materials. *Indian Perfumer,* 34(1), 75–86.

Raghavan, B., Abraham, K.O., Shankaracharya, N.B. and Shankaranarayana, M.L. (1991) Cardamom – studies on quality of volatile oil and product development. *Indian Spices,* 28(93), 20–24.

Rao, B.S., Sudborough, J.J. and Watson, H.E. (1925) Notes on some Indian essential oils, *J. Indian. Inst. Sci.,* 8A, 143.

Rizvi, S.S.H., Daniels J.A., Benado, A.L. and Zollweg, J.A. (1986) Supercritical fluids extraction-operating principles and food applications. *Food Technol.,* 40(7), 57.

Rosengarten, Jr. F. (1969) *The Book of Spices,* Livingsten Publishing Co., Wynwood. Pensylvia.

Russo, J.R. (1976) Cryogenic grinding. *Food Eng. Int.,* 1(8), 33.

Salzer, U.J. (1975) Analytical evaluation of seasoning extracts (oleoresins) and essential oils from seasonings. *International flavours and Food additives,* 6(3), 151.

Sankarikutty, B., Narayanan, C.S., Rajaraman, K., Sumathikutty, M.A., Omanakutty, M. and Mathew, A.G. (1982) Oils and oleoresins from major spices. *J. Plantation Crops,* 10(1), 1–20.

Sankarikutty, B., Sreekumar, M.M., Narayanan, C.S. and Mathew, A.G. (1988) Studies on microencapsulation of cardamom oil by spray drying technique. *J. Fd. Sci. Tech.,* 352–356.

Schultz, W.G. and Randall, J.M. (1970) Liquid carbondioxide for selective aroma extraction. *Food Technol.,* 24, 1282.

Shankaracharya, N.B. and Natarajan, C.P. (1971) Cardamom – Chemistry, Technology and uses. *Indian Food Packer*, 25(5), 28–36.

Shultz, T.H., Flath, R.A., Black, D.R., Guadagni, D.G., Schultz, W.G. and Teranishi, R. (1967) Volatiles from delicious Apple Essence – Extraction methods. *J. Food Sci.*, 32, 279–283.

Suresh, M.P. (1987) Value added products – better scope with cardamom, *Cardamom*, 20(1), 8–11.

Survey of India's export potential of spices (1968) The Marketing Research Corporation of India Ltd, New Delhi. Vol. 2A, B59.

Udayasankar, K. and Manohar, B. (1988) Extraction of essential oils using supercritical carbondioxide. In *Proc. International Sym. Supercritical Fluids*, NICE, France, p. 807.

Udayasankar (1989) Supercritical carbondioxide extraction of spices. The technical advantages. *Proc. Recent trends and Developments in Post-harvest Technologies for Spices*, CFTRI, Mysore, India, pp. 56–78.

Verghese, J. (1985) On the husk and seed oils of *Elettaria cardamum* Maton. *Cardamom*, 18(10), 9.

Wijesekera, R.O.B. and Nethsingha, C. (1975) *Compendium on Spices of Sri Lanka. 1-Cardamom*, Ceylon Institute of Scientific and Industrial Research and National Science Council of Sri Lanka, Kularatne & Co. Ltd, Colombo, p. 10.

Wistreich, H.E. and Schafer, W.F. (1962) Freeze-grinding ups product quality. *Food Eng.*, 34(5), 62.

11 Cardamom economy

M.S. Madan

1 INTRODUCTION

Cardamom is an important spice commodity of international commerce ever since the ancient Greek and Roman period. Until 1979–80 India was the largest producer and dominated the International trade in cardamom earning valuable foreign exchange for the nation. More than 90 per cent cardamom of international commerce, both small as well as large, originated in India. However, during the past two decades, Indian cardamom is facing serious threat in the world market from Guatemala, which has slowly and steadily encroached into the traditional Indian export markets. Currently this Central American country with an average annual production of more than 13,000 mt has emerged as the top producer and exporter of cardamom in the world and India has been relegated to the second position.

The cost of production of cardamom in India is relatively high compared to that in Guatemala, mainly due to poor yield and low productivity. India's highest productivity level in years of good crop is three times lesser than the yield per ha in Guatemala. Senility and poor unselected varieties, prolonged drought and over-dependence on monsoon, predominance of small holdings, problems of land tenure (lease), inadequate management practices, poor disease management, faulty post harvest practices etc. are some of the reasons responsible for the low yield (Anonymous, 1996). The situation warrants a critical evaluation of the above limiting factors. To make cardamom production competitive, the first step is an in-depth analysis of the economics of production, marketing and other aspects affecting cardamom economy of the country. In this chapter, an attempt is made to analyze the above aspects holistically.

1.1 Historic development

Table 11.1 shows the growth of cardamom production during the past 25 years. India and Guatemala are the major players in world economy of cardamom. Other small producers are Tanzania, Sri Lanka, Papua New Guinea, Honduras, Costa Rica, El Salvador, Thailand and Vietnam. India accounted for nearly 65 per cent of the world production in the early 1970s, but only 28 per cent in 1997–98. Guatemala on the other hand, stepped up its production from the middle of 1960s contributing 21.5 per cent of world production in the early 70s, but now her share is more than 65 per cent. Unlike India, Guatemala has negligible local demand, and the entire production goes for export. Making use of the 'law of comparative advantage' in cardamom production,

Table 11.1 Production in major cardamom producing countries

Period	Per cent share in total by			World production (mt)
	India	Guatemala	Others*	
1970–71 to 1974–75	65.4	21.5	13.1	4678
1975–76 to 1979–80	53.7	34.5	11.8	6628
1980–81	42.9	48.8	8.3	10,250
1984–85	31.9	60.3	7.8	12,220
1985–86 to 1989–90	26.5	67.5	6.0	14,392
1990–91 to 1994–95	28.4	65.6	6.0	19,470
1995–96 to 1997–98	29.8	64.2	6.0	24,953

Sources: Cardamom Statistics, 1984–85, Govt. of India, Cardamom Board, Cochin.
Spices Statistics, 1991, Govt. of India, Spices Board, Cochin.
Spices Statistics, 1997, Govt. of India, Spices Board, Cochin.
All India Final Estimate of Cardamom – 1997–98, Govt. of India, Ministry of Agriculture

Note
* Estimated figures (actual figures are not available).

Guatemala could increase its share in the world market from mere 30 per cent to over 90 per cent in the recent past and has captured all traditional markets for cardamom.

Cardamom is grown under the regulated shades of tropical evergreen forest of the Western Ghats. Now in the traditional cardamom growing regions of Kerala, Karnataka and Tamil Nadu, the cropping pattern in plantations (coffee–pepper) also included cardamom as a mixed/inter crop. With the development of large-scale commercial cultivation of cardamom in the last several decades, the area under as well as production and export have shown significant increase. The estimated area under the crop, was about 40 thousand hectares in the 1930s, increased to slightly over one lakh hectares by the beginning of 1980s and then started declining before crossing the one lakh hectare mark again in early 1990s. Guatemala's entry in a big way and imposition of new regulations on quality, microbial and other contaminants and pesticide residues in the produce by the importing countries went against the interest of Indian cardamom growers. The weak infrastructure supports, insufficient credit and marketing facilities in the cardamom belt also affected the prospects for cardamom adversely. A vicious circle of "low price-less production-high price-more production-low price" came into operation. There was a fall in production and consequent higher price in the domestic market in 1982–83. It was at this stage of recession in production and export that the industry was badly hit by severe drought in 1983–84, destroying substantial portion of yielding area, which was gradually made good by replanting/gap filling in subsequent years. An analysis of the performance of cardamom industry in terms of area under the crop, production, productivity, export, export earnings, prices etc. in the recent years would be relevant for planning the future programmes for its revival and development.

2   TRENDS IN AREA, PRODUCTION AND PRODUCTIVITY

2.1   The data base

Official statistics on area, production and productivity of cardamom in India are conflicting and are of doubtful reliability. There is wide disparity between official estimates of the Ministry of Agriculture and the Ministry of Commerce of Government of India. Trade estimates of production by the Indian Pepper and Spice Trade Association, Kochi gives a third figure. The official estimates have always been considerably lower than the trade estimates of production. Bias in the official estimates arise mainly out of:

(a) Inadequate sampling and estimation procedures which do not take into account the perennial nature of the crop and regional variations in cultivation.
(b) Exclusion of encroached forestland and unregistered smallholdings from the purview of estimation.

Despite such limitations, an attempt is made here to analyze the available information on area, production and productivity of cardamom with a view to get some broad indications of the possible changes that have been taking place in the crop economy during the last 25 years and future prospects for the immediate 5 years.

2.2   The emerging trends

The time series data on area, production and productivity of cardamom along with growth index worked out for the period from 1970–71 to 1999–2000 are presented in Table 11.2. A perusal of the period-wise performance indicates that significant decline in production was recorded in 1972–73, 1976–77, 1982–83, 1983–84, 1987–88 and 1996–97. As in the case of other agricultural commodities, climate exerts great influence on cardamom production, and productivity. In India, cardamom is generally grown under rain-fed conditions, and is affected by ecological changes of the forest habitats and moisture stress. Hence, along with the year-to-year fluctuations in rainfall, both its quantum and distribution – the output and productivity of cardamom have shown considerable fluctuations. The severe drought that prevailed in certain years not only affected the yield during those years but also in the subsequent years. Recently, introduction of irrigation during summer for the cardamom crop in large coffee/cardamom plantations has stabilized the yield to some extent in certain areas.

2.2.1   Area

Changes in area under the crop over the period from 1970 to 1971 onwards can be grouped as follows:

(1) 1970–71 to 1977–78 – Period of no change in the area under cardamom;
(2) 1978–79 to 1988–89 – Period of increasing trend;
(3) 1989–90 to 1997–98 – Period of decline.

While the area under the crop remained unchanged during the first period for about 8 years, there were year-to-year fluctuations in quantity produced, reflecting the climatical effect on productivity. A sudden dip in cardamom area during 1989–90 is not an actual change. It is merely a correction (change) in existing statistical figures on record. During this year the report of the survey for assessment of area under cardamom in India conducted by Spices Board was published (Spices Board, 1991).

Table 11.2. Area, production and productivity in cardamom

Year	Area (ha)	Growth index	Production (mt)	Growth index	Productivity (kg/ha)	Growth index
1970–71	91,480	100.00	3170	100.00	34.65	100.00
1971–72	91,480	–	3785	119.40	41.38	119.42
1972–73	91,480	–	2670	84.23	29.19	84.24
1973–74	91,480	–	2780	87.68	30.39	87.71
1974–75	91,480	–	2900	91.48	31.70	91.49
1975–76	91,480	–	3000	94.64	32.79	94.63
1976–77	91,480	–	2400	75.71	26.24	75.73
1977–78	91,480	–	3900	123.02	42.63	123.03
1978–79	92,760	101.40	4000	126.18	43.12	124.44
1979–80	93,950	102.70	4500	141.96	47.90	135.64
1980–81	93,950	102.70	4400	138.80	46.83	135.15
1981–82	93,950	102.70	4100	129.33	43.64	125.95
1982–83	93,950	102.70	2900	91.48	30.87	89.09
1983–84	93,950	102.70	1600	50.47	17.03	49.15
1984–85	1,00,000	109.31	3900	123.03	39.00	112.55
1985–86	1,00,000	109.31	4700	148.26	47.00	135.64
1986–87	1,05,000	114.78	3800	119.87	38.00	109.67
1987–88	1,05,000	114.78	3200	100.95	30.48	87.96
1988–89	81,113	88.67	4250	134.07	40.48	116.83
1989–90	81,113	88.67	3100	97.79	38.22	110.30
1990–91	81,554	89.15	4750	149.84	58.24	168.08
1991–92	81,845	89.47	5000	157.73	61.09	176.31
1992–93	82,392	90.06	4250	134.07	51.58	148.86
1993–94	82,960	90.69	6600	208.20	79.56	229.61
1994–95	83,651	91.44	7000	220.82	83.68	241.50
1995–96	83,800	91.60	7900	249.21	94.27	272.06
1996–97	72,520	79.27	7290	208.99	100.52	290.10
1997–98	69,820	76.32	7150	225.55	102.40	295.53
1998–99	72,135	78.85	7170	226.18	135.00	389.61
1999–2000	72,451	79.20	9290	293.06	173.00	499.28

Source: Data from various issues of 'Spices statistics', Spices Board, Cochin and Agricultural production statistics, Ministry of Agriculture, Government of India, Delhi.

Accordingly, the actual enumerated area of 81,113 hectares replaced hitherto reported area of 1,05,000 ha which was based on the area registered by the state government. In the third 'period of decline', though there was a marginal improvement during 1992–95, the area under the crop continues to be below the lower confidence level (LCL) of 84,818.39 ha (Fig. 11.1). The estimated growth index towards end of this period is negative (−23.68 per cent) in 1997–98 over the base year 1970–71 (Table 11.2).

2.2.2 *Production*

A significant feature of cardamom production in India is the cyclical fluctuations in yield i.e. after a continuous increase of production and productivity for 2–3 years a trend of decline sets in and continues before climbing up again. The climatic variations occurring in a cyclical nature give rise to this phenomenon in the case of this delicate shade-loving plant. There were cyclical fluctuations during 1973–76, 1978–81 and

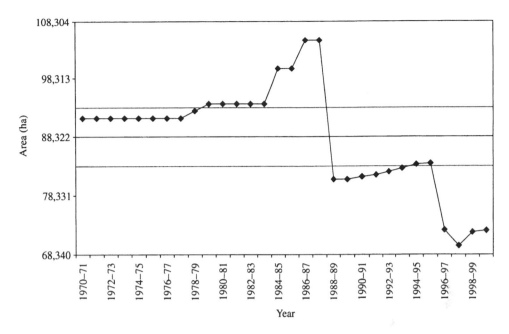

Figure 11.1 Area under cardamom in India (1970–2000).

1985–87. It can be observed in Fig. 11.2 that the peaks were achieved after gradual increase for 2–3 years, then there was a sudden dip. India's production had been showing a consistently increasing trend from 4250 tons in 1992–93 to 7900 tons in 1995–96, but declined to 7150 tons in 1997–98. But, the rate of decline was not as fast as in 70s and 80s. This may be due to improvement in productivity, using improved varieties and better production technology. There is an increasing trend in production after 1997–98 crop year.

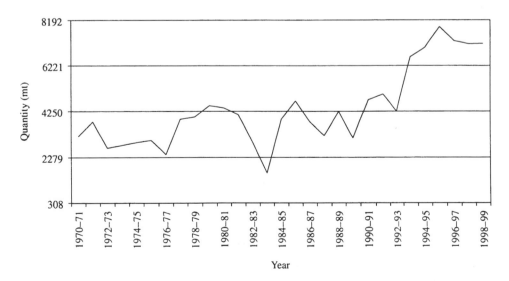

Figure 11.2 Production trend for cardamom in India (1970–2000).

2.2.3   *Productivity*

The yield level that was around 34.65 kg/ha during 1970–71 has not shown much improvement till the end of 1980, except for occasional fluctuations towards higher side (up to 48 kg/ha during 1979–80). It appears that, the yield increase during this period does not seem to have contributed to the increase in production; the entire increase in production being accounted for by area increase. However, the productivity level has improved from 1990 onwards and reached 102.4 kg/ha during 1997. Productivity registered during 1997–98 is almost 3 times the productivity of 1970–71. The estimated growth rate in production is more than 225 per cent over the base year. Record productivity of 173 kg/ha was achieved during 1999–2000 owing to better varieties and improved management practices followed (Spices Board, 2000).

2.2.4   *Growth estimates*

In order to get summary measures of long-term trends in area, production and productivity of cardamom in India, semi logarithmic growth equations are estimated and the results are presented in Table 11.3. It can be seen from the table, that the overall trend in area under cardamom registered an average annual negative growth rate of -0.6 per cent, while production has grown at the rate of 3.1 per cent per annum. In order to examine the cyclical fluctuation in cardamom production and area, decadal growth analysis was done. The decadal growth rate shows that there was a positive growth rate of 0.2 and 0.1 per cent respectively in 70s and 80s, it was -1.9 in the 90s as regards area. Production had a positive growth rate in all the three decades, but it was the highest (8.3 per cent) in the 90s. The figures for productivity are a matter of major concern, as it has direct bearing on the cost efficiency and profitability of cardamom cultivation. The estimated negative growth rate in area and positive growth rate in production during the 90s indicates the improvement in productivity i.e. with less area under the crop more quantity is produced.

2.3   State-wise area, production and productivity

Cardamom belt of India is located in the Western Ghats regions of Kerala, Karnataka and Tamil Nadu. Table 11.4 presents the area, production and productivity of cardamom in different States of India. It is clear from the table that Kerala accounts for

Table 11.3 Estimated growth equations for area and production of cardamom in India (1970–71 to 1997–98)

Characteristic	Estimated growth equation	R^2	Growth rate (%)
Area (1970–97)	ln At = 11.4956 − 0.0063T	0.278	−0.63
Area 1	ln A_1t = 11.4167 + 0.0020T	0.466	0.20
Area 2	ln A_2t = 11.4581 + 0.0012T	0.002	0.12
Area 3	ln A_3t = 11.7617 − 0.0194T	0.451	−1.94
Production	ln Pt = 7.8362 + 0.0813	0.480	3.13
Production 1	ln P_1t = 7.9181 + 0.0306	0.202	3.05
Production 2	ln P_2t = 7.9202 + 0.0161	0.029	1.61
Production 3	ln P_3t = 6.6699 + 0.0833	0.626	8.33

Notes
1 = period 1 from 1970–71 to 1978–79.
2 = period 2 from 1979–80 to 1980–81.
3 = period 3 from 1981–82 to 1997–98.

Table 11.4 State-wise area, production and productivity of cardamom in India

Year	Variables	Kerala		Karnataka		Tamil Nadu		India
		Actual	(%)	Actual	(%)	Actual	(%)	
1970–71	Area (ha)	55,190	60.33	28,220	30.81	8070	8.81	91,480
	Production (t)	2130	67.19	805	25.39	235	7.41	3170
	Productivity (kg/ha)	38.59		28.53		29.12		34.65
1980–81	Area (ha)	56,380	60.01	28,220	30.03	9350	9.95	93,950
	Production (t)	3100	70.45	1000	22.73	300	6.82	4400
	Productivity (kg/ha)	54.98		28.22		32.09		46.83
1990–91	Area (ha)	43,826	53.74	31,605	38.75	6123	7.51	81,554
	Production (t)	3450	72.63	800	16.84	500	10.53	4750
	Productivity (kg/ha)	78.72		25.31		81.66		58.24
1997–98	Area (ha)	43,050	61.66	21,410	30.66	5360	7.68	69,820
	Production (t)	5430	75.94	1240	17.34	480	6.71	715
	Productivity (kg/ha)	126.13		57.92		89.55		102.41
1999–2000*	Area (ha)	41,522	57.31	25,882	35.72	5047	6.97	72,451
	Production (t)	6550	70.51	1950	20.99	790	8.50	9290
	Productivity (kg/ha)	213		103		205		173

Source: Spices statistics (various issues) Government of India, Spices Board, Cochin.

Notes

Yield is estimated by dividing total production with area.

* Midterm estimate.

the major share of area and production of cardamom in India, and that this remained more or less unchanged over the last three decades. Karnataka State stands second, followed by Tamil Nadu. Over the years productivity per unit area has gone up both in Kerala and in Karnataka but has gone down in Tamil Nadu.

2.4 Production constraints

The major reasons attributed for low productivity of cardamom in India are:

1 Recurring climatic vagaries, especially drought in the absence of irrigation practice.
2 Absence of regular replantation activities – under the mixed cropping system farmer is happy with the additional income from the aged cardamom plants, where as re-planting with improved high-yielding varieties would have given better yield.
3 Deforestation and resultant changes in the ecological conditions prevailing in the growing area – leading to conversion of cardamom land to other competing crops like pepper.
4 Lack of eagerness among planters to adopt high production technologies – though better varieties and practically proven package of practice is available to enhance yield level up to 600 kg/ha.
5 Problems of pests and diseases.

6   Remote location of plantations.
7   System of land tenure – does not allow long-term planning for improvement by the actual producer who works on the land (Anonymous, 1996; George, 1976; Cherian, 1977).

2.5   Comparative performance of area and productivity of cardamom with other plantation crops

Coffee, pepper and arecanut are the other plantation crops raised in cardamom growing areas of India. Depending on the prevailing market price, one crop is neglected in favour of other. So, it would be interesting to examine the growth of area, production and productivity of cardamom with that of other crops. Table 11.5 presents the quinquennial average of growth indices of area; production and productivity of coffee, and pepper along with cardamom for the period from 1960–61 to 1994–95 with 1960–61 as the base year (Radhakrishnan, 1993). Some of the growers in both Karnataka and Kerala may have taken up coffee either by replacing existing cardamom plants or replacing a few rows of cardamom with coffee adopting a multiple-cropping pattern to assure economic stability (Anonymous, 1996). The table indicates that coffee has shown steady growth in area, production and productivity. But, the spice crops, cardamom and pepper, did not achieve the comparable growth rate. However, cardamom has shown a phenomenal improvement in area and production in all the quinquennial period, while there was a marked improvement in the quinquennial period from 1985–86 to 1989–90.

3   COST OF CULTIVATION

Productivity and cost of cultivation plays crucial roles in deciding the competitiveness of product in the global market. Compared to India, the cost of production of cardamom in Guatemala is much lower. This arises due to two factors: higher productivity and lower wages. Productivity per unit area in India was mere 47 kg/ha during the 1980s, when it was 91 kg/ha in Guatemala. At present, Guatemala obtains more than 200 kg/ha (dry), whereas, in India, it is only around 120 kg/ha. This in itself gives to Guatemala a cost advantage of above 225 per cent. Guatemala has an advantage over India from cost of

Table 11.5 Quinquennial averages of the indices of area, production and productivity of major commercial crops (1960–61 = 100)

Period	Rubber			Coffee			Cardamom			Pepper		
	A	P	Y	A	P	Y	A	P	Y	A	P	Y
1960–61 to 1964–65	111	131	117	106	101	95	100	113	113	100	93	93
1965–66 to 1969–70	139	258	185	131	157	126	126	113	91	99	85	86
1970–71 to 1974–75	165	448	270	128	209	163	140	120	84	117	94	80
1975–76 to 1979–80	187	569	304	166	266	159	165	160	97	109	91	84
1980–81 to 1984–85	236	647	274	194	293	151	185	273	146	106	97	91
1985–86 to 1989–90	276	930	336	213	388	182	211	320	150	144	152	105
1990–91 to 1994–95							148	173	167	181	182	101

Source: Radhakrishnan (1993).

Notes
A = Area; P = Production; Y = Productivity.

cultivation point of view also, production cost was a mere Rs. 30–40/kg during 1980s when it was Rs. 70–90/kg in India (Bossen, 1982); which adds another 225 per cent cost advantage to that country over India. Consequently, Guatemala has been able to compete successfully with India in the world market on price front. The price of cardamom from Guatemala in recent years has been about US$ 5–7/kg lower than that from India (Table 11.6). Besides, the quality of cardamom produced in Guatemala is comparable in characteristics with the exports from India. Such advantages helped Guatemala exporters to penetrate into the prime markets for Indian cardamom. Recent advances in cardamom production technology helped India to increase productivity per unit area. The highest yield achieved is 2475 kg/ha in Idukki district of Kerala by a particular farmer (1999–2000) and many others have achieved more than 1400 kg/ha (Korikanthimath, 1992; John, 2000). However, the labour component in the production cost, which accounts up to 60 per cent during the establishment stage and more than 40 per cent thereafter, in the total cost makes Indian cardamom much costlier in the international market. Studies have shown that expenditure on labour has positive correlation with yield/ha (Mahabala *et al.*, 1991). The present estimated cost in India ranges from Rs. 150

Table 11.6 Auction, wholesale and export prices of cardamom in India

Year	Auction price		Wholesale price		Export price		International spot price US$/kg		
	Rs/kg	Growth index	Rs/kg	Growth index	Rs/kg	Growth index	India	Guatmala	Differ-ence
1971–72	27.51	100.00	41.48	100.00	37.40	100.00	–	–	–
1972–73	45.46	165.25	51.89	125.10	49.47	132.27	–	–	–
1973–74	59.05	214.65	64.36	155.16	63.72	170.37	–	–	–
1974–75	74.01	269.03	87.31	210.49	81.94	219.09	–	–	–
1975–76	86.45	314.25	90.83	218.97	99.86	267.01	–	–	–
1976–77	156.75	569.79	109.81	264.73	157.13	420.13	–	–	–
1977–78	134.41	488.59	98.77	238.11	175.30	468.72	14.01	9.32	4.69
1978–79	166.42	604.94	199.98	482.11	202.90	542.51	12.65	10.39	2.26
1979–80	134.88	490.29	110.46	266.30	184.21	492.54	12.50	10.36	2.41
1980–81	98.41	357.72	60.45	145.73	148.20	396.26	12.41	9.77	2.64
1981–82	116.02	421.74	69.98	168.71	129.88	347.27	12.00	9.65	2.35
1982–83	161.08	585.53	119.04	286.98	158.61	424.09	12.03	8.65	3.38
1983–84	370.49	1346.75	328.12	791.03	210.94	564.01	16.91	9.92	6.99
1984–85	199.91	726.68	233.96	564.03	271.95	727.14	24.46	13.17	11.29
1985–86	132.80	482.73	137.75	332.09	163.39	436.87	15.50	10.51	4.99
1986–87	118.32	430.10	122.30	294.84	127.82	341.76	11.17	5.94	5.23
1987–88	140.62	511.16	140.64	339.05	125.94	336.74	10.10	5.55	4.55
1988–89	133.91	486.77	145.15	349.93	131.81	352.43	10.01	4.25	5.76
1989–90	266.73	969.57	245.76	592.48	174.51	466.60	11.03	3.82	7.21
1990–91	251.67	914.83	326.09	786.14	286.70	766.58	17.90	4.67	13.23
1991–92	287.92	1046.60	259.97	626.74	286.29	765.48	17.41	3.94	13.47
1992–93	465.37	1691.64	395.90	954.44	395.04	1056.26	16.02	4.25	11.77
1993–94	340.06	1236.13	490.18	1181.73	375.93	1005.16	13.79	4.59	9.20
1994–95	255.38	928.32	256.00	617.16	296.74	793.42	17.39	7.66	9.73
1995–96	201.58	732.75	202.00	486.98	349.33	934.04	9.58	6.33	3.25
1996–97	366.00	1330.43	366.00	882.35	384.32	1027.59	10.00	6.12	3.88
1997–98	257.32	935.37	342.00	824.49	358.15	957.62	13.78	5.48	8.30
1998–99	541.19	1967.25	–	–	541.19	1447.03	11.67	8.50	3.17

Sources: Spices statistics (various issues) Spices Board, Government of India, Cochin; Spices Statistics, Directorate of Spices and Arecanut Development, Ministry of Agriculture, Government of India.

to 200/kg depending upon the cropping system followed. Thomas *et al.* (1990) have concluded that the low productivity and high cost of production vis-à-vis stiff competition in the international market rendered Indian cardamom less competitive and subsequently unremunerative for the planters. Due to non-availability of skilled labour for harvesting and post-harvest handling (including on-farm processing), the employment of unskilled labourers resulted in less recovery of 17–19 per cent only against the desirable 25 per cent recovery. Thus, an avoidable post-harvest loss of around 6.8 per cent is also responsible for reduction in productivity.

4   DOMESTIC MARKET STRUCTURE AND PRICES

Cardamom trade in India is what may be called a regulated trade. Cardamom (Licensing and marketing) Rules 1987 was introduced to streamline the system of marketing in general and bringing about control in the form of restricting the entry of persons into the different functional categories, namely exporters, dealers and auctioneers. The declared purpose of such regulation is to ensure a fair price for the product and the timely payment of the sale proceeds. Export marketing of cardamom is regulated by the Spices Board (Registration of Exporters) Regulation 1989. The Spices Board issues the following certificates/licenses:

1   Cardamom dealer licenses.
2   Cardamom auctioneer licenses.
3   Certificate of Registration as exporter of spices.
4   Registration-cum-membership certificates to exporters (RCMC).

Market Intelligence Officers have been posted in important marketing/auction centers to collect reports on crop purchase, sales, movement and price trends.

An important aspect of the market structure is the existence of an efficient auction system, which ensures fair prices to the larger planters who take their produce to the auction centre. There are at present 11 auction centers in India. Though the auction system is efficient, the quantity flowing through the auction centers is only about 70 per cent of the production. The licensed auctioneers conduct weekly auction during the harvesting seasons in the production tracts/assembly centers on particular days as approved by the Board and as per the conditions/directions issued from time to time. A sizeable quantity (about 30 per cent) flows from the producer to dealers outside the auction system. The reasons being: (1) non-availability of auction centers in growing regions and (2) because of the fixed quantity of (500 gm/lot) the auction sales is not profitable for lots less than 16 kg. A study by Joseph (1985) has revealed that the average price obtained tends to increase with the increasing size of the lot. The price obtained by the largest lot is 12–87 per cent higher than that obtained by the smallest lot depending upon the season and year of sale.

4.1   Price analysis

Analysis of the structure and behaviour of farm prices is of considerable interest in the context of finding ways and means for increasing production and productivity. Prices often act as a guide to indicate the change in production decisions.

Cardamom is a moderately storable export commodity. Long-term storage is not possible as in the case of pepper. This necessitates market clearance within the crop year, thereby ruling out speculations. Within these limits the formation of prices in the domestic market takes place in the following manner. Depending upon the length of the summer, severity of drought, pre-monsoon showers and the quantum of rainfall during the June–July period, the well-experienced traders forecast the crop prospects for the forthcoming season. This is aided by the fact that many of the dealers and exporters are also plantation owners. If the expected production is much lower than the normal production, a significantly higher price than that ruling in the previous year is set at the beginning of the season. If on the other hand the expected production is much higher than the normal production, a much lower price is set (Nair *et al.*, 1989). Peak prices (prices ruling during the peak sales season) do not deviate very much from the opening prices in most of the years, but do deviate in abnormal years when production is low and forecast go wrong. Table 11.6 brings out this fact. It may be seen that the wide variation in prices (especially in an upward direction) is associated with sharp decrease in production.

Analysis of relationship between price, lot size and quality reveals that, during the peak season, quality (68 per cent) explains the price variation across different lots and the changes in lot size explain the price variation during the slack season (Joseph, 1985). Accordingly, the formation of auction prices occurs.

Export value of cardamom usually depends on its major quality aspect of colour. Traders are keen to acquire as much as possible of the output in the peak harvest season as high quality harvest (with good colour) comes in the middle of the season. This is what makes for the peak prices in the peak-harvesting period, which in turn, becomes the peak sales period. This period broadly falls during September–December. The keenness of the traders is further borne out by the rather high percentage of peak purchases in years when the production was lower as for instance, 1976–77, 1981–82, 1982–83 and 1983–84. A study by Joseph (1985) indicates that the export price leads the domestic price with a lag of about 1-month. But according to Nair *et al.* (1989) though there is trend synchronization between export prices and auction prices, a month-to-month correspondence does not hold. However, there exists an asymmetry, that a rise in the export price is not always paralleled by a corresponding increase in the domestic price, whereas a fall in the export price is transferred entirely to the domestic price. At times the domestic prices of both wholesale and farm used to be more than the export price indicating the strong domestic market. Therefore, an attempt is made to analyze the trends in auction, wholesale and fob prices of cardamom during the period from 1971–72 to 1997–98. As can be seen from the Fig. 11.3 all the three price variants of cardamom shows an overall upward trend with cyclical variations and short-term fluctuations.

Further, the estimated growth equation indicated that while the auction prices have registered an average annual growth rate of 6.8 per cent, both the wholesale and export prices increased at the rates of 6.4 per cent and during the entire period from 1971 to 1997–98. The closeness of the estimated values of growth rates of the three price variants is indicative of the high degree of market integration at different levels of trade.

4.2 Price transmission

Cardamom being an export-oriented commodity, the hypothesis is that farm and wholesale prices are determined by the export price. To test this hypothesis linear regression models of the following forms are estimated for the data relating to

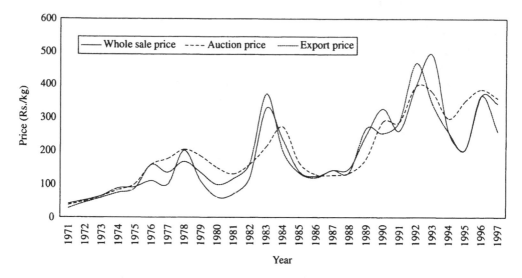

Figure 11.3 Trend in prices of cardamom (1971–97).

cardamom prices from 1971–72 to 1997–98. This analysis is based on the premise that wholesale price is determined by the export price. The validity of this premise can be tested by regression analysis of the data on cardamom prices for the last 25 years. For this the following regression models are tested:

$$P_a = \alpha_1 + \beta_1 p_e + u_1 \tag{1}$$

$$P_w = \alpha_2 + \beta_2 p_e + u_2 \tag{2}$$

$$P_a = \alpha_3 + \beta_3 p_w + u_3 \tag{3}$$

Where,

P_a = Auction/farm price
P_w = Wholesale price
P_e = Export price
u_1, u_2 and u_3 are error terms with usual econometric assumptions.

The summary results of the regression analysis are presented in Table 11.7.

To test the efficiency of price transmission, the null hypotheses formulated are:

$$H_o : \beta_1 = 1$$
$$\beta_2 = 1$$
$$\beta_3 = 1$$

If the null hypotheses are not rejected (i.e. the slope coefficients are not significantly different from unity), the commodity market is relatively efficient in transmitting price change from one market to the other.

The summary results presented in Table 11.8 provide enough evidence to accept all the three null hypotheses. Therefore, cardamom marketing appears to be efficient in

Table 11.7 Estimated growth equations of auction, wholesale and export (fob) prices of cardamom from 1970–71 to 1997–98

Characteristic	Estimated growth equation	R^2	Estimated annual growth rate (%)
Auction	In P_{at} = 4.0152 + 0.6983T	0.661	6.83
Wholesale price	In P_{wt} = 401818 + 0.0639T	0.676	6.39
Export price	In P_{et} = 4.1648 + 0.0641T	0.900	6.41

Table 11.8 Estimated price transmission models

Model no.	Estimated equation	R^2
1.	Pa = 10.0360 + 0.8766pe	0.742
2.	Pw = −12.645 + 1.0019pe	0.784
3.	Pa = 28.8991 + 0.8336pw	0.8710

transmitting price information from one market level to other. The first model shows that a rupee change in export price is estimated to have given rise to 0.88 rupee change in farm prices. The second model shows that a rupee change in export price is estimated to have given rise to one rupee change in wholesale price and third model shows that a rupee change in wholesale price has given rise to 0.83 rupee change in farm price.

4.3 Cyclical movement in price

As regards the trend, the prices have been increasing from 1950s, the increase being about ten-fold in 30 years by middle of 1980s. The compound growth rate of prices between 1954 and 1986 is 9.36 per cent per annum, while it is 7.2 per cent between 1970 and 1997.

A perusal of the period-wise movement of cardamom prices brings out an important aspect of cyclical fluctuation of prices, which seems to occur in the following manner: Prices remain stable or tend to increase during a certain period followed by a sharp fall. They remain low for the next few years, then start moving up and continue to increase or remain stable for another period; the process of decline and subsequent increase repeats itself. Period of this cycle is worked out to be around 11 years (Nair *et al.*, 1989). These cyclical fluctuations have a significant bearing on the conditions of supply side (area and production).

When the log transformed data on production and farm price for the period from 1971 to 1997 is graphed together as in Fig. 11.4; it can be observed that both curves were moving inproportionately opposite to each other. The pattern observed in graph can be interpreted as either the price response to supply, or the supply response to price. When the supply swindled in 1983 and 1992, peak of the price cycle was achieved. But in certain years the price response is much sharper than the supply response to price. As regards supply response to price, it is not immediate, but reflected in jumps over a period causing alternate occurrence of over and short supply. The upswings and downswings in prices are clearly related to these gaps and excess in supply. The length of the upswing is about 6 years. Here, what needs to be remembered is that once new planting or replanting is done, the plant may start yielding by the third year. The subsequent

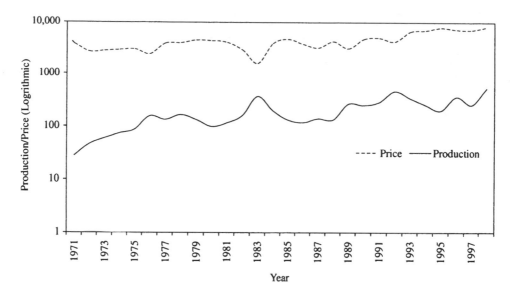

Figure 11.4 Trend in cardamom production and farm price (1997–98).

few years mark the peak yielding stage. Then the jumps in production from one cluster to the next may be interpreted as increases in area taking place about 5 or 6 years prior to that period. With such an interpretation, it may be seen that the jumps in production correspond to the early phase of the upswing in prices. The significance of the above correspondence is that, once the farmers ascertain the upswing, they respond to it by bringing new area under the crop by replanting or new planting. It takes another five to six years for these to get reflected as increases in output. These responses must have been massive; otherwise the output would not have jumped at such high rates.

5   EXPORT PERFORMANCE

Historically cardamom began to be cultivated in India primarily as an export crop. Till the end of 1960s, cardamom was in the sellers market in the world trade and India was the leader in production and export of this commodity. Guatemala stepped up its production from mid 1960s and began capturing India's traditional markets in the Middle East. Consequently export from India declined; in 1977–78, India could export only 2763 mt, whereas Guatemala exported 3610 mt reversing the position earlier existed. With further increase in production in Guatemala, India's exports have been seriously affected. Table 11.9 shows the export of cardamom as per cent total production for both India and Guatemala. The export trend depicted in the form of graph in Fig. 11.5 reveals the fact that exports was in proportion to the production till 1985–86. Thereafter, the export figure remained low with consecutive fall in export during 1986–87 and 1987–88. Though the level of production went up, quantity exported remained low. The gap between production and export started widening and the trend still goes on. At present cardamom production in the country is mainly sustained on local markets. During this period, there was a heavy flow of cardamom from Guatemala into the world market; their overall export quantity has more than doubled. So, India

Table 11.9 Cardamom exports from India (1970–98)

Year	Quantity exported (mt)	Growth index	Export value ('000 Rupees)	Growth index	Export as per cent total production
1970–71	1705	100.00	1,12,000	100.00	53.79
1971–72	2141	125.92	80,000	71.43	56.57
1972–73	1384	81.17	68,000	60.71	51.84
1973–74	1813	106.33	115,000	102.60	65.22
1974–75	1626	95.36	133,000	118.76	56.07
1975–76	1941	113.84	194,000	173.21	64.70
1976–77	893	52.38	140,314	125.28	37.21
1977–78	2763	162.05	484,363	432.47	70.85
1978–79	2876	168.68	583,536	521.01	71.90
1979–80	2636	154.60	485,581	433.55	58.58
1980–81	2345	137.54	347,539	310.30	53.30
1981–82	2325	136.36	301,969	269.62	56.71
1982–83	1032	60.53	163,690	146.15	35.59
1983–84	258	15.13	54,423	48.59	16.13
1984–85	2383	139.77	648,653	579.15	61.10
1985–86	3272	191.91	534,599	477.32	69.62
1986–87	1447	84.87	184,953	165.14	38.08
1987–88	270	15.83	34,003	30.36	8.44
1988–89	787	46 16	103,736	92.62	18.52
1989–90	180	10.56	30,668	27.38	5.81
1990–91	379	22.23	102,224	91.27	7.98
1991–92	544	31.91	155,741	139.05	10.88
1992–93	190	11.14	75,057	67.02	4.47
1993–94	387	22.70	145,483	129.89	5.86
1994–95	257	15.07	76,261	68.09	3.67
1995–96	375	21.99	131,000	116.96	4.75
1996–97	226	13.26	86,967	77.65	3.10
1997–98	297	17.42	106,371	94.97	4.15
1998–99	475	27.86	252,121	225.10	6.62
1999–2000	550	32.26	276,035	246.46	5.92

Source: Various issues of 'Spices Statistics', Spices Board, Government of India.

Note
* Midterm estimate by Spices Board.

could not compete with its competitor. Export promotion schemes such as airfreight subsidy to Middle East countries, exemption of cess and export assistance of Rs. 35/kg etc., could not help much to enhance export quantity.

As it can be seen from the Fig. 11.5 domestic market has absorbed the entire production. This domestic demand was estimated to be around 3500 tons during 1988–89 (i.e. 82.4 per cent of the total production) and during 1997–98 the domestic market has consumed more than 95 per cent of the production. World export demand has been rising at a steady rate and is expected to be around 12,000 tons in the year 2000–2001. Domestic demand in the country is also increasing fast. In the absence of household consumption data to estimate the quantum of domestic consumption, deducting the quantity exported from total production gives the quantity consumed within the country. Thus derived domestic demand during 1997–98 is around 6850 mt, which is more than half of the world demand. The estimated demand was around 7200 tons in 1998 (Anonymous, 1998).

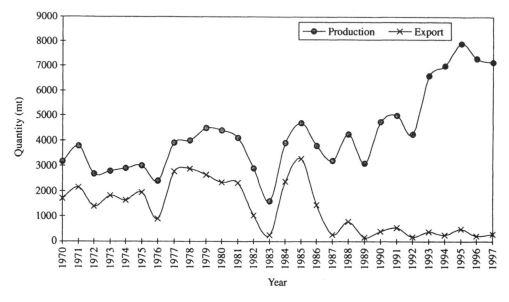

Figure 11.5 Trend in production and export of cardamom in India.

Spices Board's campaign for increasing domestic consumption during 1986–87 had a positive impact. Consequently, the enlarged domestic market could sustain the cardamom industry during the years in which international prices were at low levels. Occasionally, the domestic demand has exceeded the level of production and prices went up above the international export price. Though it is a good sign for the cardamom industry to depend on the domestic market, there is yet another problem involved in it. Since export accounts for only a small portion of the production, the prices are influenced more by the strong internal demand than by the demand abroad. In 1990s the higher domestic price paved way for smuggling of an estimated 2000 tons of cardamom into India every year (Anonymous, 1998). Thus, the country's ambitious plan to capture at least 25 per cent of the market share in the international trade of cardamom by the turn of this century gets a set back.

The growth co-efficient estimated for the period from 1970–71 to 1997–98 showed a negative figure of −8.96 per cent. The period-wise analysis revealed that the first period (1970–71 to 1978–79) registered positive growth rate of 4.5 per cent, but the subsequent two periods (1979–80 to 1988–89 and 1989–90 to 1997–98) registered negative growth rates. Though the production went up at the rate of 8.3 per cent in the third period, the export went down by −5.4 per cent. Comparatively less fall of −3.4 per cent in export value during the same period indicates the prevailing better prices in the world market for cardamom which has registered an over all growth rate of 6.4 per cent for the entire period under study.

5.1   Direction of Indian exports

Till the end of 1980s more than 80 per cent export of Indian cardamom was concentrated in the Middle East market, which accounts for over 50 per cent of the world import. Because of the high degree of quality differentiation of the product in the Middle East

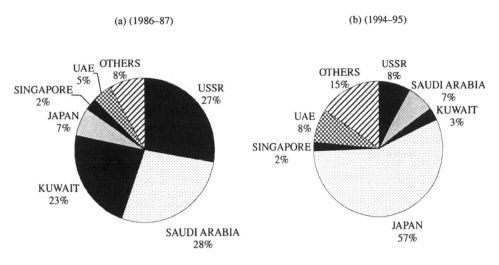

Figure 11.6 Direction of Indian export of cardamom during 1986–87 and 1994–95.

markets, the higher grade of cardamom produced in India was finding a ready market in the Arab world. The demand in this market segment was increasing in the 80s because of the enormous increase in the per capita income of these countries, consequent upon the oil boom. But, in recent years, (i.e. 90s) the share of Middle East market in Indian export in general has shown a decreasing trend. Further, the recession set in the region has slowed down the prosperity also. The result was a change over from quality consciousness to price sensitivity. Cardamom from Guatemala, which was, priced $4–5/kg less than the Indian cardamom, was readily accepted. The result was, India has lost its traditional market. India has lost its East European market also, after the collapse of erstwhile USSR. Japan is the only consistent market for Indian cardamom in recent years. In 1995–96 more than 60 per cent of the total export (226 mt) from India was to Japan. India tops the suppliers of cardamom to Japan with 66 per cent, followed by Guatemala and Vietnam. Japan buys the second grade of Alleppy Green cardamom from India. It is also worth noting that Japanese buyers maintained consistency in purchase of Indian cardamom even in years of high price. Saudi Arabia and Kuwait are the other markets. Export of cardamom during 1996–97 was to the tune of 240 tons valued Rs. 9.21 crores compared to 500 tons valued Rs. 12.40 crores of 1995–96. The two major countries to which cardamom was exported during the year have been Japan – 163 tons and Saudi Arabia 45 tons. Direction of Indian export of cardamom during the years 1991–92 and 1995–96 is presented in Figs 11.6a and 11.6b. Guatemala usually produces 12,000–14,000 tons. With almost negligible domestic requirement, their entire produce is sold in the international markets at a very competitive price. As it can be seen from Table 11.10 at times Guatemala has exported more than their production.

5.2 India's competitive position in the international market

Competitive power of a country in the export of a commodity depends crucially on its relative price over the competing countries. Table 11.9 presents the price movements of cardamom in respect of India and Guatemala in New York market during 1970 and 1997.

Table 11.10 Production and export of cardamom from Guatemala

Year	Cultivated Area (ha)	Production (mt)	Productivity (kg/ha)	Export (mt)	Export as per cent of production
1985	32,336	7348.32	90.89	6173.50	84.02
1986	38,333	8845.20	92.33	7978.82	90.21
1987	41,418	10,591.56	102.29	11,489.69	108.48
1988	42,656	10,432.80	97.83	11,303.71	108.35
1989	43,000	11,340.00	105.49	11,076.91	97.68
1990	43,000	11,340.00	105.49	11,113.20	98.00
1991	43,000	12,201.84	113.51	13,163.47	107.88
1992	43,000	12,474.00	116.04	13,240.58	106.15
1993	47,472	12,927.60	114.57	14,442.62	111.72
1994	45,133	14,969.80	126.13	13,213.37	88.27
1995	47,472	15,603.84	131.48	13,920.98	89.22
1996	47,472	16,329.60	137.59	14,500.00	130.16
1997	1,19,540	16,692.48	139.64	14,020.78	83.99
Total		11,576.70		12,491.79	

The price of Indian cardamom in general has been higher than that of Guatemala. Indian cardamom has never become cheaper, indicating the weak competitive position of India in the international market for cardamom. The competitive power, which depends on relative price, is also closely related to relative productivity and cost of production. Productivity estimates of Indian cardamom is very low compared to that of other producing countries. Therefore, attempts to increase production and productivity, which reduce the cost of production, must assume prime importance.

6   DEMAND AND SUPPLY

Cardamom production is influenced by national as well as international factors. As we have seen earlier, the prices of cardamom are subject to change from time to time depending on supply and demand. While the demand side is influenced by many other factors including the overall economic development, the supply side is influenced not only by economic factors but also by agro-climatic, biotic, and abiotic stress factors in the growing region. The response to price change gets reflected in the form of change in supply after 5–6 years only. Thus, there are multitudes of factors, which are to be considered for forecasting the future of cardamom. The kind of data available with us does not permit to go for sophisticated forecasting models, which may give a correct picture. What we have with us is only the historic data for area, production and prices. Suitable model, which can give reasonable prediction with these data, were identified and fitted and this must be seen as a step on the road towards a more sophisticated modelling analysis based on superior data later.

6.1   Model identification

A variety of statistical forecasting techniques are available ranging from very simple to very sophisticated. All of them try to capture the statistical distributions in the data provided and presents the future uncertainty quantitatively. Lack of quality data forced us to choose methodologies, which forecast the future by fitting quantitative models to statistical patterns from historic data for several years. Therefore, univariated methodologies based solely on the history of the variable (one at a time) were tried. There are three such models:

1   Simple moving average models.
2   Exponential smoothing models.
3   Box-Jenkins models.

To identify the right model the data's have to be explored first.

6.1.1   *Exploring the data*

The time series data on cardamom for area, production and prices were plotted/graphed to identify their specific characteristics in order to select an appropriate model. The characteristics observed in the time series data for cardamom can be listed as follows:

1   There is an overall positive trend (i.e. the trend cycle accounted nearly 80 per cent).
2   Non-seasonal in nature i.e. not consistently high or low and the repetitive annual pattern.
3   The time series is non-stationary in both mean and variance.

The classical decomposition of the time-series data also revealed the fact that the trend cycle accounted for about 78 per cent and above, while the irregularity accounted for the rest. Thus the forecasting model should account for trend, non-seasonality and also the non-stationary factor. This rules out the use of simple moving average technique as well as Box-Jenkins model. However, after transforming the data to stationarity, the Box-Jenkins model was also tested and used wherever it is possible. Finally, the models of exponential smoothing were selected, as these models are built upon clear-cut features like level, trend and seasonality.

6.1.2   *Model selection*

The exponential smoothing works as its name suggests, extracts the level, trend and seasonal index by constructing smoothed estimates of these features, weighing recent data more heavily. It adapts to changing structure, but minimizes the effects of outliers and noises. Three major exponential smoothing models tried were:

1   Simple exponential smoothing.
2   Holt exponential smoothing.
3   Winters exponential smoothing.

An appropriate model was selected based on minimum BIC (Bayer's information criterion). The model that minimizes the BIC is likely to provide the most accurate forecasts. Finally, the Holt exponential smoothing model was selected as the best and the forecasting was done for all the three variables i.e. area, production and price. However for forecasting the demand and quantity exported the Box-Jenkins model was used with log transformed data.

6.1.3   The model

Holt's (1957) exponential smoothing model uses a smoothed estimate of the trend as well as the level to produce forecasts. The forecasting equation is:

$$Y(m) = S_t + mT_t \tag{4}$$

The current smoothed level is added to the linearly extended current smoothed trend as the forecast into the indefinite future.

$$S_t = \alpha Y_t + (1 - \alpha / ie(1 - \alpha))(S_{t-1} + T_{t-1}) \tag{5}$$

$$T_t = \gamma(S_t - S_{t-1}) + (1 - \gamma)T_{t-1} \tag{6}$$

Where,

m = Forecast lead time
Y_t = Observed value at time t
S_t = Smoothed level at end of time t
T_t = Smoothed trend at end of time t
γ = Smoothing parameter for trend
α = Smoothing parameter for level of series.

Equation (5) shows how the updated value of the smoothed level is computed as the weighted average of new data (first term) and the best estimate of the new level based on old data (second term). In much the same way, equation (6) combines old and new estimates of the one period change of the level, thus defining the current linear (local) trend.

6.2   The forecast

Before embarking on the presentation of figures for likely future developments, it needs to be stressed again, that the present forecast is not an accurate exercise based on systematically collected elaborate data. However, the forecast based on historic data helps us to understand the overall direction in which the supply (area and production) will move and price fluctuates. Forecast is produced with upper and lower confidence limits. The upper confidence limit is calculated for 97.5 per cent and the lower for 2.5 per cent i.e. the actual should fall inside the confidence band 95 per cent of the time.

6.2.1   Demand

The major markets for cardamom are Saudi Arabia, Kuwait, Jordan, Qatar, UAE, USSR and Western Europe. Other important importers include West Germany, Pakistan, UK, Japan and Iran. The highest consumption of cardamom takes place in the Middle East where it is used in the preparation of their traditional drink 'gahwa'. This market accounts for 80 per cent of the total world consumption (UNCTAD, 1985). In Europe, Scandinavian countries use cardamom to flavour bread and pastries. Cardamom is imported in raw and ground forms for use in food manufacturing and special blends. Among the producing countries, India consumes the largest quantity in the world.

There was sharp increase in world demand in the late 70s and which has remained more or less stagnant during recent years because of declining purchasing power in the Middle East. Though the market has shifted to lower qualities supplied by

Guatemala due to higher cost of good quality Indian cardamom; the quantity did not change. With the main deciding factors of consumption (i.e. ageing and income) on the increase, the demand for cardamom in countries like Japan has gone up. Unlike in the past, new uses for cardamom in the food and industrial sector are the reason for increased demand in the world market; and it is growing almost at the rate of global population growth. The world demand was estimated around 9000 tons in 1985–86 excluding India's domestic consumption. Keeping this as the base, the demand potential is projected on the basis of an average growth rate of 2 per cent per annum, which is proportionate to the growth rate of population. Demand was accordingly estimated for 2000–2001 as 12,000 tons. According to Spices Board (1990) with an estimated annual per capita consumption of 8.5 g, the total requirement in India would be 8700 mt by 2000 AD and with an estimate of 9 g, the total requirement would be 10,800 mt by 2010 AD. As far as India is concerned, the demand for cardamom is increasing more than the population growth rate. Recent domestic consumption trends in India indicate a sharp increase. As against 1500 tons in 1985–86 domestic consumption have gone up to more than 6850 tons during 1997–98. If this level is maintained, the world demand (including India's) at the turn of the century may be more than 15,000 tons. When actual data is not available, deducting the quantity exported from total production gives an approximate consumption in the domestic market. The growth equation fitted for the consumption trend in the country is:

$$\text{In } Dt = 6.7110 + 0.0725T \quad R2 = 0.815$$

Accordingly, the estimated growth rate is 7.3 per cent per annum. Thus, the growth rate in demand is much more than the growth rates in production. Under the circumstances, it is unlikely that India can regain her position as the world's largest producer and exporter because an increasing percentage of production will be consumed domestically leaving nothing much for export. Though there is a shift in the consumption pattern of Middle East countries from high quality Indian cardamom to cheaper Guatemala cardamom, there is no reduction in quantity. Further, new markets are emerging for this commodity. While the household consumption sector remains intact, rather increasing along with the population growth, the recent development of new uses in industrial sector (food and non-food) are also expected to enhance the demand for the commodity. The global import demand for cardamom is expected to be 20,000 mt in 2005.

6.3 Projections of supply

Projections for Indian cardamom, for which a model was developed for area and production are shown in Fig. 11.7 and Fig. 11.8, which presents both the historic and fitted values along with forecasted values beyond 1997. Both area and production are expected to grow slowly in the immediate future. The growth in production is expected to be more pronounced than in area indicating the improvement in yield per unit area. As per the cyclical movement discussed earlier, after the peak so achieved in 1995–96, 3-year period of decline is already over, it is the turn of increasing trend to reach the next peak in the cycle. As per the forecasted value, by the year 2000–2001 the expected production level will be between 8000 and 10,000 tons; and the area expansion is expected to touch 90,000

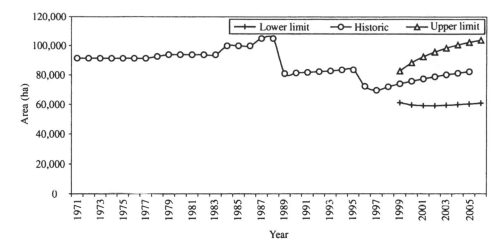

Figure 11.7 Trend in area under cardamom in India.

ha during the same period. The improvement in internal and international price will catalyze the supply to jump in the usual fashion discussed earlier.

The forecasted and the actual prices from 1990 onwards are shown in Fig. 11.9. Since the forecasted production is not sufficient to create enough export surplus, and the reports of declining production in Guatemala is already reflected in the form of less supply to the world market by that country during the 1998–99 crop year, the repercussions will be favourable to Indian farmers in the form of increased price. The prevailing higher market price is expected to continue in the near future and there is also a possibility for the price to cross the Rs. 1000/kg mark before falling down as per the usual cyclical fluctuations. Availability of less exportable surplus will have direct effect on the export. The forecasted standard scenario indicates that the trend prevailing for the last 5 years will continue for the 5 years to come i.e. the export will remain below the 300 tons mark. However, the actual estimated export is expected to lie between the upper confidence limit and actual forecast.

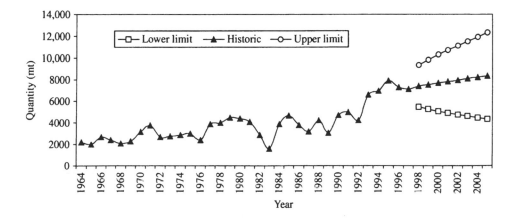

Figure 11.8 Production trend and forecast for cardamom in India.

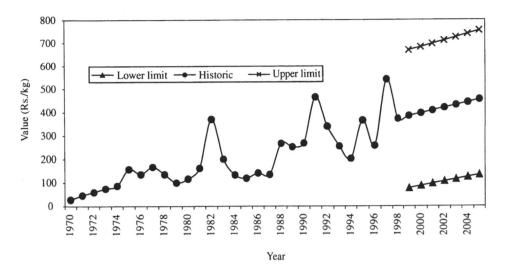

Figure 11.9 Trend in domestic price of cardamom in India.

7 CONCLUSION

The objective of this chapter is to analyze the performance of the cardamom industry in terms of area, production and productivity, export prices and their inter-relationship. In this, the movement of prices, price cycles and supply response to price variation etc. has been considered. Finally, an attempt is also made to quantify the uncertain future using forecasting models. The following are the main findings:

1   To meet the increasing demand for cardamom in the world market, supply is also rising at a fast rate. Much of the increased supply is from Guatemala. India, while consuming more than half of the total world production, contributes hardly 2 per cent to the world market.

2   While the increase in production during the 80s was mainly due to increase in area, during the 90s the increase in production is due to improvement in productivity. However, the yield gap that exists between the potential productivity level and average achieved level of productivity in the country indicates the possibility of improving yield level considerably.

3   In the export front, country has lost most of its traditional markets to Guatemala mainly because of the incompetitive price of the Indian cardamom in the international market. Guatemala derives its price competitiveness mainly from less production cost and high productivity per unit area.

4   In the changed scenario, Japan is the steady and reliable market for Indian cardamom. Due to decline in oil prices and consequent fall in purchasing power the Middle East market has changed from quality conscious to price sensitive.

5   There is a definite pattern of cyclical fluctuation in prices mainly due to the producers' response to price by new and replanting, which will yield after a certain time lags. Thus the cyclical fluctuations in prices have an implicit bearing on the condition of supply through farmer's response.

6   Finally, the forecasted future indicates that there will be a steady increase in supply (production and yield) and the price is expected to either move up or to remain steady at the present level.

The future scenario presented here can change drastically as a result of innovative product development, diversification of some entirely new application if discovered for cardamom or its products. Imaginative product development programmes have to be initiated to boost up the consumption pattern. Attractive formulations backed up by catchy advertisements can do wonders in this field.

REFERENCES

Anonymous (1996) Alternative crops for cardamom, *Report of the Expert Committee*, Spices Board, Cochin.

Anonymous (1998) Financial Express, 25 May, 1998.

Bossen, L. (1982) Plantation and labour force dissemination in Guatemala. *Current Anthropology*, 23, 3.

Cherian, A. (1977) Environmental ecology – an important factor in cardamom cultivation. *Cardamom*, 9 (1), 9–11.

George, K.V. (1976) Production constraints in cardamom. *Cardamom News*, 8 (3), 9–16.

John, K. (2000) Njallani green gold at cardamom productivity helm with precision farming techniques. In *Spices and Aromatic Plants*, Indian Society for Spices, Calicut, pp. 105–106.

Joseph, K.J. (1985) *Marketing and price formation of cardamom in Kerala*. M.Phil Thesis, Jawaharlal Nehru University, New Delhi.

Korikanthimath, V.S. (1992) High production technology in cardamom. In Y.R. Sarma, S. Devasahayam and M. Anandaraj (eds) *Black Pepper and Cardamom: Problems and Prospects*, Indian Society for Spices, Calicut, pp. 20–30.

Krishna, K.V.S. (1979) Harvesting of cardamom in Idukki district. *Cardamom*, 2, 3–8.

Mahabala, G.S., Bisaliah, S. and Chengappa, P.G. (1991) Resource use efficiency and age-return relationship in cardamom plantations. *J. Plantation Crops*, 20(suppl.), 359–365.

Nair, K.N., Narayana, D. and Sivanandan, P. (1989) *Ecology and Economics in Cardamom Development*. Centre for Development Studies, Trivandrum.

Radhakrishnan, C. (1993) Economics of pepper production in India. *Report of the Research Project Sponsored by Spices Board*. Government of India, Spices Board, Cochin.

Spices Board (1990) Status paper on spices. Spices Board, Ministry of Commerce, Government of India, Cochin, Kerala.

Spices Board (1991) Report of the survey for assessment of area under small cardamom in India. Spices Board, Ministry of Commerce, Government of India, Cochin, p. 22.

Spices Board (2000) Draft Annual Report (1999–2000). Spices Board, Ministry of Commerce, Govt. of India, Cochin, p. 22.

Thomas, E.K., Indiradevi, P. and Jessy Thomas, K. (1990) Performance of Indian cardamom – an analytical note. *Spices India*, 3(6), 19–23.

UNCTAD/GATT (1985) *Spices: A Survey of the World Market*, International Trade Centre, Geneva.

12 Properties and end-uses of cardamom

K.K. Vijayan, K.J. Madhusoodanan, V.V. Radhakrishnan
and P.N. Ravindran

1 INTRODUCTION

Spices and herbs used in food seasoning often have a mild, broad spectrum of anti-microbial activity. Many crude drugs are used medicinally because of their volatile oil content or other chemical constituents that possess biological activities. Cardamom is very popular as a spice and food additive because of its delicious flavour. The constituents of its volatile oil are responsible for the flavour and fragrance. It also possesses carminative, stomachic and antimicrobial actions. These biological activities bring about many advantages to the seasoned and prepared foods. Apart from this, cardamom finds application in the indigenous systems of medicine.

2 PHARMACOLOGICAL PROPERTIES

In *Ayurveda* and *Sidha* systems of medicine cardamom finds application as a component of several therapeutic formulations. *Charakasamhita*, the ancient Indian medical text, describes the use of cardamom as an antidote for food poisoning. This forms a constituent of *Bhrahmi rasayana*, which is used as a treatment for inflammations. Also used as a component of many balms, ointments and therapeutic oils used against cramps, rheumatic pain, inflammations etc. In Ayurvedic texts the properties of cardamom seeds are described as aromatic, acrid, sweet, cooling, stimulant, carminative, diuretic, cardiotonic and expectorant. Cardamom is used as an ingredient in preparations used for the treatment of asthma, bronchitis, hemorrhoids, renal and vesicle calculi, cardiac disorders, anorexia, dyspepsia, gastropathy, debility and vitiated conditions of *vata*. But no pharmacological investigations were carried out to validate the above properties. An aqueous extract of seeds is given to nursing mother to treat ringworm infection of child (Aloskar *et al.*, 1992). Roasted seeds are boiled with betel leaves and the extract is used to treat indigestion and worm trouble. However, such uses of cardamom in the indigenous system of medicine have not been evaluated pharmacologically.

2.1 Carminative action

Cardamom seeds and oil have carminative action. Tincture cardamom is used in many medicinal preparations, which are used as carminative, stomachic and to relieve colic pain (British Pharmacopoeia, 1993). Tincture cardamom and compound tincture of cardamom are included as an official preparation in the *British Pharmaceutical Codex* 1963, *British Pharmacopoeia* 1993 and in the pharmacopia's of China, Hungary and Japan. Martindale (1996) *The Extra Pharmacopoeia*, also describes preparations of cardamom fruits, cardamom oil and cineole, the major constituent of cardamom oil, as carminative and flavouring agent (Martindale, 1996). Cineole has been used as a counterirritant in ointments and in dentrifices. It has also been used in nasal preparations. Jain *et al.* (1994) have shown that cardamom essential oil containing preparation such as *Brahmi rasayana*, suppresses castor oil induced diarrhoea in rats.

2.2   Antimicrobial activity

The terpenoid constituents are responsible for the antifungal and anitbacterial effects. Mishra *et al.* (1990, 1991) studied the effect of cardamom on *Aspergillus flavus*, the fungus which produces the deadly aflatoxin B1. Mycostatic activity was observed at 400 ppm level. This was found to be as potent as synthetic antifungals commonly used (Hirasa and Takemasa, 1998). The flavour components also showed antibacterial effects against several food-born microorganisms (Kubo *et al.*, 1991). Another study proved that growth of *Morgenella morganii*, was moderately inhibited by the application of cardamom oil or powder (Shakila *et al.*, 1996). This organism is a potent histamine-producing bacterium growing on stored fish.

The minimum inhibitory concentration (MIC per cent) of cardamom extracts for bacteria and fungi in comparison with a few other common spices are given in Tables 12.1 and 12.2.

2.3   Anticarcinogenic activity

Banerjee *et al.* (1994) has found that cardamom oil enhances glutathione-s-transferase enzyme and acid soluble sufhydril activities. These enzymes mediate the oxidation and detoxification of xenobiotics. Cardamom oil was fed by gavage at 10 μl/day for 14 days and hepatic microsomal enzymes were measured. GST and acid-soluble sulfhydril were found to be significantly elevated ($P < 0.1$, $P < 0.001$ and $P < 0.05$ respectively).

Table 12.1 MIC (%) of cardamom and other spice extracts

	PH	BS	Sa	Ec	St	Sm	Pa	Pv	Pm
Cardamom	7	2.0	2.0	4.0<	4.0<	4.0<	4.0<	4.0<	4.0<
	5	0.1	0.5	4.0<	4.0<	4.0<	4.0<	4.0<	4.0<
Cinnamon	7	4.0	2.0	4.0	4.0	4.0	4.0	2.0	4.0
	5	0.5	2.0	4.0	4.0	4.0	4.0	1.0	2.0
Clove	7	1.0	1.0	1.0	1.0	1.0	2.0	1.0	1.0
	5	0.5	2.0	1.0	1.0	1.0	1.0	0. 5	0.5
Mace	7	0.2	0.05	4.0<	4.0 <	4.0<	4.0<	4.0<	4.0<
	5	0.1	0.5	4.0<	4.0<	4.0<	4.0<	4.0<	4.0<

Source: Hirasa and Takemasa (1998).

Notes
BS: *Bacillus subtilis*; Sa: *Staphylococcus aureus*; Ec: *Escheritia coli*; St: *Salmonella typhimurium*; Sm: *Salmonella marcescens*; Pa: *Pseudomonas aeruginosa*; Pv: *Proteus vulgaris*; Pm: *Proteus morganii*.
MIC (%) for fungi is given in Table 12.2.

Table 12.2 MIC (%) for certain fungi

	Sc	*Cp*	*Ck*	*P* sp.	*Ao*
Cardamom	4.0	4.0	4.0<	4.0<	4.0<
Cinnamon	1.0	1.0	1.0	1.0	1.0
Clove	0.5	0.5	0.5	0.5	0.5
Mace	4.0<	4.0<	4.0<	4.0<	4.0<

Source: Hirasa and Takemasa (1998).

Notes
Sc: *Saccharomyces cerevisiae*; *Cp*: *Candida parakrusei*; *Ck*: *Candida krusei*; *P* sp.: *Penicillium* sp.; *Ao*: *Aspergillus oryzae*.

Hashim *et al.* (1994) reported that cardamom oil suppresses DNA adduct-formation by aflatoxin B1 in a dose dependent manner. It appeared to be modulated through the action of microsomal enzymes.

2.4 Anti-inflammatory activity

Yamada *et al.* (1992) reported that cardamom showed potent complement system activating property. Complements represent the humoral arm of natural immunological host-defense mechanism and are essential for survival. Once activated this kills certain bacteria, protozoa, fungi and virus as well as cells of higher organism. Thus complement activation forms a major part of natural defense affording a range of mediators possessing immuno-inflamatory potency. Jain *et al.* (1994) have shown that the drug *Brahmi rasayana* containing cardamom (together with cloves and long pepper), exhibited a dose-dependent anti-inflammatory activity in the case of carrageenan induced rat paw oedema. This drug also inhibited nystatin-induced inflammation in rats. Al-Zuhair *et al.* (1996) have shown that cardamom oil when administered at 175 and 280 μl/kg of body weight inhibited the growth of carrageenan-induced paw oedema in rats by 69.2 and 86.4 per cent respectively. The anti-inflammatory activity of cardamom oil is comparable to that of indomethacin (indometacin). El-Tabir *et al.* (1997) investigated the pharmacological action of cardamom oil on various animal systems, such as the cardio-vascular system of rats, nictitating membrane of cats, isolated rabbit jejunum, isolated guinea-pig ileum and the frog sciatic nerve. The essential oil (5–20 μl/kg iv) decreased the arterial blood pressure in rats and heart rate in a dose-dependent manner. These effects are antagonistic to the treatment with cyproheptadine (1 mg/kg) for 5 min.). Atropine was also antagonistic to the cardamom-induced bradycardia. The oil was not having any effect on isolated, perfused rat heart, and did not affect electrically-induced contractions of the cat nictitating membrane. At concentrations of less than 0.08 μl/ml the oil induced contractions of the jejunum; but larger doses relaxed it. Larger dose of the oil was antagonistic to the action of acetylcholine, nicotine and $BaCl_2$ on the rabbit jejunum. The oil at concentrations of 0.01–0.04 μl/ml, induced contractions of the isolated guinea pig ileum; this effect was suppressed by atropine and cyproheptadine. Exposure of frog sciatic nerve to 0.2–0.4 μl/ml cardamom oil suppressed the frog limb withdrawal reflex, exhibiting a local anaesthetic effect (El-Tabir *et al.*, 1997). Al-Zuhair *et al.* (1996) have shown that cardamom oil also exhibited analgesic properties and inhibited spontaneous and acetylcholine induced movements of rabbit intestine *in vitro* in a dose-dependent manner.

2.5   Other pharmacological studies

From these pharmacological investigations the beneficial effects of cardamom and its oil were established. It is not a mere flavouring agent. It imparts carminative, fungicidal and bactericidal effects. It activates the complement system thereby the immunological defense mechanism of the human body is enhanced. Two other studies reported with extracts of cardamom show another aspect of its therapeutic utility. Extracts of cardamom enhance the percutaneous absorption of medicament. Yamahara *et al.* (1989) studied the dermal penetration of prednisolone using mouse skin model and reported that terpineol and acetylterpineol are the active constituents in cardamom extract, which facilitates the absorption. Huang *et al.* (1993) used rabbit skin model and *in vivo* and *in vitro* studies were conducted. They observed that the extract of cardamom enhanced the skin penetration both *in vivo* and *in vitro*. Hence addition of cardamom extract or terpineol or its acetate in balms and ointments enhances the absorption of medicaments through skin. Terpineol and bornyl acetate exhibits disinfectant and solvent properties and hence used with other volatiles for cough and respiratory tract disorders. Cineole is an ingredient along with other volatile substances for the treatment of renal and biliary calculi (Martindale, 1996).

Yaw Bin *et al.* (1999) also investigated the effect of cardamom extract on transdermal delivery of indometacin. The permeation of indometacin was significantly enhanced after pretreatment with cardamom oil both in the *in vitro* (rat, rabbit and human skin) and *in vivo* (rabbit) studies. The indometacin flux decreased as the length of the pretreatment increased. Both natural cardamom oil and a cyclic monoterpene mixture composed of the components of the oil showed similar enhancement of indometacin permeation, indicating cyclic monoterpenes are the predominant components altering the barrier property of stratum corneum. This study also showed that three minor components in cardamom oil (α-pinene 6.5 per cent, β-pinene 4.8 per cent and α-terpineol 0.4 per cent) had a synergistic effect with 1,8-cineole (eucalyptol) and D-limonene to enhance the permeation of indometacin.

2.4   Toxicity

No toxicity was reported for cardamom. The main use of cardamom is as a spice, and as a flavourant. When flavour substances are added to foods no health hazard should arise at the concentrations used, as they are used in small doses only, usually not exceeding 10–20 ppm of the total quantity of the food item. Higher concentrations could not be used because of their intense smell and taste. Most of the individual components of cardamom oil were studied to assess their toxicological actions on experimental animal system. The studies were conducted under the auspicious of the international programme on food safety. In the series of Technical reports by the joint FAO/WHO Expert Committee on Food Additives, cardamom oil and its chemical constituents are excluded from having any toxicological effects. In the allopathic medicine the use of cardamom is only as a carminative in certain medical formulations.

2.5   Antioxidant function

Cardamom exerts only mild antioxidant function and hence is not very effective in preventing food spoilage. The antioxidant function of cardamom in comparison with a few other major spices is given in Table 12.3.

Table 12.3 Antioxidant activity of cardamom and a few other spices against lard (Con. Added 0.02%)

	Ground spice *POV (meq/kg)	Petrol ether soluble fraction POV (meq/kg)	Petrol ether insoluble fraction POV (meq/kg)
Cardamom	423.8	711.8	458.6
Pepper	364.5	31.3	486.5
Cinnamon	324.0	36.4	448.9
Clove	22.6	33.8	12.8
Turmeric	399.3	430.6	293.7
Nutmeg	205.6	31.1	66.7
Ginger	40.9	240.5	35.5

Source: Hirasa and Takemasa (1998).

Note
* POV: Peroxide value, negatively correlated with the antioxidant property.

2.6   Pharmaceutical products

Blancow's (1972) Martindale is giving the following preparations using cardamom: Aromatic cardamom tincture (BPC), (Tincture cardamom aromatic; carminative tincture). Cardamom seed 1 part in about 15 parts of strong ginger tincture, alcohol (90 per cent) and oil of caraway, cinnamon and clove.

Compound cardamom tincture (BP). (Tincture cardamom compound). Prepared from cardamom, cochineal and glycerin by percolation with alcohol (60 per cent). Often the tincture is decolorized by alkaloidal salts, bismuth carbonate, calcium ions and sodium bromide.

Compound cardamom tincture (USNF). Prepared by macerating cardamom seed 2 gm, cinnamon 2.5 gm and caraway 1.2 gm with glycerin 5 ml and diluted with alcohol to 100 ml.

2.7   Other properties

2.7.1   *Effect on stored product insect pests*

Huang *et al.* (2000) studied the contact and fumigant toxicities and antifeedent activity of cardamom oil to two stored-product insect pests (*Sitophilus zeamais* and *Tribolium castaneum*). Topical application was employed for contact toxicity studies, and filter paper impregnation was used for testing fumigant action. The adults of both insects were equally susceptible to the contact toxicity of the oil at the LD 50 values of 56 and 52 µg/mg insect for *S. zeamais* and *T. castaneum* respectively. For fumigant toxicity, *S. zeamais* adults were more than twice as susceptible as *T. castaneum* adults at both the LD 50 and LD 95; 12-day-old larvae of *T. castaneum* were more tolerant than the adults to the contact toxicity of the oil. The susceptibility of larvae to contact toxicity increased with age. Cardamom oil applied to filter paper at concentrations ranging from 1.04 to 2.34 mg/cm^2 significantly reduced the hatching of *T. castaneum* eggs and the subsequent survival rate of the larvae. Adult emergence was also drastically reduced by cardamom oil. When applied to rice or wheat, cardamom oil totally suppressed F1 progeny production of both insects at a concentration of 5.3×10^3 ppm. Feeding deterrence studies showed that cardamom oil did not have any growth inhibitory or feeding deterrence effects on either adults or larvae of *T. castaneum*. However, the oil significantly reduced all the nutritional indices of the adults of *S. zeamais* (Huang *et al.*, 2000).

2.7.2   Effect of cardamom on house dust mite

It is generally held that about 70 per cent of the allergy cases show positive antigenic reaction for house dust mites (*Dermatophagoides farinae* and *D. petronyssinus*). Yuri and Izumi (1994) studied the effect of essential oil of spices on *D. farinae*, and reported that some of the spice essential oils were effective against this mite (Table 12.4). They used a concentration of 80 $\mu g/cm^2$ on filter papers, and the mortality rate was counted after 24 h. The essential oil of cardamom exerted only very low mortality rate.

Table 12.4 Mortality rate of cardamom essential oil for *Dermatophagoides farinae* in comparison with other common spice oils

Essential oil	Mortality rate %
Cardamom	4.7
Clove	97.3
Mace	0.5
Nutmeg	–
White pepper	0.1
Anise	56.5
Garlic	72.8

Antiheliminthic activity of spice extracts was studied by Tsuda and Kiuchi (1989) and found that methanol extract of cardamom exhibited antihelminthic effect on dog round worm.

3   CARDAMOM IN TRADITIONAL SYSTEMS OF MEDICINE

In Indian systems of medicine – *Ayurveda*, *Sidha* and *Unani*, cardamom is used as a powerful aromatic stimulant, carminative, stomachic and diuretic. It also checks nausea and vomiting and is also reported to be a cardiac stimulant. Powdered cardamom seed mixed with ground ginger, cloves and caraway is helpful in combating digestive ailments. Tincture of cardamom is also made and used chiefly in medicines for windiness or as stomachic. A good nasal application is prepared by using extracts of cardamom, neem and myrobalan along with animal fat and camphor. Cardamom seeds are chewed to prevent bad smell in mouth, indigestion, nausea and vomiting due to morning sickness, excessive watering in mouth (pyrosis) etc. Gargling with the infusion of cardamom and cinnamon cures pharyngitis, sore throat, and hoarseness during infective stage of flu. Its daily gargle protects one from flu infection (Pruthy, 1979).

Powdered seeds of cardamom boiled in water with tea powder imparts a very pleasant aroma to tea and the same can be used as a medicine for scanty urination, diarrhea, dysentery, palpitation of heart, exhaustion due to overwork, depression etc. (Singh and Singh, 1996). Eating a cardamom capsule daily with a tablespoon of honey improves eyesight, strengthens nervous system and keeps one healthy. It is believed by some people that excessive use of cardamom causes impotency.

One of the main properties of cardamom is its effect on different dermatological disorders. Medicated cardamom oil and cardamom powder can retard various types of hypo-pigmentation on the face (Nair and Unnikrishnan, 1997). Cardamom powder is a safe emetic that can be used in bronchial asthma patients when excess of sputum is present in the lungs. Further it is a very good cough suppressant. Cardamom finds a place in the formulation of lozenges for the management of common cold and associated symptoms (Nair and Unnikrishnan, 1997).

In the form of tincture or powder, cardamom is used as a frequent adjunct to other stimulants, bitters and purgatives. A decoction of cardamom together with its pericarp and jaggery added is a popular home remedy, which relieves giddiness caused by bili-ousness. A compound powder containing equal parts of cardamom seeds, ginger, clove and caraway is a good stomachic in atonic dyspepsia. A powder made of equal parts of parched cardamom seed, aniseed and caraway seed is a good digestive. Cardamom is used in as many as 24 important preparations in *Ayurveda* in the form of decoctions, oils and powders as well as medicated fermented beverages like *Arishta* and *Aasava* (Sahadevan, 1965). Cardamom seeds along with saffron (*Crocus sativus*), galengal (*Alpinia galanga*) and "nealgor of the corryrium" cure cataract and other eye ailments like tumours in eyelids, fleshy growth and ophthalmia.

Cardamom fruit is an emmenagogue, the only spice to qualify for this. Cardamom, cinnamon, tejpatra (*Cinnamomum tamala*) and iron wood tree (*Mesua ferrea*) taken together is known as *Chaturjata*. They are used to flavour electuaries and to promote their actions (Warrier, 1989).

Cardamom is also a component of medicinal preparation used to cure skin diseases, poisons, cold and inflammation. Preparations such as *Eladigana* (Ela-cardamom) are a common cure for *vata* (arthritis) and *kapha* (congestion) diseases, poisons, to improve complexion and to cure itching. It is also an ingredient of mixtures for improving digestion, curing vomiting, cough etc. It is used to stimulate diuresis, particularly in the case of snakebite poison. A group of medicines known as *Ariyaru kashayam* (six grains) for skin diseases of children contains cardamom.

The Burmese (Myanmar) traditional medicine formulation – O2 (TMF-O2) consists of four basic plant ingredients one of which is cardamom (others being *Anacyclus pyrethrum*, *Glycyrrhiza glabra* and *Syzygium aromaticum*).

4 CARDAMOM AS A SPICE

4.1 Cardamom for culinary use

The major use of cardamom on a worldwide basis is for domestic culinary purposes in whole or ground form. In Asia, cardamom plays an important role in a variety of spiced rice, vegetable and meat dishes. Cardamom can add a lingering sparkle to many dishes, both traditional and modern. International trade in cardamom is dependent, however, on the demand created by specialized applications that have evolved in two distinct markets, namely the Arab countries of the Middle East and in Scandinavia.

Cardamom gives a warm, comforting feeling and it is responsible for the peculiar and exotic flavour of Bedouin coffee. In the Middle East, religious ceremonies, social functions and celebrations are not complete without serving *Gahwa* or Arab coffee (cardamom-flavoured coffee). It is believed, that this drink cools down body heat in a country where extreme heat is a regular feature of daily life. It is also believed to aid in digestion and is said to be an aphrodisiac. Cardamom is also used in cooking by the indigenous population in Arab countries. The Arab population also have adopted a large number of Indian recipes and hence, cardamom is now used in *biryani* (a popular rice dish) and other similar dishes and curries popular in the Middle East countries. In Iran, cardamom is used in confectionery, bakery items and meat products for its flavour and aroma. Cardamom invariably finds a place in the spice chest of Indian kitchens. The Indian housewife uses this in a variety of meat and vegetable dishes as well as in sweets like rice porridge.

In European countries and in North America, cardamom is used mainly in ground form by food industries as an ingredient in curry powder, some sausage products, fruit cups, green pea soups, curry-flavoured soups, spice dishes, rice, Danish pastry, buns, breads, rolls, cookies, desserts, coffee cakes, orange salad, jellies, baked apple coffee, honey pickles, pickled herring, canned fish and to a small extent in flavouring cigarette and tobacco. Cardamom cola, instant gahwa, carbonated gawha, biscuits, spanish pastries, toffees, chewing gum etc. are the other products. Various breakfast foods using encapsulated cardamom oil are new products developed in the recent past using cardamom. Cardamom is also used in spiced wine and for flavouring custard (by steeping crushed cardamom seeds in hot milk). In general, the Arabs use it in coffee; the Americans in baked foods; the Russians in pastries, cakes and confectionery; the Japanese in curry, ham and sausage; the Germans in curry powders, sausages and processed meats, and so on in countless other dishes. Cardamom is widely used in baking in Scandinavian countries. The ground cardamom is mixed with flour to flavour most baked foods, it adds an exotic taste to apple pie (Rosengarten, 1973). In Sweden, cardamom is very popular with most baked foods, where the per capita consumption is about 60 times greater than that in US. Ground cardamom is also used to flavour hamburgers and meat loaves.

Indian cardamom is low in fat and high in protein, iron, vitamin B and C (Pruthy, 1993). The nutritional composition of cardamom seed is given in Table 12.5. In India, it is used as a masticatory and also in flavouring culinary preparations. In places like Kanpur, Patna, Delhi, Bhuvaneswar, Orissa, Nagpur, Calcutta, Varanasi etc. cardamom is used in all kinds of puddings, which is an inevitable item at social and religious functions throughout India. It is used whole in the preparation of savoury rice dishes like fried rice, *pulaos, biriyanis* etc. It is also used for mild sweets such as sheers and powdered seeds are added to halwas. An aromatic tea is prepared by adding cardamom along with tea leaves and with or without lime, which is popular and refreshing (Philip, 1989). Cardamom is used in flavouring conventional tea in North India. Cardamom seeds are chewed after meals to ward off foul smell and as a mouth freshener. Cardamom flavoured hot water is supplied in many of the hotels in North India. Of late, in India, a variety of cardamom-flavoured products have come to market such as biscuits, horlicks, chocolates, milk, cheese and so on. It is also used for making garlands in India and Arab countries for special occasion to present to distinguished guests.

4.2 Suitability pattern of cardamom

The suitability pattern of cardamom (Hirasa and Takemasa, 1998) is given in Fig. 12.1. Cardamom is most suitable for Indian cooking, and does not show above average

Table 12.5 Nutritional composition of cardamom seed per 100 g

Composition	USDA Handbook 8–21[1] (Ground)	ASTA[2]
Water (g)	8.28	8.0
Food energy (Kcal)	311.00	360
Protein (g)	10.76	10
Fat (g)	6.70	2.9
Carbohydrates (g)	68.47	74.2
Ash (g)	5.78	4.7
Calcium (g)	0.383	0.3
Phosphorus (mg)	178.00	210.0
Sodium (mg)	18.00	10.0
Potassium (mg)	1119.0	1200.0
Iron (mg)	13.97	11.6
Thiamine (mg)	0.198	0.18
Riboflavin (mg)	0.182	0.23
Niacin (mg)	1.102	2.33
Ascorbic acid (mg)	–	ND
Vitamin A activity (RE)	Trace	ND

Source: Stobart (1982).

Notes
1 Composition of Foods: Spices and herbs. USDA Agricultural Handbook 8–2. January 1977.
2 The nutritional composition of spices, ASTA Research Committee, February, 1977.
3 ND – Not detected.

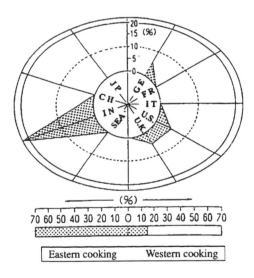

Figure 12.1 Suitability pattern for cardamom (Source: Hirasa and Takemasa, 1998).

suitability for any other national cuisines. In fact the use of cardamom is totally lacking in the cuisines of most European and American countries. Hirasa and Takemasa (1998) indicated that cardamom is more useful in meat, milk and fruit preparations. It is more useful for simmered, baked, fried, deep fried and pickled food and less suitable for steamed food.

A spice is used in food mainly for the following purposes:

(1) For the direct effect of flavouring the dish.
(2) For masking undesirable flavour notes and for deodorizing.
(3) For adding colour to the food.
(4) For adding pungency to the food.

Since a single spice cannot contribute all the above qualities often a combination of spices are used in the preparation of food. Cardamom is used mainly for its direct effect of imparting flavour. It is used for the same purpose in blended spice mixtures (curry powder, garam masala and pickle mixes).

Masking of undesirable flavour notes or odours is very important in the use of spices. According to the Weber–Fecher law, the strength of an odour perceived by the sense of smell is proportional to the logarithm of the concentration of the smelled compounds. In other words the sensational strength perceived with the five senses is proportional to the logarithm of the actual strength of these stimuli. Thus even if 99 per cent of the total smelled compounds is eliminated chemically, the sensational strength perceived is reduced only 66 per cent (Hirasa and Takemasa, 1998). Hence it is more effective and easy to mask the remaining 1 per cent through an aromatic spice. Tokita *et al.* (1984) investigated the deodorizing efficiency of various spices. The deodorizing rate (measured by the per cent of methyl mercaptan (500 mg) captured by methanol extract of the spice) of cardamom is very low (9 per cent) compared to other spices (99 per cent for thyme, 97 per cent for rosemary, 90 per cent for mint, 79 per cent for clove, 30 per cent for pepper). Ito *et al.* (1962) calculated deodorizing points for major spices in masking mutton odour. Cardamom gave a deodorizing point of 30, compared to 600 for pepper, 90 for ginger, 50 for cinnamon, 25 for celery, 23 for garlic, 5 for coriander, 4 for caraway, 3 for clove, 2.5 for thyme and 0.7 for sage. Thus cardamom is more effective in masking certain odours than many other spices, though its masking ability is poor with regard to some other flavours.

Desrosier (1978) studied the relative flavour intensities (RFI) of various spices, cardamom has a RFI of 125 (in comparison with 200 for turmeric, 260 for curry powder blend, 300 for celery seed, 400 for cinnamon, 450 for black pepper, 475 for ginger, 600 for cloves, 900 for cayenne pepper and 1000 for fresh red pepper).

4.3   Spice blends, garam masala

Ready to use spice formulations are available in market under many brand names. They are available either as dry powder form or as soluble seasonings, which are spice extracts on salt or dextrose carriers. Most common spice blends are curry powders, pickling spice mixes, fish or meat *masala* mixes and *garam masala*, and all of them are available under a variety of brand names.

Curry powders are most extensively used in Indian cuisine, and there are virtually hundreds of them for many specific purposes. The basic ingredients in most brands are coriander, cumin, turmeric for colour, chillies for pungency. Other spices such as black pepper, cardamom, cinnamon etc. are added for flavour and as taste enhancers. In many such formulations cardamom is used to impart that special flavour, but only in small quantities. The relative proportion of cardamom in a typical curry powder formulation can be understood from Table 12.6. The US Federal specification for curry powder is given in Table 12.7.

Table 12.6 Formulations for typical curry powder blends

Freshly ground Spices	US Standard Formula No. 1[a] (%)	General purpose curry formulas			
		No. 2 (%)	No. 3 (%)	No. 4 (%)	No. 5 (%)
Coriander	32	37	40	35	25
Turmeric, Madras	38	10	10	25	25
Fenugreek	10	0	0	7	5
Cinnamon	7	2	10	0	0
Cumin	5	2	0	15	25
Cardamom	2	4	5	0	5
Ginger, Cochin	3	2	5	5	5
Pepper, white	3	5	15	5	0
Poppy seed	0	35	0	0	0
Cloves	0	2	3	0	0
Cayenne pepper	0	1	1	5	0
Bay leaf	0	0	5	0	0
Chilli peppers, hot	0	0	0	0	5
Allspice	0	0	3	0	0
Mustard seed	0	0	0	3	5
Lemon peel, dried	0	0	3	0	0
	100	100	100	100	100

Source: Farrell (1985).

Notes
a From the US Military Specification Mil-C-35042A, dated 30 December 1964.
b Formula No. 2 is considered to be a mild curry, formula No. 3 a sweet curry, and formula No. 4 a hot type of curry.
c Formula No. 5 is a very hot, pungent Indian style curry more suited for use in the hot tropical areas of India.

Table 12.7 Federal Specification ES-S-631 J for curry powder

Ingredient	Limit (%)
Turmeric	37.0–39.0
Coriander	31.0–33.0
Fenugreek	9.0–11.0
Cinnamon	Not < than 7.0
Cumin	Not < than 5.0
Pepper, black	Not < than 3.0
Ginger	Not < than 3.0
Cardamom	Not < than 32.0

Source: Tainter and Grenis (1993).

Garam masala is a blend of spices having an approximate basic composition as follows (Anonymous, 1991).

Cumin seeds $-\frac{1}{2}$ oz
Cardamom $-\frac{1}{2}$ oz
Black pepper $-\frac{1}{2}$ oz
Cinnamon $-\frac{1}{4}$ oz
Cloves $-\frac{1}{4}$ oz

Table 12.8 Replacement ratios of cardamom in comparison with other spices

Spice	Replace 1# of Ground spice with		Notes
	# Oil	# Oleoresin	
Allspice	0.020	0.035	
Anise	0.020	0.050	
Basil	0.005	0.050	
Cardamom	0.030	0.015	
Caraway	0.010	0.050	
Celery	0.010	0.100	
Cinnamon	0.025	0.025	Oil use based on volatile oil
Clove	0.140	0.050	Steam. Leaf, or bud oil can be used
Coriander	0.003	0.070	
Cumin	0.020	0.040	
Dill seed	0.020	0.050	
Fennel	0.010	0.050	
Ginger	0.015	0.035	
Mace	0.140	0.070	Nutmeg oil should be used
Marjoram	0.008	0.050	
Nutmeg	0.060	0.080	Spice and oil must be from the same origin
Oregano	0.015	0.040	
Pepper, black	0.015	0.050	Oil does not provide piperine bite
Rosemary	0.008	0.040	Oleoresin is deflavoured for antioxidant
Sage	0.010	0.050	
Savory	0.005	0.065	
Tarragon	0.002	–	

Source: Tainter and Grenis (1993).

Other spices are added to impart unique tastes suitable for various dishes. Cardamom is a major component of *garam masala*.

Kalra *et al.* (1991) and recently Premavalli *et al.* (2000) analyzed the composition of garam masala brands available in Indian market. Kalra *et al.* (1991) reported the use of 11 spices, while the latter workers found 11–9 spices in different brands. The type of ingredients and the proportion are tricks of the trade. Premavalli *et al.* (2000) found that 27 spices in all are used in the preparation of garam masala out of which five are common to all brands (coriander, cumin, black pepper, cloves and cardamom). Cardamom imparts a special flavour to the masala mix and it is functioning as a taste and flavour enhancer.

4.4   Cardamom oil, oleoresin and soluble cardamom

Cardamom oil is a colourless or pale yellow oil with an aromatic pungent odour and taste. It is produced in small quantities in certain western spice-importing countries and also in India, Guatemala and Sri Lanka. The essential oil finds its main application in flavouring of processed foods, but it is used also in certain liquid products such as cordials, bitters and liquors and occasionally in perfumery. Perfumery and cosmetic industry employ the oils of many spices including cardamom in the blending of volatile and fixed oils to make dozens of alluring perfumes far in advance of the crude scents of the ancients (Bhandari, 1989).

Oleoresin of cardamom is produced in certain Western spice-producing countries and in India, and has similar applications as that of the essential oil in flavouring of processed foods but is less extensively used. Both the oil and oleoresin tend to develop "off flavours" when exposed to air for prolonged periods and their usage is generally confined to meat products and intended for other short shelf-life products such as sausages.

Cardamom oils and oleoresins are used mostly as soluble spice, mixed with a carrier, such as common salt or dextrose. They are used in processed foods industry. Soluble spices is easy to use, as it is a dry, free flowing powder compared to a liquid essential oil or oleoresin. However, much care is needed while replacing ground spice with oil, oleoresin or soluble spice. Such products often need not represent the freshly ground spice in its richness of flavour due to the loss some delicate components. Much trials need to be done in each food system to match oleoresin and oil with that of the ground spice. The replacement ratio for cardamom in comparison with other common spices are given in Table 12.8.

The Central Food Technological Research Institute (CFTRI), Mysore, India has made innovative R & D efforts for diversifying uses of cardamom, to widen the domestic and export markets for cardamom products. Cardamom flavour has been encapsulated using an innovative technology. Encapsulated spices possess unique features like free flowing nature, uniform flavour strength and convenience to use. The flavour that is encapsulated is released instantly on contact with water (Pruthy, 1993). Cardamom-cola, a fizzy drink with pleasing cardamom flavour, a *flan* mix, an instant desert mix with cardamom flavour, cardamom tea, cardamom coffee, dry powder for cardamom soft drink mix and instant pongal mix flavoured with cardamom are some other products developed by the CFTRI (Pruthy, 1993). Innovative product development programmes can help in the diversification of cardamom use leading to the creation of new dishes and food items.

Appendix 2 gives a selection of recipes where cardamom is used as a flavouring agent.

5   CONCLUSION

Cardamom is in use from ancient times for flavouring and as a component in many indigenous medicines in India. Cardamom-flavoured tea and coffee form part of the daily routine of people in the Middle East, chewing cardamom after a meal is a habit with large number of people in India, and cardamom goes for flavouring oral formulations of many medicines, bakery products and milk. Cardamom is unique because of the delicate blend of its aroma. The potentialities of this great spice have not been exploited fully. Cardamom is important only in Indian and South Asian cooking, and has little influence in continental, American or Japaneese cuisine. Diversification of products with attractive advertisement back up can create fresh demand for this important spice. The reported property of cardamom extract in enhancing the pericutaneous absorption of medicaments can be made use of in the preparation of skin ointments and balms as well as in oral formulations.

Innovative technologies are needed for increasing the global demand of cardamom. For this research need to be directed to product development including novel cardamom-flavoured dishes.

REFERENCES

Aloskar, L.V., Kakkar, K.K and Chakre, O.J. (1992) Second Supplement to *Glossary of Indian Medical Plants, Part-2.* Publication and information Directorate, New Delhi, p. 289.

Al-Zuhair, H., El-Sayeh, B., Ameen, H.A. and Al-Shoora, H. (1996) Pharmacological studies of cardamom oil in animals. *Pharmacological Res.,* **34**, 79–82.

Anonymous (1991) Try-Indian menue at home. *Indian Spices,* **28**(4), 8–9.

Banerjee, S., Sharma, R., Kale, R.K. and Rao, A.R. (1994) Influence of certain essential oils on carcinogen-metabolising enzymes. *Nutr. Cancer,* 21, 263–269.

Bhandari, N. (1989) Spices for beauty and body care. In *Strategies for Export Development of Spices,* Spices Board, Cochin, pp. 54–56.

Blancow, N.W. (1972) *Martindale. The Extra Pharmacopoeia.* The Pharmaceutical Press, London.

British Pharmaceutical Codex (1963) Pharmaceutical Press, London, p. 139.

British Pharmacopoeia (1993) General Medical Council, London, p. 138.

Desrosier, N. (1978) *Reitz Master Food Guide,* AVI Pub., Co., USA.

El-Tabir, K.E.H., Shoch, M. and Al-Shora, M. (1997). Exploration of some pharmacological activities of cardamom seed (*Elettaria cardamomum*) volatile oil. *Saudi Pharmaceutical J.,* 5(2/3), 96–102.

Farrell, K.T. (1985) *Spices, Condiments and Seasonings.* The AVI Pub., Co., USA.

Hashim, B., Aboobaker, V.S., Madhubala, R.K. and Rao, A.R. (1994) Modulatory effects of essential oils from spices on the formation of DNA adducts by aflatoxin B1 *in vitro. Nutr. Cancer,* 21, 169–175.

Hirasa, K. and Takemasa, M. (1998) *Spices Science and Technology.* Marcel Dekker, New York.

Huang, Y., Lam, S.L. and Ho, S.H. (2000) Bioactivities of essential oil form *Elettaria cardamomum* (L.) Maton to *Sitophilus zeamais* Mostschulsky and *Tribolium castaneum.* Herbst. *J. Stored Products Res.,* **36**, 167–117.

Huang, Y.B., Hsu, I.R., Wu, P.C., Ko, H.M. and Tsai, Y.H. (1993) Crude Drug enhancement of percutaneous absorption of indomethacine *in vitro* and *in vivo* penetration. *Kao Hsiung J. Haresh Ko Tsa Chih,* 9, 392–400.

Ito, Y., Miura, H. and Miyaga, K. (1962) cited from Hirasa and Takemasa (1998).

Jain, P., Khanna, N.K., Trehan, N. and Godhwani, J.L. (1994) Anti-inflammatory effects of an Ayurvedic preparation, Brahmi rasayan, in rodents. *Indian J. Expl. Biol.,* **32**, 633–636.

Kalra, C.L., Seligal, R.C., Manan, J.K., Kulkarni, S.G. and Berry, S.K. (1991) Studies on preparation, packaging and quality standards of ground spice mixes. Part 1 Garam masala. *Beverages and Food World,* **18**, 21–24.

Kubo, I., Himejima, M. and Murari, H. (1991) Antimicrobial activity of flavour components of Cardamom. *J. Agri. Food Chem.,* **39**, 1984–1986.

Martindale, The Extra Pharmacopoeia (1996) Royal Pharmaceutical Society, London, pp. 686–2, 1681–1, 1758–1.

Mishra, A.K. and Dubey, N.K. (1990) Fungitoxicity against *Aspergillus flavus. Economic Botany,* 44, 350–533.

Mishra, A.K., Dwivedi, S.K., Kishore, N. and Dubey, N.K. (1991) Fungistatic properties of essential oils of Cardamom. *Int. J. Pharmacognosy,* **29**, 259–262.

Nair, P.R.S. and Unnikrishnan, G. (1997) Evaluation of medicinal values of cardamom alone and in combination with other spices in ayurvedic system of medicine-project. Part I & II. Government Ayurveda college, Trivandrum, India, pp. 46, 51.

Philip, T.E. (1989) Spices in cookery and preservation of food. In *Strategies for Export Development of Spices,* Spices Board, Cochin, pp 40–53.

Premavalli, K.S., Manjumdar, T.K. and Malini, S. (2000) Quality evaluation of traditional products. Garam masala and puliyogere mix masala. *Indian Spices.,* 37(2), 10–13.

Pruthy, J.S. (1979) *Spices and Condiments.* National Book Trust, New Delhi, India, pp. 63–68.

Pruthy, J.S. (1993) *Major Spices of India – Crop Management and Post Harvest Technology,* ICAR, New Delhi, India, p. 514.

Purseglove, J.W., Brown, E.G., Green, C.L. and Robbins, S.R.J. (1981) *Spices*. Vol. 2. Longman Scientific and Technical, New York.

Rosengarten, F. (1973) *The Book of Spices*. Pyramid Books, New York.

Sahadevan, P.C. (1965) *Cardamom*. Farm Bulletin No. 4. Department of Agriculture, Kerala, India, p. 90.

Shakila, R.J., Vasundhara, T.S. and Rao, D.V. (1996) Inhibitory effect of spices on *in vitro* histamine production and histidine decarboxylation activity of *Morgenella morganii* and on the biogenic amine formation in market samples stored at 30 °C. *Lebensm Uters Forsh*, **203**, 71–76.

Singh, V.B. and Singh, K. (1996) *Spices*. New Age International Ltd., New Delhi, India, p. 253.

Stobart, T. (1982) *Herbs, Spices and Flavourings*. The Overlook Press, Wood stock, New York.

Tainter, D.R. and Grenis, A.T. (1993) *Spices and Seasonings*. VCH Pub. Inc., USA.

Tokita, F., Ishikawa, M., Shibuya, K., Koshimizu, M. and Abe, R. (1984). *Nippon Nogeikagaku Gakkaishi*, 58, 585 (Quoted from Hirasa and Takemasa, 1998).

Tsuda, Y. and Kiuchi, F. (1998) Koshinryo no Kinou to Seibun, Kouseikan, Tokyo. Quoted by Hirasa and Takemasa (1998).

Warrier, P.K. (1989) Spices in Ayurveda. In *Strategies of Export Development of Spices*, Spices Board, Cochin, pp. 28–39.

Yamahare, J., Kashiwa, H., Kishi, K. and Fujimura, H. (1989). *Chem. Pharm. Bull.* (Japan) 37, 855–856. (Quoted from Hirasa and Takemasa, 1998).

Yamada, C.K. (1992) Mitogenic and complement activating activities of the herbal components. *Planta Medica*, **58**, 166–170.

YawBin, H., JiaYou, F., ChenHsun, H., PawChu, W. and Yittung, T. (1999) Cyclic monoterpene extract from cardamom oil as a skin permeation enhancer for indometacin: in vitro and in vivo studies. *Biol. & Pharmaceutical Bull.*, **22**, 642–646.

Yuri, Y. and Izumi, K. (1994) *Aromatopica*, 3, 65. (Quoted from Hirasa and Takemasa, 1998).

13 Cardamom – future vision

C.K. George

1 PRODUCTION, TRADE AND CONSUMPTION

Nearly 70 years ago cardamom cultivation spread from India to Guatemala. But, India continued to be the largest producer until 1979–80. The situation has since been changed and Guatemala emerged as world's leading cardamom producer and supplier. While Indian production in the last two decades has shown a fluctuating trend, Guatemala's production has been on steady increase except in the last two years. The production of cardamom in India and Guatemala, for 12 years from 1990–2002, is given in Table 13.1 (See also Chapter 11).

World production of cardamom during 1980–81 was around 10,250 tons; in 1990–91 the production was 16,000 tons; in 2000–2001 the production was around 22,800 tons and 25,000 tons in 2001–2002, the highest so far recorded.

India has been the largest exporter of cardamom till 1980–81. The world market has been India's monopoly till Guatemalan production picked up. Practically with no domestic demand, almost the entire production of Guatemala is exported. In India, cardamom is an important spice and consumed in many households. Due to well-developed internal demand the price of cardamom in India is high. Often export is not possible because of the comparatively low international price on account of the steady supply position from Guatemala. Export of cardamom from India and Guatemala for 10 years since 1990–91 is given in Table 13.2.

India and Saudi Arabia are the largest consumers of cardamom. These two countries together require more than 50 per cent of the world's cardamom production. As cardamom is produced in India in sufficient quantities, import is not encouraged by the government to protect the domestic industry. The quantity of cardamom consumed in India and Saudi Arabia in the last 11 years is given in Table 13.3.

Saudi Arabia imports its entire requirements from Guatemala. Other important importing countries are Kuwait, Jordan, Qatar, UAE, USA, Japan, Russia, Singapore, UK, Germany and Scandinavian countries such as Sweden, Finland, Norway and Denmark. The maximum per capita consumption of cardamom is in Saudi Arabia where it is used in the preparation of the traditional drink 'Gahwa'. This drink is also popular in Kuwait, Muscat and Doha. It is reported that 'Gahwa' normally contains 30 per cent of cardamom and the rest coffee powder. It is seen that, sometimes, the ratio of cardamom seed and coffee goes up to 50:50 or even up to 60:40. 'Gahwa' is enjoyed mainly by old and conservative people in the Middle East countries (Sahadevan

Table 13.1 Cardamom production in India and Guatemala (Qty. in tons)

Year	India	Guatemala
1990–91	4750	11,500
1991–92	5000	11,120
1992–93	4250	13,500
1993–94	6600	13,500
1994–95	7000	14,200
1995–96	7900	15,300
1996–97	6625	17,000
1997–98	7900	15,000
1998–99	7170	13,000
1999–2000	9330	10,000
2000–2001	10,480	11,800
2001–2002	11,365	13,500

Sources: India — estimates by Spices Board, Guatemala – FAO, Rome, Embassy of India in Mexico.

Table 13.2 Export of cardamom from India and Guatemala (Qty. in tons)

Year	India	Guatemala
1990–91	400	11,114
1991–92	544	13,163
1992–93	190	13,000
1993–94	387	13,000
1994–95	257	14,000
1995–96	527	15,000
1996–97	226	14,500
1997–98	297	14,500
1998–99	476	12,000
1999–2000	646	8,536
2000–2001	1100	NA

Sources: India – DGCI&S, Calcutta/Shipping bills/Exporters returns. Guatemala: up to 1991–92 Banco De Guatemala and from 1992–93 estimates based on past trends.

1965). Similarly cardamom tea is also popular in Middle East as well as in India (Anonymous, 1952).

In India, cardamom is consumed not only in households, but also in industrial units and institutions. A survey conducted by the Spices Board of India on the use of cardamom in India during 1996 shows its manifold applications (Anonymous, 1997a). But the main purpose behind the use of cardamom is attributable to its cool and refreshing aroma with pleasant and sweet taste. Use of cardamom in Indian households is mainly for preparations like sweets, *payasam, pulav, biriyani*, meat dishes, flavoured tea/milk and as a mouth freshener. Cardamom is also used in many home-made medicines. The average household consumption covering both urban and rural population is about 35 g/year. The household consumption is estimated to grow at an annual compound growth rate of 3.7 per cent and attain the level of 6150 tons by 2000 AD (George and John, 1998).

Though total consumption is less, industrial units are the bulk consumers of cardamom in India. Their preparations comprise of *pan masala*, other *masala* products, herbal medicines, tobacco products, biscuits and similar items and cardamom oil. The demand from the industry was around 2050 tons in 2000 AD at a growth rate of 15 per cent per annum.

Table 13.3 Consumption of cardamom in India and Saudi Arabia (Qty. in tons)

Year	India	Saudi Arabia*
1990	4350	5272
1991	4456	6639
1992	4060	6000
1993	6213	3853
1994	6743	2709
1995	7373	7488
1996	6399	8524
1997	7603	7603
1998	6694	6569
1999	8684	6249
2000	9380	6628

Sources: Spices Board of India; 1990–2000 – UN Statistics.

Note
* Based on import statistics.

Indian institutions consuming cardamom are hotels, restaurants, bakeries, sweet-meat shops, *pan-bide* shops etc. The demand from this segment was increasing at the rate of 10 per cent per annum and reached 1250 tons in 2000 AD. The total demand for cardamom in India under different segments is thus expected to be around 9500 tons in 2000 AD registering a compound growth rate of 6 per cent per annum.

A number of measures have been undertaken in India for widening the demand base. The Regional Research Laboratory, Trivandrum and the Central Food Technological Research Institute (CFTRI), Mysore have taken up studies for developing new products from cardamom. The Indian Institute of Nutrition, Hyderabad has also conducted a study on its nutritional and medicinal values. The Ayurveda College, Trivandrum has carried out a study on the use of cardamom for developing Ayurvedic medicines for various common ailments. The Arya Vaidya Sala, Kottakkal has conducted studies on the effect of cardamom-based soaps on human skin (George and John, 1998).

The Spices Board of India has been in contact with manufacturers of various food products for promoting the use of cardamom. Certain manufacturers have begun to use cardamom flavour in their products while many have conducted tests to assess the suitability of cardamom oil as a flavouring agent in their products. As a result, a variety of new end products using cardamom flavour have been launched in the market such as cardamom-flavoured biscuits, toffee, flan and tea powder and cardamom-flavoured concentrates (George and John, 1998; Joseph and George, 1998).

2   INTERNATIONAL PRICE

In foreign markets the prices of Indian and Guatemalan cardamom are quoted regularly. Prices of Indian cardamom have always been higher than Guatemalan cardamom due to better quality, strong domestic demand for the former and comparatively higher cost of production. A study indicated that prices have been fluctuating too much in the international market. During the last 8 years while the record price of Indian cardamom in the Middle East market was in 1994–95 at US$ 17.39, for a comparable grade of Guatemalan cardamom in the same market for the same period was only US$ 7.66 per kg. The prices of Guatemalan and Indian cardamom in the Middle East market obtained from the International Trade Centre, Geneva are given in Table 13.4.

Table 13.4 Price of cardamom (Bold Green/Bold) in the Middle East market (US $/kg)

Year	Guatemala (Bold Green/Bold)	India
1994–95	7.66	17.39
1995–96	6.33	9.58
1996–97	5.69	10.31
1997–98	4.79	12.02
1998–99	9.17	13.65
1999–2000	13.10	13.96
2000–2001	17.14	16.14
2001–2002	14.92	16.58

Source: Market News Service, International Trade Centre, Geneva.

Guatemalan prices were less than half of Indian cardamom prices in certain years. There is no domestic demand for this commodity in Guatemala and in a good year of production market forces pull down the export prices. Very low cost of production, lack of understanding among exporters and non-existence of a regulatory authority for cardamom marketing as in India are other factors attributed to lower price of Guatemalan cardamom in international market (Anonymous, 1997a,b). But the price differences have been eroded in the recent years.

3 POTENTIAL APPLICATIONS

The future of any commodity depends upon the present and potential uses. Being a weak flavourant, cardamom has no strength to substitute other natural flavours like vanilla. But its certain uses especially as mouth freshener, flavouring biscuits, tea and soft drinks are likely to spread to many countries. In India, many people prefer to use cardamom seeds as an excellent mouth freshener to novelties like chewing gum. Though cardamom seed has no chewing-gum properties, because of its mild exciting taste and acceptable smell, the habit of chewing cardamom is becoming popular in many parts of India. Promotional efforts of the Spices Board of India even go to the extent of using it as a substitute for cigarette and beedi smoking. In cola drinks, the aroma of cardamom is highly acceptable and the testing of consumer preference for cardamom-flavoured cola in the capital of India has been positive (Anonymous, 1996).

Medicinal value of cardamom has not been fully studied in any country. If at all some work has been done, it is in India. *Ayurveda* is the oldest traditional system of medicine in India, as old as Vedas, which mentions the use of cardamom in certain medicinal preparations. It is a stimulant as well as carminative. Ayurvedic formulations containing cardamom is found to be effective against cough and cold. Body massage oil for head, based on cardamom has been found very soothing. But unfortunately much of the known medicinal uses of cardamom are still to be exploited commercially (Anonymous, 1952, Sahadevan, 1965).

At present consumption of cardamom in developed countries such as USA, Canada, UK, Germany, Netherlands, France and Japan is negligible. The total consumption in these countries does not exceed 1000 tons per annum. In the present trend of the movement of people to various countries as tourists or job seekers, the demand for ethnic foods is going up. Cardamom having cool and refreshing aroma and pleasant and sweet taste, the food prepared by adding it is likely to become more acceptable to more people in the coming years. However, certain amount of promotional work in the developed countries will increase the demand for cardamom products.

4   FUTURE OUTLOOK

4.1   Research and development

For successful cultivation of any crop a strong research back up is essential. Cardamom is no exception, as this crop is facing a variety of production constraints, such as drought susceptibility, virus diseases, fungal diseases, changes in the agro-ecological habitat etc. (Nair *et al.*, 1989). For production and productivity enhancement constraint alleviation together with genetic upgradation and high production technology play important roles. Research efforts in these areas can surely do wonders in pushing up the productivity of cardamom. A quantum jump in productivity should be achieved for India to compete in the global market (Anonymous, 1988).

Being a shade-loving plant, its production physiology may be different from other zingiberacious plants. Little is known about the production physiological aspects. Productivity increase can be best achieved by genetic upgradation through new gene combinations and heterosis breeding and subsequent production of hybrid seeds. Production of genetically homozygous lines for heterosis breeding is thus an urgent need. Hence an area of great importance is the production of haploids and dihaploids for hybrid seed production. This step itself can revolutionize cardamom production.

An intensive search is required to locate heat and drought tolerant lines. Heat and drought susceptibility is the most serious production constraint facing cardamom cultivation in India. Once this is achieved, then incorporation of such resistance in elite genotypes can be achieved utilizing the conventional breeding or the haploid – dihaploid – hybrid system.

Resistance to biotic stress factors especially the virus diseases is again extremely important in pushing up productivity. Survey for natural disease escapes in hotspot areas, their screening and evaluation has led to some katte and rhizome rot resistant lines in IISR, some of which are high yielding too. These lines can be exploited for the time being for planting-material production. At the same time the agronomically superior lines have to be subjected to molecular breeding for production of transgenics incorporating *katte* virus resistance either through coat protein mediated resistance or otherwise.

The human society is becoming more and more health conscious and as a result the use of agrochemicals is being discouraged in many crops. The organically grown products are slowly gaining importance at least in the developed countries. A ten-fold increase in consumption is predicted in a few years time. To catch up with this world trend a re-orientation in spices production technology is needed. Much research is needed to identify varieties suitable for organic farming and to streamline the production technology for this purpose.

Genetic engineering is going to revolutionize the future agriculture in a way that mankind has never seen before. This is going to be true of cardamom as well. The future may well witness the creation of super cardamom plants – in yield, resistance to biotic and abiotic stress and quality. New uses, novel products etc. may lead to greater demand for this noble spice necessitating higher production and productivity. Genetic engineering can help achieving this. Tailoring of cardamom plants for tropical climate (resistance to heat and drought) may well make this a crop of hot planes of South India. In due course by using the modern research tools we may be able to create cardamom plants many times superior to the present varieties in yield, quality and other agronomic characters.

4.2 Demand, production and consumption

Till now world demand has been increasing commensurate with the growth in production, and due to this there has been no slump in world market and entire production is consumed in the same year without much left over. In the last 15 years world production has increased almost $2^1/_2$ times and similarly consumption also. During this period the industry has grown many fold benefiting all those who are involved in production, processing and marketing.

It is difficult to predict the future of any non-essential agricultural produce having limited use. However, it appears that cardamom has somewhat a bright future. But any substantial jump in production by any country in a short period will upset the demand–supply position. Such a situation may lead to severe price fall making cardamom production uneconomical to farmers. A near situation of this kind has been experienced in 1997–98 for Guatemalan cardamom when its production was at its peak and price in the Middle East market was very low.

Though cardamom flavour and aroma are acceptable in many foods, consumption in the developed countries such as USA, Japan, European nations, Australia and New Zealand is very low. At present, the use of cardamom as raw material for its processed forms such as oil and oleoresin will not exceed more than 1500 tons per annum in the developed countries. This is mainly because no agency is making efforts to promote cardamom in these countries. As cardamom is produced in developing countries and the industry supports many small farmers, it is desirable to have propaganda and publicity measures implemented for increasing its consumption especially when it is a safe produce with medicinal values. There are a few well-reputed spice importers' associations such as, American Spice Trade Association (ASTA), European Spice Association (ESA) and All Nippon Spice Association (ANSA), which can be approached for promotional work. There are instances of pimento and black pepper; wherein producers and importers have jointly took up programmes in the importing countries for increasing their consumption. Such measures can be done fervently in the case of cardamom.

Other than in India there is very little or no consumption of cardamom worth mentioning in the producing countries. For sound development of agricultural industry, local consumption is a safeguard, as international demand is highly fluctuating with competition from other producing countries. Therefore, efforts are to be taken both in traditional and non-traditional countries for promoting cardamom consumption. The recent liberalization of imports of spices including cardamom into India paves the way for large scale inflow of Guatemalan cardamom into India. This may in turn destabilize the traditional cardamom production and can lead to the decline of cardamom industry. On the other hand cardamom economy can become more competitive by reducing production cost, increasing productivity, and through diversification of its uses (Joseph and George, 1998).

REFERENCES

Anonymous (1952) *The Wealth of India*. Council of Scientific and Industrial Research, New Delhi.

Anonymous (1988) *Economics of Cardamom Cultivation for Small Growers*, Spices Board, Cochin.

Anonymous (1996) *Report of the Expert Committee for Alternative Crops for Cardamom*, Spices Board, Cochin.

Anonymous (1997a) *Market Study of Small Cardamom by Dalal Consultants and Engineers Ltd.*, Madras, Spices Board, Cochin.

Anonymous (1997b) *Spices Statistics* (3rd edn) Spices Board, Cochin.

George, C.K. and John, K. (1998) Future of Cardamom Industry in India. *Spice India*, 11(4), 4. Spices Board, Cochin.

Joseph, K.J. and George, C.S. (1998) *Promotional Measures for Cardamom Cultivation*, Centre for Development Studies, Trivandrum.

Nair, K.N. *et al.* (1989) *Ecology and Economics in Cardamom Development*, Centre for Development Studies, Trivandrum.

Sahadevan, P.C. (1965) *Cardamom*. The Agricultural Information Service, Trivandrum, Kerala.

14 Yield gaps and production constraints in cardamom

P.N. Ravindran and P. Rajeev

1 INTRODUCTION

There is immense potential and technological feasibility to increase cardamom yield in India. In spite of notable increase in total output of the crop over years, the average national productivity is strikingly low. The productivity has shown only a marginal increase from 53 kg/ha in 1989–90 to about 113 kg/ha in 1994–95, and 174 kg/ha as on today. While the average productivity realised from Kerala and Tamil Nadu are slightly higher (145 kg/ha and 179 kg/ha respectively), it is only 62 kg/ha in Karnataka state. In contrast, the average productivity in Guatemala is reported to be about 350 kg/ha (Anonymous, 1996).

Cardamom economy in the Western Ghat region of India supports a sizeable proportion of small farm sector and agricultural labour and there is an urgent need to improve and sustain the production and productivity. For this the basic step is to identify the yield constraints and the yield gaps existing between average productivity and potential productivity at various sectors.

2 YIELD GAPS

Research on cardamom cultivation is being undertaken in various experiment stations for the last four decades. The results have shown high prospects for raising the yield levels. For instance, two experiments were laid out in 1980, one each under rainfed and irrigated conditions at the Cardamom Research Centre under the Indian Council of Agriculture Research located in Coorg district of Karnataka. These trials reported an average dry yield of 162 kg/ha and 650 kg/ha under rainfed and irrigated conditions respectively. The higher productivity achieved was due to the adoption of scientific cultivation practices like high yielding planting materials, timely application of manures/fertilisers, irrigation, other cultural practices and plant protection measures (Korikanthimath and Venugopal, 1989).

The above project was followed up through adaptive trials laid out in two estates in Coorg district, previously managed under two situations. In one trial, under superior management, registered a highest crop of 850 kg/ha in 1983, 2 years after planting. In contrast, the yield was only 52 kg/ha under the system of conventional management. In the second trial of a pure crop of cardamom, again under superior management, a highest crop of 1625 kg/ha was recorded, three years after planting.

Following the encouraging results obtained in experiments and adaptive trials, demonstration plots were laid out in two districts, Kodagu and Hassan in Karnataka state. The IISR Cardamom Research Centre at Kodagu provided technological back up while the Spices Board under the Ministry of commerce provided financial and extension support. The project involved demonstrations of the production technology in selected farms, training of extension workers in high production technology, training of farmers and regular follow up through farmers' meetings and field visits. A total of 42 trials were laid out covering an area of 94.1 ha in five cardamom growing taluks/blocks. An average dry cardamom yield of 494.5 kg/ha was reported in the maiden crop in 1988, 2 years after planting (Korikanthimath *et al.*, 1989).

Apart from evidences from research and demonstration trials, a recent report from the Spices Board indicated that there are large, medium and small growers whose productivity varies from 500–1000 kg/ha under situations of scientific management of the crop. The Spices Board has introduced the best Cardamom Grower Award, given to a grower annually. The yield levels achieved by the award winning cardamom growers are given below (John, 2000).

Year	Award winners yield (kg/ha)	National average yield (kg/ha)
1994–95	1035	113
1995–96	1650	128
1996–97	1087	125
1997–98	1925	149
1998–99	1875	135
1999–2000	2475	174

Such high yields are the result of superior planting material and plantation management. Thus 2475–2500 kg/ha is an achievable yield level under ideal soil, climate and management conditions. Such high yield levels are possible when plantations are raised from clonal progenies of high yielding mother clumps. In all the above award winning cases the cultivar used was *Njallani Green Gold*, a selection by a cardamom grower. On the other hand when plantations are raised from seedling progenies, which is a common practice among growers, their average performance will be much lower as in such population only about 15 per cent of the plants are found to be high yielders that contribute to about 32 per cent of the total yield. Even out of this, 4 per cent of plants only yield 500–900 g/plant, and they contribute about 15 per cent of the yield (Krishnamurthy *et al.*, 1989). To push up the average populational performance the top 4 per cent or the best 1 per cent can be multiplied clonally and used for raising plantation.

There are super clumps having very high yield potential. Some such clumps were found to give around 10 kg (fresh yield) annually (about 3 kg dry) (Korikanthimath, Personal com.). A plantation raised from such super clumps can attain an yield potential of about 6000 kg/ha. With proper micropropagation it should be possible to multiply on a large scale such super clumps for planting. If such plants are provided with high-tech management, it should be possible to reach an yield figure of 5000 to 6000 kg/ha.

High yield levels have also been reported from Papua New Guinea. Though only a few plantation exist there, some of them have attained an yield level of 2000 kg/ha (Krishna, 1997). This infact is a realisable figure under good management and irrigation.

The yield gaps existing can be represented as in Fig. 14.1. High production levels could be achieved only through a combination of (i) superior planting material, (ii) superior management and (iii) ideal combination of edaphic and climatic factors.

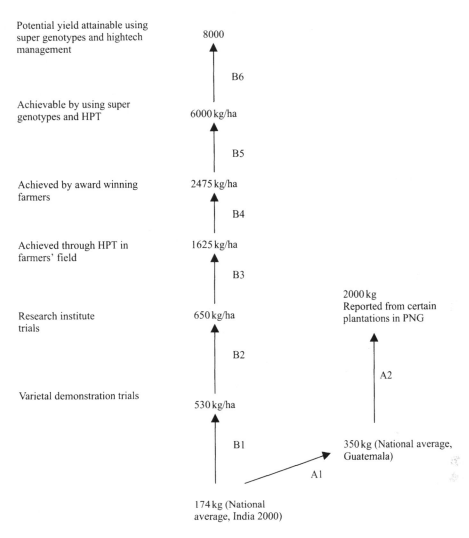

Figure 14.1 A1 and A2 are gaps existing between national average (India) and average yield in Guatemala and that of the reported yield in certain plantations in Papua New Guinea (PNG). B1–B6 are components of yield gap existing between the national average in India and the highest achievable using super genotypes and superiror production technology.

So areas where ideal edaphic – climatic combination exists should be identified for practising the input – intensive production technology. A single technology may not be suitable for all locations and conditions. So location and situation-specific production technology has to be evolved. Such production technology should approach the problem holistically, so that the constraints existing in the location can be tackled appropriately.

3   PRODUCTION CONSTRAINTS

There are wide gaps between potential, possible and actual yield levels of the crop. It is essential to identify the possible reasons for low yield in farms, before chalking out

Table 14.1 Constraints in cardamom production

Area	Constraints	Classification	Percentage of response	Rank
Planting material and nursery management	Low germination percentage, scarcity of quality planting materials, failure of tissue cultured plants	Knowledge constraint	70	3
		Infrastructure constraint	30	7
		Conviction constraint	20	9
Nutrition management	Scarcity in organic manure, quality aspects, transportation	Infrastructural constraint	65	4
Soil erosion	Leaching and gully formation	Knowledge constraint	25	8
Pest and disease management	Detection and management of *Katte*, persistence of stem borer and root grub	Knowledge constraint	75	2
Labour management	Changes in labour availability	Socio-economic constraint	80	1
Marketing	Absence of local auction centre and price fluctuation	Socio-economic constraint	60	6

a strategy to bridge the wide yield gaps. Only a few studies and reports are available dealing with the major constraints to profitable production of cardamom. A brief attempt is made here to arrive at a holistic understanding of those major production constraints. A study conducted by Rajeev and Korikanthimath (1998) brought out the major constraints faced by growers of the Coorg region (Table 14.1), which can be broadly classified under:

- Technological factors
- Socio-economic factors
- Ecological concerns.

These constrains are equally applicable to the cardamom growing regions of Kerala and Tamil Nadu as well.

3.1   Technological factors

Adoption of improved cultivation practices is a pre-requesite for realising higher yields in cardamom. Research stations located at Myladumpara and Pampadumpara (Kerala state) and Appangala and Mudigare (Karnataka state) have carried out extensive research on technology development in cardamom, with respect to selection and improvement of planting material, chemical fertiliser and manurial trials, chemical and biological control of pests and diseases, as well as post harvest technology. However, there is widespread contention that these technologies have not diffused to field level. Even now, large-scale incidences of pests and diseases, poor management and senile and poor-yielding plantations are cited as the factors responsible for low productivity.

A report from IISR Cardamom Research Centre, Coorg district (Rajeev and Korikanthimath, 1999) has identified the following reasons for low productivity.

- use of unselected planting material
- sparse plant population
- inadequate shade management
- non-adoption of timely cultural practices
- little or no use of fertilisers
- meagre plant protection measures
- over dependence on monsoon
- recurring drought and lack of irrigation
- faulty harvesting and processing methods.

A vast area under cultivation still depends upon monsoon seasons; south-west and north-east monsoons received from June–November. Cardamom is highly sensitive to micro-climatic conditions and moisture stress. When raised rainfed, the crop responds adversely to recurring climatic vagaries, erratic monsoon and prolonged drought. The data from Centre for Development Studies, Trivandrum, Kerala (Nair *et al.*, 1989) indicated sharp decrease in production and productivity during certain years (1976–77, 1982–83 and 1983–84), which were years of severe drought. In cardamom hills of Kerala and Annamalais in Tamil Nadu state, rainfall is around 300 cm, while in certain other tracts rainfall exceeds 400 cm. Under both conditions crop is exposed to prolonged dry spell from January to May. Drought during this period severely affects the critical stages of crop growth; flowering, panicle initiation and fruit set. Hence protective irrigation beginning from January and contining up to May is highly essential for ensuring high yield.

Katte, the viral disease has spread rapidly in most cardamom growing tracts. The effect of such diseases are devastating that the yield of infected clump is reduced by 99 per cent within 3 years of incidence and the plant is completely destroyed within 5 years. It is possible to keep the level of incidence below 5 per cent through effective phytosanitary measures. However, the management of the disease is found to be inadequate in majority of plantations. A recent survey carried out by Cardamom Research Centre, Appangala (Venugopal, 1999) reported that over 66 per cent of surveyed estates showed virus infections to varying degrees. In 60 per cent of estates, the incidence was reported as high as 63 per cent. In most of the estates surveyed growers did not follow any systematic phytosanitation or management measures. Similarly the *Azhukal* (rot) disease, widely occurring in Kerala, a new viral disease, *Kokka Kandu,* reported from isolated tracts in Karnataka and persistent attack of thrips are also reported to be factors of serious concern.

Thus it is amply clear that unless the pests and diseases are managed properly and systematic removal and replanting of severely virus-infected estates are taken up, it may not be possible to better the current yield levels.

3.2   Socio-economic factors

Being a cash crop, factors related to levels of production, movement of prices and variation in labour market have got a significant bearing on cardamom economy.

As cardamom is mainly raised rainfed, fluctuations in rainfall, both its quantum and distribution and conditions of drought during dry spell affect the total output. Fluctuations in production level is in turn associated with variations in prices. Being an export crop, factors operating in international market also influence prices. Fluctuation in prices has an implicit bearing on farmers' responses on crop management, new planting and replanting. For instance, the range of domestic price of cardamom in the last decade varied from Rs. 180 per kg to Rs. 700 per kg. Such frequent fluctuation in prices serves as a strong disincentive for consistent management of estates. A report from Spices Board indicated that Rs. 250 per kg was the minimum incentive price at which the farmer would invest on improved technology.

In any plantation crop, a productive labour force is essential. Cardamom is a highly labour intensive crop which requires year round employment of manual labour to carry out cultural operations like shade lopping, mulching, trashing, weeding, irrigation and harvesting. Labour wages constitute about 60 per cent of total production cost. In most of the cardamom growing tracts there is growing problem of labour shortage. Escalating labour wages, decline in conventional family labour and permanent labour system, emergence of contract labour and exploitation by middle men (labour contractors) are some of the recent changes in labour market.

In short, risk due to price fluctuations, escalating cost of production and increasing labour shortage are emerging as some of the major constraints that would influence cardamom economy in future.

3.3   Ecological concerns

There is an emerging school of thought that unfavourable changes in ecological foundations in the tropical evergreen forest ecosystem is one of the possible reasons for stagnation in productivity. The rate of deforestation due to anthropogenic pressures in cardamom hill reserves has been significantly high. A study from the Centre for Earth Science, Trivandrum has reported that 87 per cent of the tract in Idukki, a dominant cardamom tract in Kerala was covered by forests in 1905 (Chatopadhyaya and Srikumar, 1984). Within a span of 70 years the forest cover declined to a mere 33.4 per cent. The large scale removal of forest cover has adversely affected the micro-climate conducive for growth and production of cardamom. Such vast devastation led to severe aridity effect in Idukki district as a result of hot winds from hinterlands east of the Western Ghats, blowing across the district unhindered. The salubrious climate of the famed cardamom hills have given way to the present unsuitable climatic situation. As a result there was extensive damages to the cardamom plantations, year after year, making cardamom growing rather uneconomical, and subsequently leading to shifts in cropping systems. Changes in cropping pattern, human encroachment and changes in agricultural methods followed, all have influenced the ecological processes in the region. The social cost, cultivators have to bear due to such changes, is difficult to be worked out. Again, these changes have demographic, socio-economic and political dimensions.

4   STRATEGIC IMPLICATIONS

Indian cardamom industry is experiencing a phase of crisis due to low productivity and declining trend in exports. Traditional cultivation practices and escalating cost of production are the major problems of the sector. With a view to develop cardamom industry, following strategic measures have to be taken up by development agencies. Considering the technological feasibility to increase the productivity per unit area and to reduce per kg cost of production, there is great scope for reviving the sector through the following measures:

(a)   Full exploitation of irrigation potential in hill zones and enhancing area under assured irrigation.
(b)   Use of improved quality planting material.
(c)   Popularise integrated pest and disease management through extensive campaigns.
(d)   Adoption of timely cultural operations.
(e)   Wide-ranging programmes to demonstrate the profitability of improved technologies.
(f)   Organizational innovations like training and legal provisions to improve labour productivity.
(g)   Promotion of domestic market of cardamom.

The following strategies will be useful in breaking the present yield barriers existing in cardamom production.

(a)   Survey of cardamom gardens for locating super clumps that are capable of yielding above 2 kg dry capsule regularly (that is more than 6 kg fresh capsules).
(b)   Large scale multiplication of such super clumps using micropropagation or by conventional methods.
(c)   Distribution of the plantlets of super clumps for planting and for replanting programmes.
(d)   Adoption of the high production technology combined with integrated nutrient, pest and disease management.
(e)   Systematic rouguing of poor, unhealthy plants and off types, and virus-infected ones if any, and thus maintaining the plantation at a high level of yield and efficiency.
(f)   Proper post-harvest processing to achieve high quality of produce.

By adopting the above strategies it should be possible to raise the average yield level of plantations above 500 kg/ha, and the cost of production will also go down substantially. A concerted effort from the part of R & D organizations can surely achieve this target.

REFERENCES

Anonymous (1996) Report of the Expert Committee for Alternative Crops for Cardamom. Spices Board, Cochin.
Chatopadhyaya, S. and Srikumar, P. (1984) Deforestation in parts of Western Ghats region (Kerala, India). *Tech. Bull.*, Centre for Earth Science Studies, Trivandrum.
John, K. (2000) Njallani green gold at cardamom productivity helm with precision farming techniques. In *Spices and Aromatic Plants*, ISS, Calicut, pp. 105–106.
Korikanthimath, V.S. and Venugopal, M.N. (1989) High production technology in cardamom. *Tech. Bull.*, National Res. Centre for Spices, Calicut.

Korikanthimath, V.S., Venugopal, M.N. and Naidu, R. (1989) Cardamom production – a success story. *Spices India*, 11(9), 19–24.

Krishna, K.V.S. (1997) Cardamom Plantations in Papua New Guinea. *Spice India*, 10(7), 23–24.

Krishnamurthi, K., Khan, M.M., Avadhani, K.K., Venkatesh, J., Siddaramaiah, A.L., Chakravarthy, A.K. and Gurumurthy, B.R. (1989) Three decades of Cardamom Research at RRS., Mudigere. *Tech. Bull.*, Univ. Agri. Sci., Bangalore.

Nair, K.N., Narayana, D. and Sivanandan, P. (1989) Ecology or economics in cardamom development. *Tech. Bull.*, Centre for Development Studies, Trivandrum.

Rajeev, P. and Korikanthimath, V.S. (1999) Constraint analysis in cardamom production – a system approach. *IISR Annual Report* (1998) IISR, Calicut.

Rajeev, P. and Korikanthimath, V.S. (1998) In IISR Annual Report (1997–98), IISR, Calicut.

Venugopal, M.N. (1999) Natural disease escapes as source of resistance against cardamom mosaic causing *katte* disease of cardamom (*Elettaria cardamomum* Maton). *J. Spices and Aromatic Crops*, 8, 145–151.

15 Large cardamom (*Amomum subulatum* Roxb.)

S. Varadarasan and A.K. Biswas

1 INTRODUCTION

Large cardamom (L. cardamom) or Nepal cardamom (*Amomum subulatum* Roxb.) is a spice cultivated in the sub-Himalayan region of north-eastern India, especially in Sikkim since time immemorial. In the past the aboriginal inhabitants of Sikkim, *Lepchas*, collected capsules of large cardamom from natural forest, but later on these forests passed into village ownership and the villagers started cultivation of large cardamom. The presence of wild species, locally known as *churumpa*, and the variability within the cultivated species supports the view of its origin in Sikkim (Subba, 1984). Later the cultivation has spread to northern Uttar Pradesh, north-eastern States of India (Arunachal Pradesh, Mizorum and Manipur), Nepal and Bhutan. Sikkim is the largest producer of large cardamom; the annual production in India is about 3500–4000 mt of cured L. cardamom. The average productivity is 100–150 kg/ha, but in well-maintained plantations the productivity reaches 1000–2000 kg/ha. Nepal and Bhutan are the other two countries cultivating this crop with an annual production of about 1500 mt. This spice is used in Ayurvedic preparation in India as mentioned by Susruta in the sixth century BC and also known among Greeks and Romans as *Amomum* (Ridley, 1912). L. cardamom contains ca.1.98–2.67 per cent volatile oil and is mainly used in flavouring food products (Gupta *et al.*, 1984). The seeds also possess certain medicinal properties, as carminative, stomachic, diuretic, cardiac stimulant, antiemetic and are a remedy for throat and respiratory troubles (Singh *et al.*, 1978).

The commodity is mainly sold in the domestic markets of northern India. Over the past few years large cardamom is also being exported, and in 1997–98, Rs. 12.6 crores of foreign exchange was earned by exporting 1784 tonnes. Pakistan, Singapore, Hongkong, Malaysia, UK and Middle East countries are the major importers of this commodity.

2 HABIT AND HABITAT

Amomum subulatum is a perennial herb that belongs to the family Zingiberaceae under order Scitaminae. The plant consists of subterranean rhizomes and several leafy aerial shoots/tillers (leafy shoot, Fig. 15.1). Number of such rhizhomatous leafy shoots varies between 15–140 in a single plant (a clump). Height of leafy shoot ranges from 1.7 to 2.6 m depending on cultivar and possess 9–13 leaves in each tiller. Leaves are distichous, simple, linear and lanceolate, glabrous on both sides with a prominent mid rib. Inflorescence is a condensed spike on a short peduncle (Fig. 15.2). Flowers are

Figure 15.1 A clump of large cardamom in flower.

Figure 15.2 A tiller with three inflorescences (spikes).

bracteate, bisexual, zygomorphic, epigynous, and cuspinated. The yellowish perianth is differentiated into calyx, corolla and anther crest. Each spike contains about 10–15 fruits (capsules) and rarely up to 20–25 capsules, depending on cultivars. Flowering season begins early at lower altitude with peak flowering during March–April, whereas it starts at higher altitudes in May with a peak during June July. Harvesting begins during August–September at lower altitudes and in October–December at higher altitudes (Gupta and John, 1987).

The fruit is a round or oval shaped capsule, trilocular with many seeds. Capsule wall is echinated, reddish brown to dark pink. Seeds are white when immature and become

Figure 15.3 A view of a large cardamom field affected by snow fall.

dark grey to black towards maturity. The capsules formed at the basal portion of spike are bigger and bolder than others (Rao *et al.*, 1993).

L. cardamom is grown in cold humid conditions under shade of trees at an altitude between 800–2000 m above mean sea level (amsl) with an average precipitation of 3000–3500 mm, spread over about 200 days and the temperature ranging between 6 °C in December–January to 30 °C in June–July (Singh, 1988). Frost, hailstorm and snowfall are the major deleterious factors affecting L. cardamom (Fig. 15.3). The crop grows well in moist, but well drained loose soil. The depth of soil varies from a few centimeters to several meters depending upon the topography and soil formation. Soil is acidic and rich in organic matter (Mukherji, 1968).

3 CULTIVARS

All the cultivars of L. cardamom cultivated commercially belong to the species, *Amomum subulatum* Roxburgh. Out of the total 150 species of *Amomum* occurring in the tropics of old world, only about eight species are considered to be native of eastern Sub-Himalayan region viz., *A. subulatum* Roxb., *A. costatum* Benth., *A. linguiformae* Benth., *A. pauciflorum* Baker, *A. corynostachyum* Baker, *A. dealbatum* Roxb. (*A. sericeum* Roxb.), *A. kingii* Roxb. and *A. aromaticum* Roxb (Hooker, 1886). Later 18 species of *Amomum* were reported from the north-eastern Himalayan regions (Anonymous, 1950). In the Indian subcontinent itself there is another centre of diversity in the Western Ghats region in the South West India. Gamble (1925) has reported six species from this region.

There are mainly five cultivars of L. cardamom viz., *Ramsey, Sawney, Golsey, Varlangey* (*Bharlangey*) and *Bebo* (Gyatso *et al.*, 1980). They are well known. Some other sub-cultivars of the above ones (*Ramnag, Ramla, Madhusey, Mongney* etc.) are also seen in cultivation in small areas in Sikkim State. Another cultivar *Seremna* or *Lephrakey* (a *Golsey* type) is also getting importance and is spreading to more areas in lower altitudes (Upadhyaya and Ghosh, 1983).

3.1   *Ramsey*

The name *Ramsey* was derived from two Bhutia words – '*Ram*' meaning Mother and '*sey*' for Gold (yellow). This cultivar is well suited for higher altitudes, even above 1500 m. on steep slopes. Grown up clumps of 8–10 years age group possess 60–140 tillers. The tillers colour is maroonish green to maroon. Second half of May is the peak flowering season. Capsules are small, the average being 2.27 cm in length with 2.5 cm diameter, with 30–35 capsules in a spike, each containing 16–30 seeds. The harvest is during October–November. Peak bearing of capsules is noticed in alternate years.

This cultivar is more susceptible to viral diseases like *foorkey* and *chirke* especially if planted at lower altitudes. It occupies a major area under L. cardamom in Sikkim and Darjeeling district of West Bengal. Two strains of this cultivar viz., *Kopringe* and *Garadey* from Darjeeling district having stripes on leaf sheath, are reported to be tolerant to *chirke* virus (Karibasappa *et al.*, 1987).

3.2   *Sawney*

This cultivar got the name from *Sawan* in Nepali, corresponds to August by which month this becomes ready for harvest at low and mid altitudes. This cultivar is widely adaptable, especially suited for mid and high altitudes i.e. around 1300–1500 m. It is robust in nature and consists of 60–90 tillers in each clump. Colour of tillers is similar to *Ramsey*. Each productive tiller on an average produces two spikes. Average length and diameter of a spike is 6 and 11 cm. Flowers are longer (6.23 mm) and yellow in colour with pink veins. Second half of May is the peak flowering time (Rao *et al.*, 1993).

Capsules are bigger and bold and number of seeds in each capsule are more (35) than in *Ramsey*. Harvest begins in September–October and may extend up to November in high-altitude areas. This cultivar is susceptible to both *chirke* and *foorkey* viral diseases. Cultivars such as *Red Sawney* and *Green Sawney* derived their names from capsule colour. *Mongney*, a strain found in south and west districts of Sikkim, is a non-robust type with its small round capsules resembling mostly that of *Ramsey*.

3.3   *Golsey (Dzongu Golsey)*

The name has derived from Hindi and Bhutia words; '*Gol*' means round and '*Sey*' means gold. This cultivar is suitable to low altitude areas below 1300 m AMSL especially in Dzongu area in North Sikkim. Plants are not robust like other cultivars, and consist of 20–50 straight tillers with erect leaves. Alternate, prominent veins are extended to the edges of leaves (Biswas *et al.*, 1986). Unlike *Ramsey* and *Sawney*, tillers are green in colour. Each productive tiller on an average produces two spikes. Flowers are bright yellow. On an average each spike is 5.3 cm long with 9.5 cm diameter and contains an average of seven capsules. Capsules are big and bold, 2.46 cm in length and 3.92 cm in diameter and contain about 60–62 seeds. This cultivar becomes ready for harvest in August–September.

Golsey is tolerant to *chirke* and susceptible to *foorkey* and leaf streak diseases. The cultivar is known for its consistent performance though not a heavy yielder. Many local cultivars are known in different locations such as *Ramnag* from north Sikkim. *Ram* meaning 'mother' and *Nag* for black, which refers to its dark pink capsules. *Seto-Golsey* is from west district of Sikkim with robust leafy stems/tillers and green capsules. *Madhusey* with elliptic and pink coloured capsules is having robust leafy stem and has sweet seeds compared to other cultivars (Rao *et al.*, 1993).

3.4 *Ramla*

The plants are tall and vigorous like *Ramsey* and have capsule characters like *Dzongu Golsey*; the colour of tiller is pink like *Ramsey* and *Sawney*. Cultivation is restricted to a few mid-altitude plantations in north Sikkim. The capsules are dark pink with 25–38 seeds per capsule. *Ramla* appears to be a natural hybrid between *Dzongu Golsey* and *Ramsey*. They are susceptible to *foorkey* but are moderately tolerant to *chirke* disease.

3.5 *Varlangey*

This cultivar grows in low, medium and high altitude areas in South Regu (East Sikkim) and at high altitudes at Gotak (Kalimpong subdivision in Darjeeling district of West Bengal). Its yield performance is exceptionally high at higher altitude areas i.e. 1500 m and above. It is a robust type and total tillers may range from 60 to 150. Colour of tillers is like that in *Ramsey* i.e. maroonish-green to maroon towards collar zone; girth of tillers is more than that of *Ramsey*. Each productive tiller on an average produces almost three spikes with an average of 20 capsules/spike. Size of capsules is bigger and bold with 50–65 seeds. Harvest begins in last week of October. This cultivar is also susceptible to *foorkey* and *chirke* diseases.

3.6 *Bebo*

This cultivar is grown in Basar area of Arunachal Pradesh. The plant has unique features of rhizome and tillering. The rhizome rises above the ground level with roots penetrating deep into the soil and the young tillers are covered under thick leafy sheath. It is supposed to be tolerant to *foorkey* disease. The spikes have relatively long peduncle (10–15 cm) and the capsules are bold, red or brown or light brown; seeds contain low level of essential oil (2 per cent v/w) (Dubey and Singh, 1999).

3.7 *Seremna* (*Sharmney* or *Lephrakey*)

This cultivar is grown in a small pocket at Hee-Gaon in west Sikkim at low altitude and is known for its high yield potential. Plant features are almost similar to *Dzongu Golsey* but the leaves are mostly drooping, hence named as *Sharmney*. Total tillers range from 30 to 49 and is not robust in nature. On an average 2–3 spikes emerge from each productive tiller with an average of 10.5 capsules per spike, each having 65–70 seeds.

Comparative morphological characters of the four most important cultivars – *Ramla*, *Ramsey*, *Sawney* and *Golsey* are given in Tables 15.1 and 15.2.

Table 15.1 Growth performance of *Ramla, Ramsey, Sawney* and *Golsey* cultivars of large cardamom (average of 3 years)

Cultivars/Characters	Ramla	Ramsey	Sawney	Golsey	CD at 5% level
Plant height (cm)	200.83	192.08	196.00	190.05	6.54
Number of tillers per plant	59.08	42.00	40.50	39.80	12.67
Number of spikes per plant	40.25	36.00	35.00	30.00	4.70
Spike length (cm)	7.06	7.00	6.40	6.50	0.49
Spike breadth (cm)	8.60	7.00	6.00	7.20	1.47
Number of capsules per spike	16.00	12.00	13.00	14.00	1.12
Fresh capsule yield/plant (g)	375.00	185.00	190.00	216.00	14.00
Dry capsule yield/plant (g)	70.00	47.00	48.00	52.00	15.90

Table 15.2 Capsule characteristics in different cultivars of large cardamom (average of 3 years)

Cultivars/Characters	Ramla	Ramsey	Sawney	Golsey	CD at 5% level
Fresh weight of a single capsule (g)	4.00	3.50	4.00	4.50	0.56
Dry weight of a single capsule (g)	0.90	0.75	0.85	1.00	0.14
Moisture percentage	13.00	14.00	13.00	15.00	1.54
Number of seeds per capsule	38.00	36.00	35.00	40.00	3.28
Percentage of volatile oil	2.67	2.50	2.00	1.98	0.48

4   PLANT PROPAGATION

Propagation of L. cardamom is done through seeds, rhizomes (sucker multiplication) and tissue culture techniques. Cultivars suitable for specific areas, altitudes, agro-climatic conditions and mother plant/clump of known performance are selected for collection of seed, rhizome and vegetative bud.

4.1   Nursery practices

4.1.1   *Propagation through seeds*

Healthy plantation, free from viral disease in particular, is selected for seed capsules. Gardens with productivity of 1000 kg/ha or more during the past 3 years are considered. Higher number of spike bearing (reproductive) tillers per plant (bush), higher number of spikes and capsules, bold capsules, higher number of seeds per capsule etc. are some of the criteria looked into for selecting a plot for collection of seed capsule.

Spikes are harvested at maturity and seed capsules are collected from the lowest two circles in the spike. After dehusking, the seeds are washed well with water to remove mucilage covering of seeds, mixed with wood ash and dried under shade.

The dried seeds are treated with 25 per cent nitric acid for 10 min for early and higher percentage of germination (Gupta, 1989). The acid-treated seeds are washed thoroughly in running water to remove the acid residue and are surface dried under shade. The seeds are sown immediately after acid treatment.

4.2   Nursery site selection

An open area with gentle slope and having facility for irrigation is selected for nursery establishment. L. cardamom nursery is raised in two stages viz., primary nursery and secondary nursery. Seedlings raised in primary nursery (by seeds) are transplanted to the secondary nursery beds or to polybags.

4.2.1   *Primary nursery*

There are two seasons for sowing seeds: pre-winter (September–October and early November) and post-winter (late February–March). September sowing results in quicker and better germination (within 25–30 days), 30–40 per cent for acid-treated seeds (Gupta, 1989; Anonymous, 1998a).

Seedbed is prepared in a well-drained area. The soil is cut to a depth of 30 cm and exposed to sun for a week. Bed of 15–25 cm height, 90 cm width and convenient length is prepared, incorporating well-decomposed compost/cattle manure. Seeds (acid treated) at the rate of 100 gm for a bed size of 1 × 3 m are sown in furrows along the width and they are covered with a thin layer of soil. The space between the furrows is maintained at 10 cm. After sowing, the beds are covered with thick mulch (with paddy straw or dry grass) and watered regularly to keep the bed moist.

The seedbeds are examined for germination between 25–30 days after sowing. Once the germination starts the following operations are made:

(a) Overhead shading of convenient height with Bamboo poles and agro-shade net/bamboo mat (50 per cent shade, black agro-shade net is found ideal).
(b) The mulch from the bed is removed and cut into small pieces and are spread over in between the seed rows.
(c) The beds are watered regularly to keep the bed moist and weeding is attended whenever necessary.

Once seedlings in the primary seed bed reach 3–4 leaf stage (in February/March if seeds are sown in September/October or April/May if the seeds are sown in February/March) they are transplanted either to polybags or into secondary nursery beds in February/March or April/May respectively (Gupta, 1989).

4.2.2   *Polybag nursery*

Topsoil in virgin land/forest area rich in leaf mold is collected and mixed with well-decomposed cattle manure to get good potting mixture. A potting mixture of 5:1, topsoil: cattle manure is prepared and filled in polybags of 8″ × 8″ size with perforations at the base for drainage. Polybags are arranged under overhead shaded shed. Primary seedlings are transplanted in polybag (one seedling per polybag) during February/March or April/May.

Polybag seedlings are watered regularly with rose can to keep the soil moist. Care is taken to cover the collar region or exposed roots of seedlings with thin layer of topsoil, which help in better anchorage and tillering. Over watering should be avoided. Polybag seedlings attain a height of 30–40 cm with 2–4 tillers by July–August if transplanting is done in February/March. These seedlings are planted in the main plantation in

July/August. The polybags are removed and seedlings with the soil ball intact are planted. Sometimes the seedlings are maintained in polybag till April next year and planted in May–June in the main field (Anonymous, 1998a).

4.2.3   Secondary nursery

Seedlings from primary nursery are sometimes transplanted in beds. Beds of the size and nature similar to that of primary beds are prepared; and seedlings at 3–4 leaf stages are transplanted in March/April/May, maintaining a spacing of 15 cm between the seedlings. A layer of well-decomposed cattle manure is applied and incorporated in the soil. Watering is given at regular intervals to keep the soil moist. The entire secondary nursery is maintained under overhead shade (preferably with black agro-shade net). Seedlings are maintained for 10–12 months. Expected growth of seedlings is about 45–60 cm heights with 5–10 tillers each. These seedlings are transplanted in June–July in the main field (Anonymous, 1998a).

4.3   Propagation through rhizome

High yielding, disease free planting materials are selected for multiplication. Trenches of two feet width, two feet depth and convenient length are made across the slopes. Trenches are filled with topsoil, leaf mold and decomposed leaf litter. Rhizomes with one mature tiller and two young shoots or vegetative buds are planted at a spacing of three feet in the trenches during June–July. Thick mulching with dry leaf/grass is applied at the base of the rhizome and watering if required, is done regularly to keep the soil moist.

Once fresh vegetative buds appear, well-decomposed cattle manure is applied one foot around the rhizome and incorporated to the soil. The rhizome multiplication plot is maintained with 50 per cent shade, either under shade trees or under agro shade-net. When the rhizomes are planted in June/July, about 15–20 tillers are produced from each of the rhizome within 6 to 10 months. Each such clump is split into units of two to three tillers and are used for planting in the main field during June/July or used for further multiplication (Anonymous, 1998a).

4.4   Micropropagation

L. cardamom can be multiplied on a large scale through micropropagation. Protocols for micropropagation were developed at Indian Institute of Spices Research (Sajina et al., 1997; Nirmal Babu et al., 1997).

Axillary buds of 0.5–2 cm lengths from promising, virus disease free mother plants are used as e xplants. The explants are thoroughly washed in clean running water and then in a detergent solution and treated in 0.15 per cent $HgCl_2$ for 2 min, and then passed through absolute alcohol for 30 sec. These are cultured using the modified MS medium, solidified with agar and with the following adjuvant.

Step 1:   For initial bud development and its growth *in vitro* (culture period 6–8 weeks): kinetin 3–5 mg/l + IBA 1–2 mg/l + sucrose 20 g/l

Step 2:   For proliferation of the auxiliary bud rhizome (6–8 weeks): BAP 2 mg/l + NAA 3–5 mg/l + sucrose 20 g/l

Step 3:   For rooting and establishment of plantlets (6–8 weeks): IBA 1–2 mg/l + KN 3–5 g/l + sucrose 20 g/l.

Nirmal Babu *et al.* (1997) and Sajina *et al.* (1997) accomplished both multiple shoots and rooting in the same medium, in MS basal + BAP (1 mg/l) and IBA (0.5 mg/l) with 3 per cent sucrose and gelled with 0.7 per cent agar at pH 5.8 and 12 h photoperiod at a light intensity of 2500 lux. This combination produced 8–12 shoots per culture and roots per shoots.

After the plantlets attain 4–5 leaf stage with good rooting, they are removed from culture flasks, thoroughly rinsed in distilled water and sprayed with Bavistin or Dithane M-45 (0.25 per cent) to check fungal infection, and planted in micropots (8 cm) filled with sterile peat moss and vermiculite (1:1 v/v) or 1:2:1 mixture of sterile sand, peat and perlite. Liquid fertilizer in very low concentration is given after 4–5 days and there after at 15 day intervals. Plantlets are maintained under shade (5000–8000 lux) and high humidity (>90 per cent RH at 25 °C). After about two months plantlets can be transferred to polybags filled with potting mixture or to secondary nursery beds and maintained under shade (10,000–15,000 lux).

Each axillary bud (explant) gives 12–18 plantlets in one subculture; 3–4 subcultures are possible with the initial explant to produce normal TC plantlets, which perform well under field condition. The *in vitro* method is also useful in the conservation of L. cardamom germplasm through slow growth. The protocol for the *in vitro* conservation developed at Indian Institute of Spices Research (IISR), Calicut consists of *in vitro* generated plantlets maintained in 1/2 MS + 10 g/l sucrose + 10 g/l mannitol (Nirmal Babu *et al.*, 1997).

5   PLANTATION MANAGEMENT

5.1   Soil condition, preparation of land and shade development

L. cardamom is grown in forest loamy soils having soil depth a few inches to several feet. Colour of soil ranges from brownish yellow to dark brown; in most cases from dark yellowish brown to very dark grayish brown. Texture is sandy, sandy loam, silty loam or clay. In general, soil is acidic having pH ranging from 5 to 5.5 or more, and with 1 per cent or more organic carbon (Bhutia *et al.*, 1985; Biswas *et al.*, 1986). On an average, these soils are high in available Nitrogen and medium in Phosphorous and Potassium. The mean nutrient concentrations reported from one study (mg/g of soil) are: organic carbon 23.87, total nitrogen 3.30 and total phosphorous 0.75 (Sharma and Sharma, 1997). As the terrain is gentle to deep slope, chances of water logging is less, however, water-logged conditions are not suitable and adequate drainage is quite essential for better stand of the crop (Singh *et al.*, 1998).

In general L. cardamom is cultivated on hill slopes, and often in terraced lands (earlier under paddy cultivation), after raising adequate shade trees. In case of land under gentle slope, cardamom is planted on the slopes and in case of medium and steep slopes, the slopes are cut into terraces before planting. L.cardamom is a shade-loving crop. It grows well under dense shade (60–70 per cent of full day light

Figure 15.4 A view of a large cardamom field under natural forest; note the shearing of leaves caused by hail storm.

interception) to light shade (about 30 per cent full day light interception) condition (Singh *et al.*, 1989). The daylight intensity required for optimum growth of L. cardamom ranges from 5000 to 20,000 lux. Therefore, in virgin forests it is necessary to clean the under-growth; over-head shade regulation is essential in such a way that at least 50 per cent shade is maintained in the area (Fig. 15.4). In other areas having insufficient shade, planting shade tree saplings of different species is done in June–July.

The most common shade trees are *Utis* (*Alnus nepalensis*, 600–2000 m above mean sea level (amsl)); *Chilaune* (*Schima wallichi*, 550–1515 m amsl); *Panisaj* (*Terminalia myriocarpa*, 400–1000 m amsl); *Pipli* (*Exbucklandia populnea* 900–2000 m amsl) *Malato* (*Macaranga denticulata*, 670–1515 m amsl); *Asarey* (*Cole brookianum* 850–2000 m amsl); *Gogun* (*Saurauvia nepalensis*, 1400–2000 m amsl); *Karane* (*Symplocos ramosissima* Wall 1500–2400 m amsl); *Bilaune* (*Maesa chesia* 670–1515 m amsl) etc. It is advisable to plant more than one species of shade trees commonly grown in a particular locality. In case of bare land *Utis* is the first choice as it is quick growing, capable of fixing atmospheric nitrogen and has faster rates of nutrient cycling. Roots of *Alnus* species are nodulated with *Frankia* as an endophyte, and are efficient in biological N_2 fixation. Monoculture plantation of *A. nepalensis* is known to fix 29–117 kg N/ha/year. A study on dry matter production and nutrient cycling in agro-forestry system of L. cardamom grown under N_2-fixing *Alnus* and mixed tree species (non-N_2 fixing) indicates that the stand total biomass, tiller number, basal area and biomass of L. cardamom crop was much higher under the influence of *Alnus*. The agronomic yield of L. cardamom is increased by 2.2 times under the canopy of *Alnus*. Litter production and its disappearance rate is higher in the *Alnus*–L. cardamom stand. The L. cardamom based agro-forestry system under the influence of *Alnus* was more productive with faster rate of nutrient cycling (Sharma and Ambasht, 1984, 1988).

Depending on the altitude, planting of shade tree saplings in a row with a distance of 9–10 m is ideal. While planting the sapling, the course and direction of the sun movement and the slope of the hill are generally considered. Usually the tree rows are run along the southwest direction inside the plantation.

5.2   Planting

For planting of L. cardamom, pits are opened at spacing suitable for the variety/cultivar. In case of robust variety/cultivar such as *Sawney, Varlangey, Ramsey* etc., spacing followed is 150 × 150 cm while a spacing of 120 × 120 cm is used for *Golsey (Dzongu)*. Pits are opened in April–May. The size of pits usually is 30 × 30 × 30 cm.

After the receipt of a few showers, pits are filled at least 15–20 days before planting, with top soil, decomposed cattle manure or compost or leaf mold mixed well with the top soil along with 100 g rock phosphate. The ideal season for planting is June–August depending on rains. While planting, too much soil is avoided on the base of seedling, which may otherwise cover the collar region that leads to rotting. Staking is very essential for better anchorage during the initial stage of establishment.

5.3   Mulching

The plant base should be mulched with dried leaves, weeds and trashes. Mulching is done immediately after planting as well as in October–December in the existing plantation. This practice helps to cover the exposed roots, to conserve soil moisture in the ensuing dry months and helps in recycling of nutrients.

5.4   Plant nutrition

In L. cardamom, much of the nutrients are removed by leafy shoots and very less by capsules and spikes. It is observed that robust varieties like *Ramsey, Sawney* etc. remove almost double the quantities of nutrients as compared to non-vigorous ones like *Dzongu Golsey*. It is estimated that for producing about 100 kg dry L.cardamom, the robust types remove (in kg) 10.33 N: 1.95 P: 26.24 K: 19.10 Ca and 11.9 Mg; whereas *Dzongu Golsey* removes only about 5.74:0.99:3.54:9.18 and 5.86 of NPK, Ca and Mg respectively. Old leafy shoots, removed during harvesting are recycled as soil mulch. As this crop is grown under forest cover, manuring and application of fertilizer is not usually practiced. Being a low-volume and less nutrient-exhausting crop, it has a degree of sustenance in terms of nutrient cycling. However, to get sustainable high yield fertilizer application is necessary.

Application of NPK fertilizer in three splits, once in April–May after the first summer showers and second split in June and third in September–October before monsoon ceases increases yield. Fertilizer is applied along a circular band at a distance of 30–45 cm from the clump, with mild forking. After application of fertilizer in September/October, the fertilizer is covered with mulch and soil from the adjacent areas, so as to form a mulch-soil base. It helps to improve the physical make up of soil, with more moisture retention capacity, cover the exposed roots and help the plant to withstand the ensuing dry winter season (Anonymous, 1998a).

A fertilizer schedule of 18.4, 6 and 18.6 g NPK/clump produced more number of tillers and spikes/clump in *Sawney*. However *Golsey* did not respond to fertilizer application (Parthasarathy and Govind, 1995). Soil application of NPK at the rate of 20:30:40 kg/ha (in two splits: April–May and September) was also found to increase growth and yield characters (Anonymous, 1998a). Soil application of Potassium chloride shows significant effect on growth and yield characters.

Foliar application of urea (0.5 and 1 per cent), DAP (0.5 per cent) and Muriate of potash (0.5 and 1 per cent) during February, April and October enhanced yield of *Ramsey*. Potassium chloride and potassium sulphate at the rate of 15 g/clump were found to be effective in increasing the yield of Pink *Golsey*.

L. cardamom cultivation, is a natural farming with least external inputs and interference with the soil-ecosystem. Organic recycling of nutrients is very essential for sustained productivity.

5.5   Weeding

About 51 species of weeds have been recorded in L.cardamom plantations (Anonymous, 1984). Depending on the intensity of weed growth, two to three rounds of weeding are required in a year. First weeding is done in February–April, before application of first split of fertilizer and just before flowering. Second and third round of weeding is done before harvesting in August–September–October along with removal of dried leaves, unproductive tillers etc. The weeded materials are used for mulching.

5.6   Shade regulation

Most of the shade tree species are deciduous in nature and hence frequent shade regulation is not required. However during early years of shade establishment and also at 2–3 year intervals, the under growing side branches are cut to encourage straight growth and to allow the branches to spread at least 3–4 m above ground level so that moderate shade is maintained (Gupta, 1986).

5.7   Irrigation

Yield performance is better in plantations where irrigation is given during dry winter and summer months. Watering during November–March is found essential to maintain a sustainable good yield in the plantation. Water is tapped through pipes and is provided to plants through surface channel/hose in different directions.

5.8   Rouging and gap filling

One of the reasons for poor productivity of L.cardamom (as low as 50–150 kg/ha) is that most of the plantations have become senile and unproductive. The two viral diseases *Foorkey* and *Chirke* – have not only spread widely reducing the yield, but also made the plantations highly unproductive and uneconomical. Regular rouging of diseased and senile plants and replanting with high yielding, disease-free seedlings/suckers/TC plantlets are necessary for improving and sustaining the yield in the years to come.

6 CROP IMPROVEMENT

6.1 Flowering and pollination

L. cardamom, is essentially cross-pollinated, insect pollination is the rule. The flower morphology is adapted for such a mode of pollination. Each spike consists of about 40–50 flowers, which open in the acropetal sequence over a period of about 15–25 days. Flower opening starts at early morning hours, i.e. 3–4 a.m., anthers dehisce almost simultaneously whereas stigma receptivity starts an hour later and lasts for 24 h. The stigma was found to be receptive even after 36 h from the time of flower opening during rain-free days (Gupta and John, 1987; Karibasappa, 1992).

Bumble bees (*Bombus* sp.) are the main pollinators, though a variety of honeybees and other insects do pollinate the flowers (Varma, 1987). Bumble bees are effective pollinators due to its compatible size with the flowers and is having brush-like hairy structures on its dorsal thorax which helps in carrying pollen mass and depositing it on stigmatic surface while entering the flowers. The highest foraging activity of bumble bees is seen during 6–7 a.m., but becomes slow or dull in rainy conditions.

6.2 Genetical studies

Studies on coefficient of variation was undertaken by Karibasappa *et al.* (1987c, 1989) in five varieties of large cardamom (*Sawney, Pink Golsey, Ramsey, Ramnag* and *Madhusey*) for mature tillers per clump, panicles per clump and capsule–panicle ratio. This study indicated high heritability coupled with high genetic advance for characters such as length of mature tiller, panicles per clump, panicle weight and capsule yield. The capsule yield was directly correlated with clump girth, panicle weight, panicle per clump, mature tillers and capsule–panicle ratio. Correlation studies (Karibasappa *et al.*, 1989) also indicated that mature seed index, total soluble sugars (TSS) of seed mucilage and 1000 seed weight were associated positively with oleoresin and negatively with cineol contents.

6.3 Clonal selection

Explorations for collection of germplasm of L. cardamom was carried out by Indian Cardamom Research Institute (ICRI), Regional Station, Gangtok at various tracts and a gene bank consisting of 180 accessions is established at Pangthang. Rao *et al.* (1990) reported a promising selection of Barlanga cultivar from high altitudes, having desirable characters like high ratio of mature tillers to productive spikes (1:3.6) and capsules having very bold size (with 50–80 seeds per capsule). Based on a preliminary evaluation, four selections (SBLC-5, SBLC-42, SBLC-47 and SBLC-47A) having high yield potential have been identified by the ICRI, Regional station, Gangtok. They are multiplied in large numbers using micropropagation technique and distributed among the farmers for cultivation.

7 INSECT PEST MANAGEMENT

More than 22 insect pests are known to be associated with L. cardamom, and only a few of them cause substantial damage to the crop (Bhowmik, 1962; Pangtey and Azad Thakur, 1986; Azad Thakur, 1987; Azad Thakur and Sachan, 1987).

7.1   Leaf caterpillar

Leaf eating caterpillar (*Artona chorista* Jordon, Lepidoptera: Zygaenidae), is a major pest of L. cardamom in Sikkim and West Bengal (Yadava *et al.*, 1992; Singh and Varadarasan, 1998). Its outbreak was recorded in 1978 in Sikkim where about 2000 acres of L. cardamom plantations were severely defoliated (Subba, 1980). The leaf caterpillar is first recorded as *Clelea plumbiola* Hampson on large cardamom by Bhowmik (1962).

A. *chorista* occurs sporadically in epidemic form in Sikkim and West Bengal every year. Usually the incidence of the pest is observed from June to July and October–March in the field. Severe damage was recorded in Lower Dzongu, Phodong, Ramthung Basti (north Sikkim), Soreng, Hee, Chako (west Sikkim), Kewizing (south Sikkim); Assamlinzey, Dalapchand and Rongli (east Sikkim) and Gotak (Darjeeling Dist. of West Bengal).

7.1.1   *Nature and extent of damage*

The leaf caterpillars are monophagous and highly host specific. The caterpillars are gregarious in nature (60–200 caterpillars/leaf) and feed on chlorophyll contents underneath the leaf, leaving transparent epidermis and veins (skeletonization). The damaged portion of the leaf becomes brownish and which can be identified easily (Fig. 15.5). The mature larvae completely defoliate the plant leaving the midrib of the leaves. Defoliation of the plant by the pest affects the yield indirectly. The area of a medium-sized cardamom leaf is about 160–170 sqcm and a mature larva consumes about 2.12 sqcm cardamom leaf in a day (Singh and Varadarasan, 1998).

Figure 15.5 Damage caused by leaf caterpillar.

7.1.2 Biology

The adult is a moth, black and very small in size (10–15 mm). The male moth can easily be distinguished from female with its bushy antennae, white bands on the abdomen and smaller size. The sex ratio is recorded as 1 female : 2.2 male. Adults copulate in tail-to-tail position during night; they survive for 10–12 days in captivity. The female moth lays 250–300 eggs in clusters on underside of cardamom leaves. Eggs are smooth, round in shape, small, translucent and cream in colour. The eggs hatch in 20–22 days in April–May. The immature larvae are greenish-yellow and mature larvae are dull yellow and measure 20–25 mm in length (Table 15.3). The larvae bear small setae and dark brown longitudinal stripes along the dorsal side.

The larval period is completed in 58–62 days during monsoon i.e. May–July in first generation and 170–190 days (about 6 months) from October to March in the second generation. The larvae diapause from the middle of December to February by covering under the web on dry cardamom leaves and then pupate in pale brown silken cocoon (for about 25–30 days) in July–August and 35–40 days in March–April. Moths emerge in April–May and August–September).

There are two generations of *A. chorista* in a year. First generation takes about 4 months i.e. from May to August (egg to adult) and second generation about 8 months i.e. from September to April.

7.1.3 Management

Mechanical control: The larvae are gregarious in nature and feed underneath the cardamom leaf; the infested leaf can easily be identified from a distance and these can be collected along with larvae and destroyed in June–July and October–December. A demonstration trial in planters' fields at Phodong (north Sikkim) where 97 per cent and 86 per cent infestation was recorded in 1996 and 1997, has given 58 per cent and 67 per cent control over the initial infestation in both years (Table 15.4). Group approaches of mechanical control by all the farmers of the locality, when the outbreak is observed, totally suppress this pest in an area within a few years.

Biological control: A species of predatory pentatomid bug was recorded on the leaf caterpillar in the field. It kills 1–3 larvae/day by sucking their body fluid. Two larval-pupal

Table 15.3 Life cycle of leaf caterpillar, *Artona chorista* on large cardamom

Stages and other parameters	I Generation (April–July)	II Generation (August–March)
Incubation period (days)	20–22	24–26
Larval duration (days)	58–62	170–190
Pupal duration (days)	24–28	35–40
Adult longevity (days)	10–12	10–12
Total life period (days)	112–124 (About 4 months)	239–268 (About 8 months)
Minimum temp. range (°C)	8–17	2–18
Maximum temp. range (°C)	20–30	12–28

Table 15.4 Pre- and post-treatment observations on demonstration trial on mechanical control of leaf caterpillars, *Artona chorista*

Particulars	1996–97		1997–98	
	Pre-treatment observation	Post-treatment observation	Pre-treatment observation	Post-treatment observation
Clumps infested	95.0%	36.92%	86.0%	19.0%
Tiller infested	33.21%	4.38%	22.38%	1.61%
Damage on leaves/clump	1–37	1–11	1–23	1–6
Mean leaves damaged/clump	B.14 ± 0.72	1.20 ± 0.30	5.39 ± 0.59	0.36 ± 0.09
No. of caterpillars/leaf	66–192	16–136	22–236	8–86
Mean no. of caterpillars/leaf	127.91 ± 11.80	67.52 ± 10.14	97.47 ± 13.15	33.40 ± 7.84

parasitoids, *Medina* sp. and genus of *Bactromyra* (F. Tachinidae) were recorded. Single black colour adult fly emerges from a cocoon of leaf caterpillar after 22–24 days of pupation. There was 7.08 per cent field parasitization. Two larval-pupal hymenopteran hyper parasitoids (*Venturia* sp. and *Mesochorus* sp. (F. Ichneumonidae) were also recorded. *Venturia* sp. is black and 6–8 mm in size whereas *Mesochorus* sp. is brown and 5–7 mm in length. Yadava *et al.* (1992) have reported two hymenopterans viz., *Apanteles* sp. and *Dolichogenedea* sp. (F. Braconidae) an endolarval parasites on *A. chorista*. These natural enemies kill the larvae and pupae of the pest and reduce the pest population considerably in the field.

Chemical control: Even though insecticides like Quinalphos and Endosulfan (0.05 per cent) are effective, growers are advised to avoid use of such chemicals, which kill the natural enemies also. This crop is not exposed to any chemical insecticides (Subba, 1979, 1980, 1984). Therefore naturally-occurring parasitoids keep the pest population under check. If at all required any one of the following insecticide can be sprayed:

(a) If caterpillars are in early stage (skeletonization of the leaves)
    Quinalphos 25 EC 0.05 per cent (200 ml/100 l of water).
    Endosulfan 35 EC 0.05 per cent (143 ml/100 l of water).
(b) If the caterpillars are in the later stage (defoliation of the plant)
    Quinalphos 25 EC 0.1 per cent (400 ml/100 l of water).
    Endosulfan 35 EC 0.1 per cent (286 ml/100 l of water).
(c) If the infestation is very severe, second round of insecticide spray after 1 month can be given.

During rainy season, 50–100 ml wetting agent per 100 l of water is added with insecticide solution.

7.2   Hairy caterpillars

Hairy caterpillars are a group of defoliators infesting L. cardamom. Severe infestation by hairy caterpillars affects the crop yield. Among the hairy caterpillars, *Eupterote* sp. are the predominant ones found in cardamom plantations. The major species are: *Eupterote fabia* Cramer and *Eupterote* sp. (Lepidoptera: Eupterotidae).

Incidence of hairy caterpillar is recorded during post-monsoon period i.e. from August to December. *E. fabia* and other species are sporadic and polyphagous and feed on the leaves of cardamom causing defoliation. Sometimes, *E. fabia* causes severe damage of the leaf.

7.2.1 *Biology and management*

The adult moth of *E. fabia* is large (10.8 cm across the wing) and yellow in colour. The female moth lays eggs on the underside of leaf in clusters of 20–140 eggs. The hatching period lasts for 19–21 days. The mature larvae measure 7.4 cm in length. The larval development is completed in 83–97 days. The pupation takes place in soil in a silken cocoon. Adult comes out in 120–180 days (Azad Thakur and Sachan, 1987). Hairy caterpillar is a minor pest.

7.3   Aphids

The aphids cause more damage as a vector rather than as a pest. The aphids are associated with the transmission of viral diseases (*Foorkey* and *Chirke*) of L. cardamom. The aphid population is recorded high during summer months at lower altitudes. The major species are:

1   *Pentalonia nigronervosa f. caladii* (Goot) (Hemiptera: Aphididae).
2   *Micromyzus kalimpongensis* (Hemiptera: Aphididae).
3   *Rophalosiphum maidis* (Fitch) (Hemiptera: Aphididae).
4   *R. padi* (Lin.) (Hemiptera: Aphididae).

P. nigronervosa f. caladii and *M. kalimpongensis* are known to be the vectors of *Foorkey* or virus yellow disease. The aphids colonize at the base (rhizome) of the clump and if population is more, they move to the aerial portion of the clump. Two to six aphids/tiller were recorded from the *Foorkey* infected plants during summer months. *P. nigronervosa f. caladii* is also reported as the vector of *Katte* viral disease in small cardamom. These aphids are dark brown in colour, small and measure 1–1.5 mm in length. They remain mostly inside the soil close to rhizomes and suck the sap from the pseudostem. The alate (winged) and apterous (wingless) forms complete life cycle in 20–30 days.

Maize aphids, *R. maidis* and *R. padi* are reported on the lower surface of the leaves of large cardamom, congregating near the mid-rib and veins. These aphids are known to be the vector of another viral disease, mosaic streak or *Chirke* (Raychaudhary and Chatterjee, 1965).

The removal and destruction of diseased plants is helpful in control of further spread of disease and in reduction of aphid population. Spraying of 0.03 per cent Dimethoate or Phosphomidon after removal of *Foorkey* and *Chirke* affected clumps in March–April, gives adequate control of aphids.

7.4   Shoot fly

Shoot fly, *Merochlorops dimorphus* Cherian (Diptera: Chloropidae), recorded as a major pest of L. cardamom is damaging young shoots. Low to moderate damage by shoot fly is recorded in L. cardamom plantation in Sikkim and West Bengal. In the main field, more damage is recorded at higher altitudes than in the lower. Kumar and Chatterjee (1993) have reported another species of shoot fly, *Bradysia* sp. (Diptera: Sciaridae), damaging large cardamom. The pest is recorded throughout the year in L. cardamom growing tract. The incidence was as high as 56 per cent of new shoots.

The tip of the shoot becomes brown and later whole shoot dries up causing 'dead heart' symptom. Single, pale glossy white larva bores the young shoot and feeds on the central core of pseudostem from the top to the bottom resulting in its death. The mature larva measures 8–10 mm long and pupates inside the infested stem. The pupal period lasts for 20–24 days and adult survives for 4–6 days in the laboratory condition. The adult fly is smaller (5–6 mm) in size and brownish yellow in colour.

For managing this pest, infested young shoots should be removed at ground level and destroyed.

7.5   Stem borer, *Glypheterix* sp. (Lepidoptera: Glyphiperidae)

It is a minor pest and is specific to L. cardamom. Eventhough Azad Thakur and Sachan (1987) recorded 19 per cent infestation of stem borer in 1978–79, the incidence is very low in main cardamom plantations but sometimes it causes considerable damage in nurseries. Stem borer is recorded from March to November at Ghotak (West Bengal), Pangthang, Khasay, Gamdong (east Sikkim), Kabi and Mangan (north Sikkim). The larvae feeds on the central portion of the shoot and as a result the terminal leaf of the plant gets dried up and this symptom is known as 'dead heart'. Infestation of this pest is also indicated by the presence of entry holes plugged with excreta. The intensity of infestation has been found higher at lower altitudes (about 5 per cent) in Sikkim on seedlings and main plantations. It is a minor pest and can be controlled by removing infested shoots along with caterpillars.

7.6   White-grubs *Holotrichia* sp. (Coleoptera: Melonthidae)

It is a polyphagous white grub infesting roots and rhizomes of L. cardamom. The infested plant shows yellowing of leaves and withering symptoms. The grubs are white and 'C' shaped with brown head. The incidence was recorded at Panthang (East Sikkim) and Kabi (North Sikkim) in September–December. This is a minor pest, hence no control measures are adopted.

7.7   Leaf folding caterpillar

This feeds on chlorophyll contents inside the folded cardamom leaf (Fig. 15.6). It's damage is very minor in nature, and was observed during July–December in east and north Districts of Sikkim. The caterpillar pupates inside the folded leaf during winter. Adult is a medium sized (24–28 mm across the wing), brownish yellow moth. *Cotesia euthaliae* Bhatnagar (Hymenoptera: Braconidae), was recorded as larval parasitoid. For control infested leaves along with caterpillars are collected and destroyed.

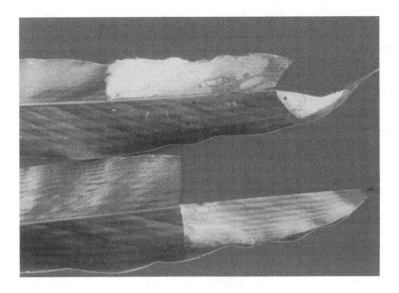

Figure 15.6 Leaf folding caused by leaf folding caterpillar.

7.8   Minor pests

7.8.1   *Mealy bugs*

These occur on underground rhizome of the L.cardamom plants. They feed on the roots and rhizome part of the plant and cause yellowing during summer months. The pest is recorded at Neem (East Sikkim), Tarku (South Sikkim), Chawang (North Sikkim) and Singling (West Sikkim) in March–October.

7.8.2   *Leaf thrips*

Heliothrips haemorrhaidalis Bouche (Thysanoptera: Thripidae), is a minor pest on leaves, infests the undersurface of leaves and suck the sap. The damage is more on seedlings and are recorded throughout the year. *Rhipiphorothrips cruentatus* Hood (Thripidae: Thysanoptera) is also reported as a minor pest of L. cardamom on seedlings (Azad Thakur and Sachan, 1987). The leaf thrips population is more on lower surface compared to the upper surface. The infested leaves turn brown and wither out gradually. For management, infested leaves are removed and destroyed. If infestation is severe it can be controlled by spraying Monocrotophos or Quinalphos at 0.025 per cent (Singh *et al.*, 1994).

7.8.3   *Lacewing bug*

Stephanitis typica (Distant) (Hemiptera: Tingidae), is a minor sucking pest on large cardamom leaves. Severe infestation was recorded in 1997 in north Sikkim where about 1000 plants in an isolated patch was damaged. The infested area was open without any shade trees. The damage is recorded in main field where shade was thin during pre- and post-monsoon period. The infestation is recognizable even from a distance due to its grayish yellow feeding spots on the leaves. The bugs suck the sap on lower surface of the leaves. So, in case of severe infestation, plant growth and yield gets affected (Singh *et al.*, 1994).

7.8.4   Grasshoppers

Grasshoppers are of minor importance and present in main field and nurseries. Both nymphs and adults feed on the leaves. The major species are: *Mazarredia* sp. Bolivar (Tettigonidae) and *Chrotogonus* sp. (Acrididae)

7.8.5   Bagworm

Acanthopsyche sp. (Lepidoptera: Psychidae), is a minor pest. It's larvae cause small holes in the cardamom leaves. The insect was recorded throughout the year in most of the plantations but damage is negligible.

7.8.6   Fruit borer

The grub of Scolytid beetle (*Synoxy* sp.) makes a hole in the immature capsule and feeds on the seeds inside and pupates inside the capsule. The pest was recorded at Hee Gaon (West Sikkim).

7.8.7   Scale insects

The scale insects colonize near the mid vein on lower surface of the leaf. They suck sap, which results in brownish spots on leaves. Minor infestation of scales was recorded in seedlings at Cardamom Nursery, Mallipayong (South Sikkim).

7.8.8   Rhizome weevil

Adult is about 1.5 cm brownish weevil feeding on rhizome by making tunnel. It is a minor pest and was recorded at Kabi (North Sikkim) in April.

7.8.9   Leaf beetle

The adult, *Lema* sp. (Coleoptera: Chrysomelidae) are greenish brown in colour and 8–10 mm in size and make irregular holes in tender leaves.

7.8.10   Green beetle

Green colour beetle, *Basilepta femorata* Jacoby (Coleoptera: Eumolpidae), is recorded in April and October on L. cardamom leaves. It's exact nature of damage on large cardamom is not yet known. The grub of *Basilepta fulvicorneis* is a major pest on cardamom (*Elettaria cardamomum* Maton) (Varadarasan *et al.*, 1991). However, Azad Thakur and Sachan (1987) reported that adult beetles are very destructive as they nibble and eat away fresh leaf buds.

7.8.11   *Nematodes*

Root knot nematode, *Meloidogyne incognita*, is found to infect seedlings/plants in nurseries and plantations causing considerable damage. The affected seedlings/plants show stunted growth, the leaves become narrow and arrange in a rosette due to reduction in internodal area. The root system shows excessive branching with galls.

Deep digging and exposing the soil to sun (solarization) before preparation for nursery may reduce the nematode intensity. Use of the same site for nursery maintenance is avoided. Farmers are encouraged to change nursery sites every year and to raise seedlings in polybags containing good potting mixture. This practice not only reduces nematode incidence but also prevents movement of nematode through soil from one location to another.

7.8.12   *Other pests*

Rodents, squirrels and wild cats damage the fruits before harvesting. Black cats (*Kaala*) a nocturnal mammal is known to cause heavy losses as they are voracious feeders of near-maturing capsules.

7.8.13   *Storage pest*

A reddish brown caterpillar of cardamom moth, bores into the capsule and feeds on the mucilaginous seed coat; however the hard seeds are unaffected; the appearance of capsules as well as seed quality is adversely affected.

8   DISEASES

The crop is susceptible to a number of diseases, which are mainly viral and fungal in origin. Large cardamom productivity is affected seriously by viral diseases. However fungal diseases are not major constraints. There are two viral diseases on large cardamom causing severe damage to the plantations. *Chirke* is serious as far as rate of spread is concerned; *Foorkey* is serious as far as yield loss is concerned. Among fungal diseases, flower rot, clump rot, leaf streak and wilt are known to cause damage to the plant and ultimately reduce the crop yield.

8.1   Chirke

This virus disease is characterized by mosaic with pale streak on the leaves. The streaks turn pale brown resulting in drying and withering of leaves and plants. The flowering in diseased plants is extensively reduced and only one to five flowers develop in one inflorescence, as against 16–20 in an inflorescence of healthy plants (Raychaudhary and Chatterjee, 1958, 1965) and by the end of third year of crop the loss is around 85 per cent. The cultivar *Kopringe* is resistant to *chirke* while the perennial weed, *Acorus calamus* L. was found to be highly susceptible (Raychaudhary and Ganguly, 1965). The disease is readily transmitted by mechanical sap inoculation and in field it is spread by aphids, *Rhopalosiphum maidis* Fitch., within a short acquisition feeding period of 5 min.

Primary spread of diseases from one area to another area is through infected rhizomes and further spread within the field is by aphids (Raychaudhary and Chatterjee, 1958, 1965). Rapid serological method was developed to locate *chirke* diseased plants under field condition in the manner as described by Bradley and Munro (Ganguly, 1966).

8.2   Foorkey

This disease is characterized by dwarf tillers with small, slightly curled pale green leaves. The virus (spherical particles of 37 nm diameter, Alhwat *et al.*, 1981) induces remarkable reduction in size of leafy shoots and leaves of the infected plants and also stimulates proliferation of large number of stunted shoots arising from the rhizome. The spikes/inflorescence are transformed into leafy vegetative parts and fruit formation is altogether suppressed. The diseased plants remain unproductive and gradually degenerate. *Foorkey* symptom appears both on seedlings and grown up plants (Varma and Capoor, 1964).

Unlike *Chirke*, *Foorkey* virus is not transmitted through sap but by the aphids *Pentalonia nigronervosa* Cog. and *Micromyzus kalimpongensis* Basu (Basu and Ganguly, 1968; Sharma *et al.*, 1972). The primary spread of disease from one area to another is through infected rhizomes and further spread within the plantation by aphids. Infected rhizomes can be killed by injecting Agroxone-40.

8.3   Management of chirke and foorkey

The following methods are adopted to considerably minimize these two viral diseases in the affected plantations (Chattopadhyay and Bhowmik, 1965):

(i)   Regular rouging of diseased plants, if any.
(ii)   The diseased plants are uprooted and destroyed as and when they are traced.
(iii)   Uprooted plants are taken to an isolated place, chopped into small pieces buried in deep pits for their quick decomposition.

8.3.1   Preventive measures

(a) Use of healthy and disease free planting materials preferably seedlings; (b) suckers as planting material from diseased area are avoided for replanting and (c) raising nurseries in the vicinity of infected plantations are avoided.

8.4   Leaf streak disease

It is a fungal disease caused by *Pestalotiopsis royenae* (D. Sacc) Steyaert., and is a serious disease among foliar diseases and is prevalent round the year. The disease symptom is the formation of numerous translucent streaks on young leaves along the veins. The infection starts from emerging folded leaves; infected leaves eventually dry up causing loss of green part and reduce the yielding capacity of the plant. *Dzongu Golsey* is found more susceptible to leaf streak (Srivastava and Varma, 1987).

Three rounds of 0.2 per cent spray of copper oxychloride at 15 days interval, two schedules in a year i.e. February–March and September–October can control this disease.

8.5 Flower rot

A fungal disease caused by *Fusarium* and *Rhizoctonia* sp. The affected flowers turn dark brown and fail to develop into capsules, if infection takes place before or at the time of fertilization. If infection occurs after flowering or during fruit set, the affected fruit or capsule loses colour and odour (Srivastava, 1991). The disease can be managed by avoiding: (a) accumulation of leaf mass or mulch over the inflorescence/spike during rainy season and, (b) soil spills over spikes.

8.6 Wilt

A fungal disease caused by *Fusarium oxysporum* is prevalent in swampy and open areas. Early symptom is chlorosis of the older leaves commencing from the petiole region and progressing inwardly towards the young leaves. As the infection progresses, the pseudostem also gets rotten, thereby blocking the vascular bundles, and pseudostem is collapsed as the infection progresses. Ultimately the plant dries up.

Drenching 0.5 per cent Dithane M-45 or Thiram would help in checking further spread of disease in nurseries as well as in main plantation. Planting in swampy or dry area should be avoided.

9 HARVESTING AND POST HARVEST TECHNOLOGY

First crop comes to harvest about 2–3 years after planting of sucker or seedling. However stabilized yields are obtained only from the 4th year up to 10–12 years. Sustainable yield depends on proper plantation management like regular rouging coupled with replanting, weeding, mulching plant bases, winter/summer irrigation, shade regulation etc. Harvesting season starts in August/September in low altitudes and continues up to December at high altitudes. Usually harvesting is done in one round and hence the harvested produce often contains capsules of varying maturity. Harvesting is done when the seeds of top capsules in the spike attain dark gray colour. A special type of knife, locally known as *Elaichi chhuri* is used for harvesting. The stalk of the spike is cut very close to the leafy shoot. After harvest, individual capsules are separated manually. Capsules after harvest are cured to reduce moisture level to 10–12 per cent. The traditional curing is called *Bhatti* curing system (direct heat drying). L. cardamom is also cured by flue pipe curing system (indirect heat drying).

9.1 Bhatti system

Drying of cardamom is generally done in kilns locally called *Bhatti*. A *Bhatti* consists of a platform made of bamboo mats/wire mesh, laid over a four-walled structure made of stone pieces with a V shape opening in the front for feeding fire wood. Capsules are spread over the platform and are dried by direct heat generated from the firewood. About 70–80 kg of firewood is required for curing 100 kg of green cardamom in this traditional kiln (Sundriyal *et al.*, 1994). Both green and dry woods are used and, as a result, huge volume of smoke is generated that passes through cardamom (Singh, 1978). Depending on the thickness of the cardamom spread, it takes 60–72 h for curing (John and Mathew, 1979). Colour of cardamom cured under this system is dark brown or black. If smoke percolates through cardamom, it loses original colour, gets smoky smell and fetches low price (Karibasappa, 1987).

9.2   Portable curing chamber

This is a prototype of 'Copra dryer' developed at CPCRI (ICAR), Kasaragod, Kerala. The unit comprises of an air heating chamber and a furnace cylinder with chimney. The whole unit is fitted in a detachable angular iron frame. The air-heating chamber is enclosed by asbestos sheets on all the four sides leaving the top open for drying. Some space is left below the walls for air to enter into the heating chamber. Fresh cardamom is spread on the platform (wire mesh of size 3–5 mm) to a thickness of about 15–20 cm layer. Firewood is burnt inside the furnace cylinder. Air around the furnace gets heated up quickly and convected upwards and passes through the produce there by drying it. The temperature of the heating chamber is regulated by the rate of burning of firewood and also by regulating the chimney valves. It takes about 20 h for curing of about 50 kg raw capsules (Annamalai *et al.*, 1988).

9.3   Flue pipe curing system

In order to overcome the drawbacks of the *Bhatti* curing, a system known as "flue pipe system" has been introduced by the Spices Board of India for L.cardamom curing (see Chapter 9). This method of curing requires a construction with corrugated tin sheets as roof and fitted with furnace, flue pipes, chimney ventilation system etc. Flue pipes made of galvanized iron sheets of 20 gauge, of 25 cm radius are provided in the room which runs from one end to the other along the center of the room below the wire gauge fittings. One end of the pipe is connected to the furnace, the firing place and the other end to a chimney to lead the smoke up through the roof into the air. Ventilators are provided in the room to lead water vapour out in the process of drying. An arrangement is made to keep a thermometer for reading the inside chamber temperature, from outside and accordingly ventilation as well as charging of fuel is regulated.

Cardamom is spread over the wire mesh floor and shelves in one or two layers after the chamber is kept at 40 °C. As soon as cardamom spreading is over, the curing room is kept closed and hot air is passed through the furnace into the flue pipes, bringing the room temperature to 45–50 °C. This temperature is maintained for about 3–4 h. At this stage, sweating of capsules takes place and moisture is released. Ventilators are then opened for sweeping out the accumulated moisture completely from the chamber. Ventilators are then closed and temperature is again maintained at 40 –45 °C. For uniform drying, cardamom is stirred one or two times. It is important to note that curing once started should be a continuous process till the drying is

Figure 15.7 Large cardamom capsules after polishing: left – cured by *bhatti* system; right – flue pipe cured.

over. The whole process of curing takes about 28 to 29 h. Cardamom capsules thus cured are immediately collected and rubbed in trays or processed in Cardamom Polishing Machine for removing the tail (Anonymous, 1998b) (Fig. 15.7). Clean produce is then packed in polythene lined gunny bags and stored in wooden boxes. Cured cardamom on an average gives 25 per cent by weight of fresh cardamom (John and Mathew, 1979).

Cardamom dried in this method of curing has following advantages over those of the *Bhatti* system (Karibasappa, 1987).

(i)   Original colour (pink) and flavour (sweet camphor aroma) is retained.
(ii) Fetches better market price.
(iii) Curing expenses are low. Firewood consumption is less.
(iv) Total curing takes only 28–29 h.
(v)   Uniform drayage is ensured.

9.4   Natural convection dryer

A dryer similar to the one described above was designed by CFTRI (Joseph *et al.*, 1996). The dryer consisted of (i) a furnace, (ii) flue ducts, (iii) wire mesh tray for charging capsules and (iv) supporting structures. The furnace is fabricated using 8 mm thick MS sheets. A brick lining inside the furnace provides insulation. Flue ducts are made of 1.6 mm GI sheets, and the ducts were arranged in two tiers, one over the other, with sufficient space between them. Two crimped steel wire mesh trays with border were placed one over each duct. Firewood is burned in the furnace. The hot flue gases passing through the ducts set up convection currents in the air between the duct wall and wire mesh trays. The convection currents pass upward through the mesh and the bed of cardamom capsules on the mesh are subjected to drying.

This dryer has thermal efficiency of 5.6, better than the conventional flue curing kilns. It can dry 300 kg large cardamom at a time and for drying to a level of 10 per cent moisture the time required is 24 h, much better than the conventional method (Joseph *et al.*, 1996).

9.5   Gasifier system of curing

It is an up gradation of the *Bhatti* curing system, developed by TATA Energy Research Institute, New Delhi. The solid fuel i.e. firewood through biomass gasification and thermo chemical reaction is converted into gaseous fuel by partial combustion. A mixture of producer gas consisting of CO (Carbon monoxide), H_2 (Hydrogen), CH_4 (Methane), CO_2 (Carbon dioxide) and N_2 (Nitrogen) is obtained, which is combustible and is used to burn. Through updraft type of biomass gasifier, air enters the gasifier from bottom and producer gas is taken out from top for curing cardamom.

A prototype unit, which is fitted in the existing *Bhatti* curing system, is successfully field tested in Sikkim and an improved quality large cardamom with better appearance and more volatile oil content is obtained (Anonymous, 1998b).

Gasifier system has definite advantages over the traditional *Bhatti* curing, the uncontrolled burning of wood logs in *Bhatti* system results in (a) loss of volatile oil; (b) exposure to smoke imparts a smoked-smell to the volatile oil and (c) charring of capsules due to localized over heating. But the controlled burning in a gasifier system of curing helps in (a) retaining more volatile oil; (b) better quality of volatile oil without any burnt or smoking smells due to clean burning of gaseous fuel and (c) the cured capsules retain the natural pink colour of large cardamom (Anonymous, 1998b).

9.6   Chemical composition

Large cardamom seeds (dried) on analysis gave the following results: moisture (8.5 per cent), protein (6 per cent), volatile oil (2.8 per cent), crude fiber (22 per cent), starch (43.2 per cent), ether extract (5.3 per cent), alcohol extract (7 per cent) and (4 per cent). They are found to contain (for 100 g seeds), calcium 666.6 mg, magnesium 412.5 mg, phosphorous 61 mg, fluoride 14.4 ppm. The seeds contain the glycosides petunidin 3,5-diglucoside, leucocyanidin 3-o-β-D-glucopyranoside and subulin, an aurone glucoside. Cardamomin – a chalcone, alpinetin – a flavanone, are also reported (Shankaracharya *et al.*, 1990).

The powdered seeds on steam distillation yield 1–3.5 per cent of a dark brown, mobile essential oil. A study has shown that the volatile oil content was 2.44 per cent in *Sawney*, 2.42 per cent in Pink *Golsey*, 2.25 per cent in *Ramnag* and 1.66 per cent in *Ramsey* (Shankaracharya *et al.*, 1990). The oil has the following properties: sp. gr (20 °C): 0.9142, ref. index (26 °C); 1.46, optical rotation: −18 °3; acid value 2.9; saponification value 14.53; and saponification value after acetylation 40.2.

Large cardamom oil is characterized by flat cineol odour, harsh aroma and inferior flavour as against the warm spicy, aromatic odour of cardamom (Table 15.5). The

Table 15.5 Composition of volatile oil of large cardamom in comparison with the oil of cardamom

Constituents	Large cardamom (range) (%)	Cardamom (range) (%)
α-terpinene	0.5–11.13	0.37–2.5
α-pinene	2.0–3.11	1.10–13.00
β-pinene	2.4–3.67	0.2–4.9
Sabinene	0.2–9.10	2.5–4.9
Camphene	0.44	0.02–0.13
ν-terpinene	0.2–16.2	0.04–11.2
Limonene	6.38–10.3	0.12–2.1
ρ-cymene	0.20–0.30	0.40–0.70
1,8-cineole	63.3–75.27	23.4–51.30
Linalool	0.41	2.1–4.5
Geraniol	0.12	0.25–0.38
α-terpineol	4.9–7.2	0.86–1.90
Terpinen-4-ol	1.42–2.0	0.14–15.3
Nerlidol	0.12–1.0	0.23–1.60
Nerlacetate	0.14	0.02–0.09
α-terpinyl acetate	5.10	34.60–52.5
α-bisabolene	1.3–3.6	0.07–0.83
β-terpineol	0.8	0.70–2.10

Source: Shankaracharya *et al.* (1990), Balakrishnan *et al.* (1984).

L. cardamom oil is almost devoid of α-terpinyl acetate, and is very rich in 1,8-cineole (Balakrishnan *et al.*, 1984; Gurudutt *et al.*, 1996).

10 PROPERTIES AND USES

Large cardamom is used as an ingredient as well as a flavouring agent with masala and curry powders; in flavouring sweet dishes, cakes, and pastries; as a masticatory and for medicinal purposes. The seeds are used for chewing along with betel quid (betel leaf, areanut lime, with or without tobacco). In gulf countries large cardamom is used as a cheaper substitute for spicing tea in place of cardamom. In the Indian systems of medicines – *Ayurveda* and *Unani*, it is used as preventive as well as curative for throat troubles, congestion of lungs, inflammation of eyelids, digestive disorders and even in the treatment of pulmonary tuberculosis (Kirtikar and Basu, 1952). The seeds are fragrant adjuncts for other stimulants, bitters and purgatives. The seeds have a sharp and good taste and are tonic to heart and liver. The pericarp is reported to be good for alleviating headache and heals stomatites (Anonymous, 1950). Decoction of the seeds is used as a gargle in afflictions of teeth and gums. With melon seeds they are used as a diuretic in cases of gravel of the kidneys. They promote elimination of bile and are useful in congestion of liver. They are also used in the treatment of gonorrhea. In large doses with quinine, they are used in neuralgia. The seed oil is applied to the eyes to allay inflammation.

The direct uses of large cardamom are in pickles, pulao, meat and vegetable dishes. It is also used in industrial sector for flavouring toothpastes, sweets, soft drinks, toffees, flavoured milk and alcoholic drinks. The ripe fruits are eaten raw by people of Sikkim and Darjeeling and are considered a delicacy (Gyatso *et al.*, 1980).

11   CONCLUSION

Large cardamom is a crop of the north-eastern Himalayan tracts, the largest producer is the state of Sikkim in India. It is used extensively as a spice in South Asian countries, and as a substitute for true cardamom in the Middle East regions. It is also important in tribal and indigenous medicine.

Research inputs on this crop has been meager. Efforts are yet to be made in evolving superior genotypes combining high yield and quality. A search for aroma quality is essential to locate lines of superior flavour and quality composition. More important is the management of diseases and pests, especially developing resistant or tolerant lines against the two serious virus diseases – *foorkey* and *chirke*. The rich genetic diversity found in the centre of origin (Sikkim and adjoining areas) has to be screened for locating natural resistance against diseases and pests, and for locating superior genotypes. A search for types adaptable to lower elevation will be a boon to take the cultivation of this crop to the lower hills. A good tissue culture protocol for large cardamom is available and large scale multiplication of some of the elite lines will provide disease free superior planting material to the growers. Appropriate research back up is also needed to provide and popularize an efficient drying technology that will be suitable to the location and acceptable to small and medium growers.

REFERENCES

Alhwat, Y.S., Raychaudhuri, S.P., Yora, K. and Dot, Y. (1981) Electron mircroscopy of the virus causing Foorkey disease of large cardamom, *Amomum subulatum* Roxb. *Natl. Acad. Sci. Letters*, 4(4), 165.

Annamalai, S.J.K., Patel, R.T. and John, T.D. (1988) Improved curing method for large cardamom. *Spice India*, 1(4), 5.

Anonymous (1950) *Amomum. Garden Bull.*, 13, 192–214.

Anonymous (1984) Large cardamom–package of practices. Cardamom Board, Gangtok, Sikkim.

Anonymous (1998a) Spices Board's Annual report (1997–98). Spices Board, Sugandha Bhawan, Cochin.

Anonymous (1998b) Design, development and field-testing of an advanced cardamom-curing prototype for Sikkim. Project Report, Tata Energy Research Institute, New Delhi.

Azad Thakur, N.S. (1987) Seasonal incidence of insect pests of Large cardamom (*Amomum subulatum* Roxb.) in Sikkim. *Research Bulletin No. 21*, ICAR Res. Complex for N.E.H. Region, Shillong, Meghalaya, India.

Azad Thakur, N.S. and Sachan, J.N. (1987) Insect pests of large cardamom (*Amomum subulatum* Roxb.) in Sikkim. *Bull. Ent.*, 28, 46–58.

Balakrishnan, K.V., George, K.M., Mathulla, T., Narayana Pillai, O.G., Chandran, C.V. and Verghese, J. (1984) Studies in Cardamom-1. Focus on oil of *Amomum subulatum* Roxb. *Indian Spices*, 21(1), 9–12.

Basu, A.N. and Ganguly, B. (1968) A note on the transmission of 'foorkey disease' of large cardamom by the aphid, *Micromyzus kalimpongensis* Basu. *Indian Phytopathology*, 21, 127.

Bhowmik, T.P. (1962) Insect pests of large cardamom and their control in West Bengal. *Indian J. Ent.*, 24, 283–286.

Bhutia, D.T., Gupta, R.K. and Biswas, A.K. (1985) Fertility status of the soils of Sikkim. *Sikkim Science Society Newsletter*, 5.

Biswas, A.K., Gupta, R.K. and Butia, D.T. (1986) Characteristics of different plant parts of large cardamom. *Cardamom*, 19(3), 7–10.

Chattopadhyay, S.B. and Bhowmik, T.P. (1965) Control of 'foorkey' disease of large cardamom in West Bengal. *Indian J. Agri. Sci.*, 35, 272–275.

Dubey, A.K. and Singh, K.A. (1999) Large cardamom – a spice crop of India. *Indian Farming*, 49(2), 17–18.

Gamble, J.S. (1925) *The Flora of the Presidency of Madras*. Bot. Sur., India, Calcutta (Reprint).

Ganguly, B. (1966) A rapid test for detecting Chirke affected Large cardamom (*Amomum subulatum* Roxb.) plants in field. *Science and Culture*, 32(2), 95–96.

Gupta, P.N. (1983) Export potential in large cardamom. *Cardamom*, 16(1), 3–9.

Gupta, P.N. (1986) Shade regulation in Large cardamom. *Cardamom*, 19(8), 5, 7.

Gupta, P.N., Nagvi, A.N., Mistra, L.N., Sen, T. and Nigam, M.C. (1984) Gas chromatographic evaluation of the essential oils of different strains of *Amomum subulatum* Roxb. growing wild in Sikkim. *Sonderdruck ans Parfumeric and Kodmetik*, 65, 528–529.

Gupta, U. and John, T.D. (1987) Floral Biology of Large Cardamom. *Cardamom*, 20(5), 8, 15.

Gupta, U. (1989) Studies on germination of seeds of large cardamom. *Spice India*, 2(3), 14.

Gurudutt, K.N., Naik, J.P., Srinivas, P. and Ravindranath, B. (1996) Volatile constituents of large cardamom (*Amomum subulatum*). *Flavour and Fragrance J.*, 1, 7–9.

Gyatso, K., Tshering, P. and Basnet, B.S. (1980) Large cardamom of Sikkim. *Publication of Dept. Agric.*, Govt. of Sikkim, 2(4), 91–95.

Hooker, J.D. (1894) *The Flora of British India*, Vol. VI, Reeves and Co, London.

John, J.M. and Mathew, P.G. (1979) Large cardamom in India. *Cardamom*, 3(10), 13–20.

Joseph, J., Raghavan, B., Nangundaiah, G. and Shankaranarayana, M.L. (1996) Curing of large cardamom (*Amomum subulatum* Roxb.), Fabrication of a new dryer and a comparative study of its performance with existing dryers. *J. Spices and Aromatic Crops*, 5, 105–110.

Karibasappa, G.S. (1987) Post harvest studies in large cardamom. (*Amomum subulatum* Roxb.), *Sikkim Sci. Soc. News Letter*, 6(3), 2–10.

Karibasappa, G.S., Dhiman, K.R., Biswas, A.K. and Rai, R.N. (1987) Variability and association among quantitative characters and path analysis in large cardamom. *Indian J. Agric. Sci.*, 57, 884–888.

Karibasappa, G.S., Dhiman, K.R. and Rai, R.N. (1989) Half sib progeny analysis for variability and association among the capsule characters and path studies on oleoresin and its cincole content in large cardamom. *Indian J. Agri. Sci.*, 53, 621–625.

Karibasappa, G.S. (1992) Phenology of flowering in large cardamom at Gangtok. *Indian J. Agric. Sci.*, 59, 205–209.

Kirtikar, R.P. and Basu, B.D. (1952) *Indian Medicinal Plants*, Vol. 4, Balu, L.M., Pub., Allahabad (Rep.).

Kumar, A. and Chatterjee, S.V. (1993) Record of new pest *Bradysia* sp. and its predator, *Phaonia simulans* on large cardamom. *J. Appl. Zool. Res.*, 3(1), 103–104.

Mukhurji, M.K. (1968) Large cardamom cultivation in Darjeeling district of West Bengal. *Cardamom News*, 2 (2), 1–8.

Nirmal Babu, K., Ravindran, P.N. and Peter, K.V. (1997) *Protocols for Micropropagation of Spices and Aromatic Crops*, Indian Institute of Spices Research, Calicut.

Pangtey, V.S. and Azad Thakur, N.S. (1986) Insect pests of Large Cardamom. *Indian Farming*, 35(12), 17–19.

Rao, Y.S., Gupta, U., Kumar, A. and Naidu, R. (1990) Phenotypic variability in large cardamom. *Proc. Nat. Symp. on New Trends in Crop Improvement of Perrennial Species*, Rubber Research Institute of India, Kottayam.

Rao, Y.S., Kumar, A., Chatterjee, S., Naidu, R., and George, C.K. (1993) Large cardamom (*Amomum subulatum* Roxb.) – a review. *J. Spices & Aromatic Crops*, 2(1,2), 1–15.

Raychaudhary, S.P. and Chatterjee, S.N. (1958) A preliminary note on the occurance of a new virus disease of the large cardamom (*Amomum subulatum* Roxb.), in Darjeeling Dist. *Proc. Mycological. Res.* Workers conference, ICAR, 174.

Raychaudhary, S.P. and Chatterjee, S.N. (1965) Transmission of chirke disease of large cardamom by aphid species. *Indian J. Ent.*, 27, 272–276.

Raychaudhary, S.P. and Ganguly, B. (1965) Further studies on 'chirke' disease of large cardamom (*Amomum subulatum* Roxb.). *Indian Phytopathology*, 18, 373–377.

Ridley, H.N. (1912) *Spices*. McMillan & Co. Ltd., London.

Sajina, A., Mini, P.M., John C.Z., Nirmal Babu, K., Ravindran, P.N. and Peter, K.V. (1997) Micropropagation of large cardamom (*Amomum subulatum* Roxb.). *J. Spices & Aromatic Crops*, 6, 145–148.

Shankaracharya, N.B., Raghavan, B., Abraham, K.O. and Shankaranarayana, M.L. (1990) Large cardamom – Chemistry, technology and uses. *Spice India*, 3(8), 17–25.

Sharma, D.C., Raychaudhuri, S.P. and Capoor, S.P. (1972) Countdown on the cardamom virus. *Farm and Parliament*, 7(12), 13–15.

Sharma, E. and Ambasht, R.S. (1984) Seasonal variation in nitrogen fixation by different ages of root nodules of *Alnus nepalensis* plantation in the eastern Himalayas. *J. Appl. Eco.*, 21, 265–270.

Sharma, E. and Ambasht, R.S. (1988) Nitrogen accretion and its energetics in Himalayan alder plantations. *Functional Ecology*, 229–235.

Sharma, H.R. and Sharma, E. (1997) Mountain farming system, Mountain Agricultural Transformation processes and sustainability in the Sikkim Himalayas. International Centre for Integrated Development (ICIMOD), Kathmandu, Nepal. Discussion paper series, No. MFS 9712.

Singh, G.B. (1978) Large cardamom. *Cardamom*, 10(5), 3–13

Singh, G.B., Gupta, P.N. and Pant, H.G. (1978) Large cardamom, a foreign exhange earner from Sikkim. *Indian farming*, 3, 7–8.

Singh, K.A., Rai, R.N. and Bhutia, D.T. (1989) Large cardamom (*Amomum subulatum* Roxb.) plantation: on age old agroforestry system in eastern Himalayas. *Agrofrestry System*, 9, 241–257.

Singh, B.P. (1988) Large Cardamom in Sikkim. *Spices News letter*, 22(10), 21.

Singh, J., Sudharshan, M.R. and Kumaresan, D. (1994) Cardamom foliar pest, lace-wing bug (*Stephanitis typica* (Distant) (Heteroptera: Tingidae). *Ann. Entomol.*, 12(2): 81–84.

Singh, J., Srinivas, H.S. and John, K. (1998) Large cardamom cultivation in Uttar Pradesh Hills. *Spice India*, 11(4), 2–4.

Singh, J. and Varadarasan, S. (1998) Biology and management of leaf caterpillar, *Artona chorista* Jordon (Lepidoptera: Zygaenidae), a major pest of large cardamom (*Amomum subulatum* Roxb.). *PLACROSYM-XIII*, Coimbatore, 1998, pp. 46–47 (Ab).

Srivastava, L.S. (1991) Occurrence of spike, root and collar rot of large cardamom in Sikkim. *Plant Diseases Res.*, 6, 113–114.

Srivastava, L.S. and Varma, R.N. (1987) *Amomum subulatum*, a new host for *Phakospora elettariae* (Roxb.) Cumins from Sikkim. *Curr. Sci.*, 56, 544.

Subba, J.R. (1979) Efficacy of commonly used insecticides for the control of hairy caterpillar, *Clelea plumbiola* Hmp. on large cardamom, *Amomum subulatum* Roxb. *Krishi Samachar*, 1(3), 17–19.

Subba, J.R. (1980) Field problems of large cardamom cultivation in Sikkim. *Cardamom*, 12, 3–5.

Subba, J.R. (1984) *Agriculture in the Hills of Sikkim*. Sikkim Science Society, Gangtok, Sikkim, p. 286.

Sundriyal, R.C., Rai, S.C., Sharma, E. and Rai, Y.K. (1994) Hill agroforestry, systems in South Sikkim, India. *Agroforestry systems*, 26, 215–235.

Upadhyaya, R.C. and Ghosh, S.P. (1983) Wild cardamom of Arunachal Pradesh. *Indian Hort.*, 27(4), 25–27.

Varma, P.M. and Capoor, S.P. (1964) 'Foorkey' disease of large cardamom. *Indian J. Agri. Sci.*, 34, 56–62.

Varma, S.K. (1987) Preliminary studies on the effect of honey bees on the yield of greater cardamom. *Indian Bee J.*, 49, 25–26.

Yadava, S., Kumar, A. and Naidu, R. (1992) Biology of *Artona chorista* Jordon (Lepidoptera: Zygaenidae) on large cardamom. *Proc. Bioecol. and Control Ins. Pests*, pp. 73–78.

16 False cardamoms

P.N. Ravindran, M. Shylaja and K. Nirmal Babu

1 INTRODUCTION

The cardamom of commerce (also known as small cardamom) is the true cardamom, the 'queen of spices', and is obtained from the plant botanically known as *Elettaria cardamomum* Maton, a member of the family Zingiberaceae. In addition, there are many other plants belonging to the genera *Amomum* and *Aframomum*, both belonging to the cardamom family, producing aromatic seeds. Among them the most important and the one that is being grown commercially is *Amomum subulatum* Roxb., the Nepal cardamom or large cardamom (see Chapter 15). The others are clubbed together as false cardamoms. They do not have much of commercial importance except in the case of *Aframomum corrorima* (*A. melegueta*, the Korarima cardamom or 'grains of paradise'), which is cultivated on a small scale in some West African countries. But most of these false cardamoms are important locally as spices and flavourents and as remedies for various ailments. Detailed studies on these species are lacking, though a few have been subjected to chemical analysis. Brief notes on various false cardamoms are given in this chapter.

2 ELETTARIA

2.1 Sri Lankan wild cardamom *Elettaria ensal* (Gaertn.) Abheywickrme (*E. major* Thawaites)

This is the Sri Lankan wild cardamom. Morphologically similar to the true cardamom, but a more robust plant, bearing erect panicles and much elongated fruits (3–5 cm) (see Chapter 2, section 1.2). Some controversy still exists regarding the taxonomic entity of this species. Burtt (1980), as well as Burtt and Smith (1983) did not treat this as a separate species, but included under *E. cardamomum* only. Abheywickreme (1959) treated this as a separate species because of its more robust nature, distinctly different fruit size and of course the chemical composition is very different. Burtt is of the opinion that these characters are not sufficient enough to separate the Sri Lankan wild cardamom into a different species. The fruits contain more seeds and the taste and flavour are distinctly different from the true cardamom. Oil separated from seeds through the cold expression process was analyzed by Bernhard *et al.* (1971). The two most important constituents of cardamom oil, 1,8-cineole and α-terpinyl acetate, are present only in traces in Sri Lankan wild cardamom. Bernhard *et al.* (1971) carried out comparative chemical analysis of the different cardamoms and reported the following constituents in the oil of Sri Lankan wild cardamom:

Trans-sabinene hydrate (22.2 per cent), α-terpinenol (15.3 per cent), α-pinene (13 per cent), α-terpinene (11.2 per cent), sabinene (4.9 per cent), β-pinene (4.9 per cent), methyl heptenone (4.1 per cent), linalool (3.7 per cent), 1,8 cineole (3.3 per cent), myrcene (2.5 per cent), D-limonene (2.5 per cent), *cis*-citral (1.3 per cent), 4-terpinyl acetate (1.7 per cent), geranyl acetate (1.5 per cent), α-terpineol (0.86 per cent), citronellol (0.7 per cent), *trans*-nerolidol-(0.44 per cent), *cis*-nerolidol (0.37 per cent), linalyl acetate (0.31 per cent), α-terpinyl acetate (0.14 per cent) and traces of farnesol, ascaridole, neryl acetate, DL-borneol etc.

Both 1,8-cineole and α-terpinyl acetate, which are predominant in true cardamom, are present only in negligible quantities in *E. ensal*. Similarly the most important constituent of *E. ensal, trans*-sabinene hydrate is practically absent in true cardamom; so also 4-terpinenol, α-pinene and α-terpinene.

Steam-distilled oil from the wild Sri Lankan cardamom, has shown differences from the above constituents based on the analysis of cold expressed oil. The steam-distilled oil was found to have the following major compounds (Rajapakse-Arambewela and Wijesekeara, 1979):

p-cymene (35 per cent), 4-terpineol (29 per cent), α-thugene (3.8 per cent), linalool (3.6 per cent), α-terpinyl acetate (3 per cent), *trans* sabinene hydrate (2.5 per cent) etc. No information is available on other aspects of the plant. The seeds are used by local people for flavouring, and also in tribal medicine, otherwise has little commercial importance.

2.2 MALAYSIAN AND INDONESIAN *ELETTARIA*

Elettaria longituba (Ridl.) Holtt., in Gard. Bull. Sing., 13, 238, 1950.

> *E. longituba* is a perennial, vigourous herb endemic to Malaysia, and seems to be conspecific with *E. aquatilis*, reported from Sumatra. It is a large species, its flowers appear singly at longer intervals, only very few flowers occur per cincinnus, flowering stolons (panicle) is 3–4 m long, anther dehiscing by pores without a hairy flap, fruit is long, globose or pyriform, smooth and slightly ribbed. There is no reported use (Holttum, 1950).

Recently Sakai and Nagamasu (2000) while studying the Zingiberaceae of Indonesia listed seven species of *Elettaria* from the region:

> *E. rubida* R.M.Sm., in Bot. J. Linn. Soc., 85, 66, 1982.

> Inflorescence red with orange flowers, anther ecristate, leaves obovate with attenuate base.

> *E. stolonifera* (K. Schum.) Sakai and Nagamasu, Edinb. J. Bot., 57, 227–243, 2000.

> Plant 0.8–1.5 m tall. Inflorescence 25–60 cm long, trailing on the ground, flowers white, anthers 3 mm long, anthers dehiscing in the upper half only, anther crest 3-lobed, fruit unknown.

> *E. kapitensis* Sakai and Nagamasu, Edinb. J. Bot., 57, 227–243, 2000.

> Plant about 0.7 m tall, inflorescence 40–60 cm long, flowers white, calyx 17 mm long, corolla tube and calyx free above the ovary, anther dehiscing through out their length, anther crest ca. 2mm long, fruit unknown.

E. surculosa (K. Schum.) B.L. Burt and R.M. Smith, in Notes Roy. Bot. Gard. Edin. 31, 312, 1972 (syn. *E. multiflora* R.M. Sm. in Notes Roy. Bot. Gard. Edinb., 43, 452, 1986).

Plant about 1.2–2.0 m tall, inflorescence, upto 2 m long, creeping on the ground, flowers white, anther dehisce by small pores, pore covered with a flap with long hairs, anther crest up to 2 mm long, fruit ca. 40 × 16 mm, ellipsoid, sparsely pubescent, reddish brown.

E. linearicrista Sakai and Nagamasu Edinb. J. Bot., 57, 227–243, 2000.

Plant 0.6–1.5 m tall, leaves narrowly oblong, inflorescence ca. 40 cm long, axis densely pubescent, somewhat erect, flowers white, anther dehiscing by longitudinally elongated pores at the upper middle of the thecae, fruit unknown.

E. longipilosa Sakai and Nagamasu Edinb. J. Bot., 57, 227–243, 2000.

Plant 1–1.3 m tall, leaves densely pubescent on the lower surface, inflorescence ca. 50 cm long, creeping just below the ground, flowers white, anther thecae dehiscing by a small pore just above mid-point, pore covered with a flap with very long hairs, anther crest deeply three lobed, fruit unknown.

E. brachycalyx Sakai and Nagamasu Edinb. J. Bot., 57, 227–243, 2000.

Plant about 1.0 m tall, inflorescence ca. 60 cm long, flower white, calyx ca. 6 mm long, fissured for two thirds of its length, anther thecae dehiscing by pores covered with a flap of long hairs, anther crest ca. 1 mm long, 3-lobed, fruit unknown.

3   *AFRAMOMUM* SP.

3.1   Korarima cardamom: *Aframomum corrorima* (Braun). Syn., *A. melegueta* (Roscoe) Schum. (Meleguetta pepper, grains of paradise or alligator pepper)

Korarima cardamom grows in wild and is also cultivated sporadically in Ethiopia, Nigeria and the nearby regions and this species is endemic to that area. It is a perennial aromatic herb having strong, fibrous, subterranean scaly rhizomes and with a leafy stem of ca. 1–2 m high. Rhizomes are subterete, brown–reddish brown, smooth, glabrous, scales brown, thin, deciduous. Leafy stem unbranched, formed by leaf sheaths. Leaves distichous 10–30 × 2.5–6 cm, leathery, entire, glabrous, dark green above, lighter green below. Inflorescence is short stalked (condensed spike), arise from the root stock near the base of the leafy stem, peduncle up to 7 cm long, covered by imbricate brown–purplish-brown glabrous scales, ca. 5-flowered, each flower surrounded by a bract, structure typical zingiberaceous, having tubular corolla, 3-lobed at the apex, 3.5–4.5 cm long, white–pale violet, glabrous outside; labellum obovate, having a tubular fleshy claw of up to 3 × 1.5 cm and a glabrous lobe of 3 × 3.5 cm, lobe pale violet, but yellow at throat inside, fertile stamen 1, ovary inferior, sub-cylindrical, glabrous, 3-locular with numerous ovules, placentation axial, fruit indehiscent, fleshy, shiny green when immature, turning light red at maturity, seeds glabrous, light to dark brown with thin aril covering the seed; seeds have small endosperm, copious perisperm, embryo single, 1.5 × 0.5 mm (Jensen, 1981).

The plants grow naturally at 1700–2000 m altitude in Ethiopia, and in other tropical West African countries, they flower from January to September, fruits mature in 2–3 months after flowering. Usually bees are the pollinators. Propagated both by seeds and rhizome parts. No cultivation practices have been recorded. Harvesting is done when the fruits turn red. They are then dried in the sun. The dried products are usually of poor quality due to improper drying. Dried fruits are sold in local markets and exported to other African countries and small quantities are exported to Saudi Arabia, Iran and Gulf countries. A rust disease caused by *Puccinia aframomi* Hansfood is common in this plant.

3.1.1 Composition

Seeds contain 1–2 per cent essential oil. The oil has a typical odour and is sometimes called 'nutmeg-cardamom'. Lawrence (1970) steam distilled dried, powdered fruits and obtained 3.5 per cent of pale yellow volatile oil with a flat, cineolic odour. The major components of the oil are:

1,8-cineole (35.1 per cent), limonene (13.5 per cent), α-pinene (3.2 per cent), β-pinene (6.8 per cent), sabinene (6.7 per cent), γ-terpinene (2.6 per cent), p-cymene (3.9 per cent), terpinen-4-ol (5.4 per cent), α-terpineol (3.4 per cent), geraniol (4.8 per cent) and small quantities of myrcene, α-phellandrene, α-terpinene, camphene and terpinolene. The seeds also contain hydroxy phenyl alkanones such as (6) paradol, (7) paradol and (7) gingerol. The pungency of the grains is due to the presence of paradol. Escoubas *et al.* (1995) reported in addition to the above compounds, gingerdione.

Ajaiyeoba and Ekundayo (1999) carried out a gaschromatography–mass spectrophotometry (GC–MS) study on hydrodistilled oil from samples purchased from local market in Nigeria. The oil yield is about 1.21 (w/w), 27 compounds have been identified, and the most important ones are: α-humulene (60.9 per cent) and β-caryophyllene, which together are making up 82.6 per cent of the oil. Two other sesquiterpenoid hydrocarbons (germacrene-D and δ-cadinene) occur in trace amounts (0.1 per cent). Three oxides (humulene oxide II 5.5 per cent, caryophyllene oxide (2.9 per cent, and humulene oxide I (10.6 per cent) occurred in relatively higher amounts in the volatile oil. Thus the main compositional feature of the Nigerian sample of *A. corrorima* is the dominance by α-humulene, β-caryophyllene and their oxides (91.6 per cent). This is very different from the composition reported by Lawrence (1970) In addition seven sesquiterpene alcohols (nerolidol, T-murolol, δ-cadinol, α-cadinol, humulenol and caryophylladienol I and caryophylladienol II) were also detected in trace amounts. The hexane extract of seeds contains pungent compounds such as zingerone (3.7 per cent), 6-paradol, 6-gingerol and 8-gingerol (Tackie *et al.*, 1975).

Menut *et al.* (1991) made a comparative study on the oil composition of *A. melegueta*, (*A. corrorima*) *A. danielli* and *A. sulcatum*, using GLC and GC–MS. They reported oil yields of 0.8 per cent (w/w) for *A. corrorima*, 2.3 per cent for *A. danielli* and 1 per cent for *A. sulcatum*. They had distinct odours. GC–MS analysis indicated wide variation in the chemical composition of the three oils. The major components in *A. corrorima* were α-humulene (31.3 per cent), humulene oxide (26.4 per cent), Caryophyllene oxide (17.9 per cent), β-caryophyllene (8.5 per cent), linalool (5.7 per cent) 1-Methylhexyl acetate (3.3 per cent), and lesser quantities of γ-cadinene, heptan-2-ol etc. *A. danielli* oil contained predominantly 1,8 cineole, (48.9 per cent), limonene (12.8 per cent), β-pinene (12.7 per cent), α-terpineol (7.4 per cent), and lesser quantities of α-pinene, β-caryophyllene (1.12 per cent), δ-guaiene (1.1 per cent), γ-cadinene (2.8 per cent) etc. The oil of *A. sulcatum* was different from the other two, containing mainly: α-guaiene (26.7 per cent), δ-guaiene (18.1 per cent), limonene (10.5 per cent), β-caryophyllene (5 per cent), γ-cadinene (4.8 per cent), β-selinene (7.1 per cent), β-elemene (3.9 per cent), α-selinene (2.6 per cent), β-guaiene (2.4 per cent), α-humulene (2.8 per cent) and lesser quantities of β-pinene, *trans*-β-ocimene, germacrene-D, humulene oxide etc. Monoterpene hydrocarbons were absent in *A. corrorima*, constituted 30 per cent in *A. danielli* and 13.1 per cent in *A. sulcatum*. Oxygen containing monoterpenes was in the order 5.7 per cent in the first, 56.3 per cent in the second and 0.4 per cent in the third species. The presence of sesquitenpene hydrocarbons were to the tune of 41.3 per cent in *A. corrorima*, 11 per cent in *A. danielli* and 81.1 per cent in *A. sulcatum*. The oxygen-containing sesquitenpenes constituted 48.7 per cent of oil in the first, 0.7 per cent in the second and 3.3 per cent in the third species, Oxygen containing non-terpenoid compounds were 3.9 per cent in *A. corrorima* and were absent in the other two species (Menut *et al.*, 1991).

Menut *et al.* (1991) further compared the chemical composition of *Aframonum* and *Amomum* and came up with the following classification:

Type 1
Aframomum angustifolium   Roughly equal amounts of hydrocarbons
A. corrorima   and oxygen containing compounds
A. mala
Amomum muricatum
Type 2
Amomum compactum
A. kravanh   Oxygen containing monoterpenes with
A. subulatum   1,8-cineloe as the major constituent.
Type 3
Amomum globosum (from Thailand)   Oxygen containing monoteperes different
A. villosum   from 1,8-cineloe
A. giganteum
Type 4
Amomum globossum (from China)   Sesquiterpenes, α-humulene and farnasol
Type 5
Amomum ptychloimatum   Aliphatic compounds

3.1.2 Uses

Korarima cardamom was once important in trade with Europe, and part of the West African coast was then known as the Grain Coast (now Liberia). It was in demand as a spice and stimulating carminative and for a variety of other ailments by local people. The fruit pulp around the seed is eaten especially before maturity and is chewed as stimulant. The seeds are used as a spice in food to flavour all types of dishes in tropical African countries, used in flavouring tea and coffee, and for some special kinds of breads.

In Ghana the seeds are used in enema preparations. In Lagos and Sierra Leone the seeds are a remedy for inflamed conditions of the throat, and in fevers. The tribe Susus, treat acute inflammation of the throat and tonsils with the bruised young leaves of the tree 'apil' (*Vismia leonensis*, Clusiaceae), along with the crushed fresh seeds of this cardamom in small quantities of water taken repeatedly. An extract of the seeds dropped in an infant's mouth ensures evacuation of the meconium, as well as in the treatment of yaws (iron rust boiled with lime juice and with the addition of ants and grains of paradise). Seeds when chewed with Kola produce a benumbing effect and are used to relieve toothache. For external use seeds are crushed and rubbed on the body as a counter irritant, or applied as a paste for headache, ear ache etc. and pulverized to put on wounds and sores (Dalziel, 1937). The whole plant decoction is taken as a febrifuge, the roots have a cardamom like taste, and a decoction is given for constipation. The roots are regarded as a vermifuge for tapeworm. The juice of the young leaves acts as a styptic (Dalziel, 1937).

This was used in UK and USA in the preparation of medicines for cattle and also in the preparation of the spiced wine (called hippocras), flavoured with korarima cardamom, cinnamon and ginger (Bently and Trimen, 1880).

Rafatullah *et al.* (1995) investigated the antiulcer and gastro protective activities of an ethanol extract of korarima cardamom seeds in rats using pyloric legation, hypothermic restraint stress, indomethacin (indometacin), cystamine and narcotizing agents. The extract (250 and 500 mg/kg, ip or po) exhibited dose-dependent inhibition of basal gastric secretion in pylorus-ligated rats and reduced significantly the intensity of gastric lesions induced by pyloric-legation, hypothermic-restraint stress, indomethacin and mucosal damaging agents and duodenal ulcers induced by cystamine. The concentration of non-protein sulfhydryls decreased significantly in gastric mucosa after administration of 80 per cent ethanol; treatment with the extract of 500 mg/kg (p.o) of korarima cardamom seed prevented this depletion. It has been suggested that the extract of korarima cardamom exerts antiulcer effects due to its anti-secretary activity.

Galal (1996) has studied the antimicrobial activity of 6-paradol and related-compounds present in korarima cardamom. These compounds were active against *Mycobacterium chelonei, M. intracellulare, M. smegonatis* and *M. xenopi* minimum inhibitory concentration (MIC) values of 10–15 μs/ml). The desmethyl derivative of 6-paradol was more active against *Candida albicans*. Escoubas *et al.* (1995) have shown that extract of seeds exhibited antifeedent activity against termites.

3.2 *A. daniellii* Schum

A perennial herb occurring in West Africa, growing to about 75 cm, having leaves of about 25–30 cm. Flowers red, many in an inflorescence, lip pink, yellow at base, and margin white. Fruit capsule red, smooth, seeds smooth, shining olive brown, with a turpentine like taste. Pulp is agreeably acidic, eaten by the people to refresh and relieve thirst in fever. *A. daniellii* inhibited the growth of *Salmonella enteriditis, Pseudomonas fragi, P. fluorescense, Proteus vulgaris, Streptococcus pyogenes, Staphylococcus aureus, Aspergillus flavus, A. parasiticus, A. ochraceus* and *A. niger*. The minimum inhibiting concentration determined for *Klebsiella pneumoniae* and *Pseudomonas aeruginosa* was 1 in 320, while the MIC for *S. aureus* was 1 in 8000 (Adegoke and Skura, 1994).

3.3 *A. granum – paradisi* K. Schum (black *Amomum*)

The immature fruit is dark in colour, hence known as black *Amomum*. Occur in Ghana, Gambia, Sierra Leone, S. Nigeria. Also called Anitelope's pepper. The fruit capsule is ten or more groved and succulent. The seeds are sub-globose or slightly angular having smooth and shining surface. The seeds are aromatic and possess camphoraceous taste and flavour. The seeds are used by people together with shea butter (from *Vittellaria paradoxa* Gaertner, Sapotaceae) to make body pomade. Whole plant is used medicinally as in the case of *A. corrorima*. Leaves are cooked with food for imparting flavour. Wrongly identified as grain of paradise earlier.

3.4   *A. latifolium* K. Schum (large or grape seeded Amomum)

Indigenous to the tropical African country of Sierra-Leone. The seeds are like grape stones, smooth, shining and weakly aromatic. The ripe fruit is acidic and is eaten by natives; given to refresh and to relieve thirst in fever and is said to be good for fatigue (Dalziel, 1937). A decoction of the whole plant is mixed with the leaves of *Morinda lucida* (Rubiaceae) and used by the tribe Susus as a daily bath to restore strength after fever.

3.5   *A. biauriculatum* K. Schun.

Chisowa *et al.* (1998) analyzed the rhizome of this species and identified 29 compounds in the rhizome oil and 17 compounds in the leaf oil. The major components of leaf essential oil were β-pinere (58.5 per cent), β-caryophyllene (14.2 per cent), α-pinene (7.3 per cent) and *trans*-sabinyl acetate (5.6 per cent). The rhizome oil was dominated by β-pinene (23.8 per cent), 1,8-cineole (22.5 per cent), α-terpineol (5 per cent), γ-pinene (4.9 per cent), α-pinene (4.6 per cent) and limonene (4.6 per cent).

3.6   *A. sanguineum* K. Schum

It is a native of West Africa. The seeds of this plant are used in cosmetics, to flavour tea and rice, and in traditional medicine to prevent throat infections and to treat stomachache, dysentry and snake bites. Twenty-five constituents were identified in the essential oil, important ones are 1,8-cineole (38.5 per cent), alpha terpinyl acetate (9.6 per cent) and geranyl acetate (9.4 per cent) (Hari *et al.*, 1994).

4   *AMOMUM* SP.

4.1   Bengal cardamom *Amomum aromaticum* Roxb. (Jalpaiguri cardamom)

A perennial herb, growing to about 1 m height, indigenous to the northern West Bengal, Assam, Meghalaya, Sikkim, North Bihar and northern West Bengal, extending to Nagaland and UP. Cultivated in the northeastern regions at the foot of Himalayas. Leaves oblonglanceolate, glabrous, 15–30 cm $\times$ 5–10 cm, flowers pale yellow, in small globose spikes, corolla-tube white tinged with brown, capsules rugose, 25 cm long, trigonous, oblong or globose, with 7–13 narrow membranous longitudinal wings, seeds many in each cell.

The seeds on steam distillation yield 1–1.2 per cent of an essential oil, having strong camphoraceous and cineol-like odour and taste; contains large quantity of cineole. The seeds are used as spice and are medicinal, used in place of *A. subulatum*, the large cardamom.

4.2   *A. dealbatum* Roxb.

A herbeaceous perennial with large oblong-lanceolate leaves and globose short, ped-uncled spike, distributed in Khasi hills and eastern Himalayas. The leaves on distillation give an oil (0.018 per cent), having 1-β-pinene as the main constituent, and smaller quantities of α-pinene. The seeds are feebly aromatic in taste and odour. Seeds are used as a substitute for large cardamom.

4.3   *A. longiligulare* Wu

A species occurring in Vietnam, and used locally for medicinal purposes as well as for flavouring; used by locals as a remedy for dyspepsia, vomiting, diarrhea and dysentry. The seed oil contains 47.14 per cent camphor, 39.12 per cent D-bornyl acetate and smaller quantities of α-pinene (3.84 per cent), β-pinene (1.48 per cent), D-limonene (4.47 per cent), D-borneol (2.48 per cent), camphene (0.04 per cent), linalool (0.54 per cent), caryophyllene oxide (0.13 per cent) and β-caryophyllene (0.09 per cent) (Dung and Thin, 1992).

4.4   *A. tsao-ko* Crevost Lemarim (Vietnam cardamom)

A herb, 2–3 m high, leaves oblong-lanceolate, lamina 60–70 $\times$ 18–20 cm, glabrous, spike globose, peduncle short, outer bract 2.5 cm long, ovate, corolla tube pale brown 2.5 cm long, lip pale yellow. Anther crest large, petalloid, lobes rounded. Capsules 2.5 cm long, oblong, trigonous and three valves with numerous seeds in each cell. Cultivated in South China, Vietnam especially at the forest regions of the Hoang Lien Son Mountain.

Seeds contain 1–2 per cent of essential oil, the main constituents of which are: 1,8-cineole (30–34 per cent), *trans*-2-undecenal (17.33 per cent), geranial (10.5 per cent), α-terpineol (4.34 per cent), neral (6.98 per cent), 2-phenyl 2-butenal (4.38 per cent), methyl cinnamic aldelyele (2.82 per cent) nerylacetate (1.66 per cent), *trans*-2-dodecenal (3.8 per cent) and other miscellaneous compounds (11.84 per cent). About 80 chemical constituents were reported from the essential oil (Lizhu-Qiang *et al.*, 1998).

The seeds of *A. tsao-ko* are used as spice in Vietnam and China in place of true cardamom. It is also used as a local remedy for many ailments including dyspepsia, diarrhoea, indigestion etc.

Seven antioxidative compounds were reported from *A. tsao-ko* (Fangjen and De, 1996).

4.5   *A. krevanh*

Indigenous to Thailand (some times called Thai cardamom) where it is used in traditional medicine. Kamchonwongpaisan *et al.* (1995) has isolated novel diterpene peroxide, the structure of which was elucidated by them. This compound exhibited potent activity against *Plasmodium falciparum*.

4.6   *A. xanthiodis* Wall (Tavoy cardamom)

Distributed in the forests of Indo-Burma, Malayan region. Perennial herb, leafy stem reaching 1.5–2 m; spikes globose, few flowered and shortly peduncled; fruits (capsule) echinate, oblong, trigonous, pale brown. The fruits are used in diarrhea, dysentry, cramps, hiccups, nausea, splenitis, stomachache, anodyne, antitussive, apenient, carminative and sedative. Seed is said to hasten the solution of coins, fishbones or other foreign substances accidentally swallowed. Seeds contain D-borneol, borneol acetate, D-camphor, linalool, nerolidol and terpene. Seeds are effective in relieving painful urination, and for regulating the frequency of motions. For this, powdered seed mixed with butter is administered. The Chinese and Burmese use this as an important condiment (Kirtikar and Basu, 1952).

4.7   *A. compactum* Soland (round cardamom) syn. *A. kepulaga* Sprague and Burkill

Round cardamom, also known as Siam cardamom, is reported to be antitoxic, antiemetic, carminative and stomachic. In China, it is a folk remedy for ague, cachexia, cancer, catarrh, cold, cough, cramps, dyspepsia, gout, heartburn, hepatosis, nausea, ophthalmic, rheumatism, and vomiting. Rarely used alone, more frequently in combination with other plant products. Mixed with fresh egg yolks are used during parturition. Elsewhere the plant is used in folk remedies for indurations of the liver and uterus and for cancer (Perry and Metzger, 1980).

4.8   *A. constalum* Benth

Endemic to the eastern Himalayan Forests. Perennial herb with stout leafy stem 1.5–2 m; spikes 5–7.5 cm, oblong shortly peduncled fruit ovoid, strongly ribbed, smooth, seeds many, aromatic. In Chinese medicine the seeds are employed for ailments of the stomach and for asthma, pulmonary afflictions, and general debility. Tribals use the seeds for relieving stomachache.

4.9   *A. pavieanum* Pierre and Gagnep.

A. pavieanum is endemic to the south-east Asia. An infusion of the fruits of this plant is used in the treatment of diarrhoea and general debility following certain dysentries. The rhizome is used as seasoning agent in the eastern and southern parts of Thailand. (Perry and Metzger, 1980). Scheffer *et al.* (1988) studied the chemical composition of oil of rhizome extracted by steam distillation. The oil (yield is about 0.23 per cent vol/wt) is colourless having an anise-like odour. GC analysis has identified 41 compounds in rhizome oil. The dominant compound of the oil is methylchavicol (91.6 per cent). Only four other components amounted to more than 1 per cent: α-pinene, camphene, β-pinene and camphor. The composition of *A. pavieanum* is quite distinct from other species, by the absence of 1,8-cineole as the dominant component. It seems that there is no other *Amomum* sp. having oil composition resembling that of *A. pavieanum* (Scheffer, 1988).

4.10   *A. acre* Valeton

An Indonesian species mainly found in Sulawesi region. The fruits and inner part of petioles are used for pickling, often pickled in vinegar and subsequently used as a spice and a flavorant. Very young stem is used directly, the plant itself is very poorly understood. Two forms have been distinguished – *rombo* and *kautopi* (de Guzman and Siemonsma, 1999). Information on chemical composition is lacking.

4.11   *A. ochreum* Ridley

Occur in Peninsular Malaysia. The seeds of this species are used locally as a substitute for cardamom. The plant is 3–4 m tall, produces inflorescences (spikes) from the base, flowers having obovate, 3-lobed labellum, orange-yellow with red veins and spots, anther appendage transversely oblong, faintly red-spotted (de Guzman and Siemonsma, 1999). Common in peninsular Malaysia.

4.12   *A. testaceum* Ridley

Occur in Thailand, Vietnam, Peninsular Malaysia, and Borneo; occasionally cultivated. Seeds are aromatic and locally used like true cardamom.

Rhizomatous herb with leafy shoots about 3 m tall, inflorescence an oblongoid spike emerging from the rhizome, flowers have obovate labellum, white with a broad dull yellow patch towards the apex and paler yellow median band flanked by purple lines, stamens with 8 mm long filament and a 3-lobed anther appendage (de Guzman and Siemonsma, 1999). Fruits slightly ribbed and pinkish, slightly hairy. Seeds brown, aril thin, and aromatic. Chemical composition not known.

4.13   *A. xanthophlebium* Baker (Syn. *A. stenoglossum* Baker)

Occur in peninsular Malaysia and Borneo. Flowers are used for flavouring curries. Rhizomatous herb, leaf stem up to 4–5 m tall, inflorescence are ellipsoidal spikes arising from the rhizome, flowers have obovate labellum, edges crinkled, white suffused with red stripes and spots and yellow stripes towards the apex, anther connective 3-lobed, filament 1 cm long, pink. Fruit obovoid, smooth appressed silky hairy (de Guzman and Siemonsma, 1999). Flowers are used by locals for flavouring various dishes. Seeds are used by tribals in local medicinal preparations. No information on chemical composition.

The following species of *Amomum* occur in the Western Ghat forests of South India: *Amomum involucratum* Benth., *A. hypoleucum* Thw., *A. cannicarpum* (Wight) Benth. ex Baker., *A. muricatum* Beddome., *A. ghaticum* Bhatt., *A. masticatorium* Thwaites and *A. pterocarpum* Thwaites. There is no information about their use in local or tribal medicine. No study has gone into the chemical composition of the seeds of these species.

REFERENCES

Abheywickreme (1959) A provisional checklist of the flowering plants of Ceylon. *Ceylon J. Sci. (Biol. Sci.)*, 2, 119–240.

Adegoke, G.O. and Skura, B.J. (1994) Nutritional profile and antimicrobial spectrum of the spice *Aframomum danielli* K. Schum. *Plant Foods for Human Nutrition*, 45, 175–182.

Ajaiyeoba, E.O. and Ekundayo, O. (1999) Essential oil constituents of *Aframomum melegueta* (Roscoe) K. Schum. Seeds (alligator pepper) from Nigeria. *Flavour Fragr. J.*, 14, 109–111.

Bernhard, R.A., Wijesekera, R.O.B. and Chichestr, C.O. (1971) Terpenoids of cardamom oil and their comparative distribution among varieties. *Phytochemistry*, 10, 177–184.

Bently, R. and Trimen, H. (1880) *Medicinal Plants.* Vol. IV (Reprint), J.A. Churchill, London, pp. 267–268.

Burtt, B.L. (1980) Cardamoms and other Zingiberaceae in *Hortus Malabaricus.* In K.S. Manilal (ed.) *Botany and History of Hortus Malabaricus,* Oxford and IBH, New Delhi, pp. 139–148.

Burtt, B.L. and Smith, R.M. (1983) Zingiberaceae. In M.D. Dassanayeke (ed.) *A Revised Handbook to the Flora of Ceylon.* Vol. IV, Amerind Pub., New Delhi.

Chisowa, E.H., Hall, D.R. and Farman, D.I. (1998) Volatile constituents of leaf and rhizome oils of *Aframomum biauriculatum* K. Schum. *J. Essential oil Res.,* 10, 447–449.

Dalziel, J.M. (1937) *The Useful Plants of West Tropical Africa.* Crown Agents for Overseas Dev., London, pp. 470–472.

de Guzman, C.C. and Siemonsma, J.S. (eds) (1999) *Plant Resources of South-East Asia.,* No. 13, *Spices.* Backhuys pub., Leiden.

Dung, N.X. and Thin, N.N. (1992) Some important medicinal and aromatic plants from Vietnam. In S.P. Raychandhuri (ed.) *Recent Advances in Medicinal, Aromatic and Spice Crops.* Vol. 2. Today and Tomorrows Pub., New Delhi, pp. 299–308.

Escoubas, P., Lajide, L. and Mizutani, J. (1995) Termite antifedent activity in *Aframomum melegueta. Phytochemistry,* 40, 1097–1099.

Fang Jen, W.U. and De, S.J. (1996) Isolation, purification and identification of antioxidative components from fruits of *Amomum tsao-ko. J. Chinese Agri. Chemical Soc.,* 34, 438–451.

Galal, A.M. (1996) Antimicrobial activity of 6-paradol and related components. *Intern. J. Pharmacology,* 34, 64–69.

Hari, L., Bukuru, J. and Pooter, H.L. de (1994) The volatile fraction of *Aframomum sanguineum* K. Schum. from Burundi. *J. Essential oil Res.,* 6, 395–398.

Holttum, R.E. (1950) The Zingiberaceae of Malay Peninsula. *Gard. Bull., Singapore,* 13, 1–249.

Jensen, P.C.M. (1981) *Spices, Condiments and Medicinal Plants in Ethiopia,* PUDOC, Wageningen.

Kamchonwongpaisan, S., Nilanonta, C., Tarnchompoo, B., Thebtaranonth, C., Thebtaranonth, Y., Yuthavong, Y., Kongsaeree, P. and Clardy, J. (1995) An antimalarial peroxide from *Amomum krevanh* Pierre. *Tetrahedron Lett.,* 36, 1821–1824.

Kirtikar, R.P. and Basu, B.D. (1952) *Indian Medicinal Plants.* Vol. 4. LM Balu Pub., Allahabad, (Rep.).

Lawrence, B.M. (1970) Terpenoids in two *Amomum* Species. *Phytochemistry,* 9, 665.

Lizhu-Qiang, Lei, L., Wang-Hua, D., Rong, H. and Yuan-Qing, Z. (1998) Chemical constituents of the essential oil in *Amomum tsao-ko* from Yunnan Province. *Acta Bot. Yunnanica,* 20, 119–122.

Menut C., Lamaty, G., Amvam Zollo, PH., Atogho, B.M., Abondo, R and Bessiere, J.M (1991). Aromatic plants of Tropical central Africa. V. Volatile oil components of three Zingiberaceae from Cameroon: *Aframomum melegueta* (Roscoe) K. Schum., *A. danielli* (Hook. f.) K. Schum. and *A. sulcatum* (Oliv. and Hanb.) K. Schum. *Flavour and Fragrance J.,* 6, 183–186.

Perry, l.M. and Metzger, J. (1980) *Medicinal Plants of East and South-East Asia.* MIT Press, Cam., USA.

Rafatullah, S., Galal, A.M., Al-Yahya, M.A. and Al-Said, M.S. (1995) Gastric and duodenal antiulcer and cytoprotective effects of *Aframomum melegueta* in rats. *Intern. J. Pharmacognosy,* 33, 311–316.

Rajapakse-Arambewela, L.S. and Wijesekera, R.O.B. (1979) G.L.C. study of the essential oil of wild cardamom oil of Sri Lanka. *J. Sci. Food Agric.,* 30, 521–527.

Sakai, S. and Nagamasu, H. (2000) Systematic studies of Bornean Zingiberaceae: II. *Elettaria* of Sarawak. *Edinb. J. Bot.,* 57, 227–243.

Scheffer, J.J.C., Vreeke, A., Looman, A. and Mondranondra, I-0. (1988) Composition of essential oil of the rhizome of *Amomum pavieanum* Pierre & Gagnep. *Flavour and Fragrance J.,* 3, 91–93.

Tackie A.N. *et al.* (1975) (cited from Menut *et al.* (1991)).

Appendix 1
Specification for cardamom

1 CARDAMOM CAPSULES AND SEEDS

1.1 Scope

This standard prescribes the requirements for cardamom, *Elettaria cardamomum* (L.) Maton (Chhoti elaichi) of the family Zingiberaceae.

This standard does not cover the capsules or seeds of the genus *Amomum*.

1.2 Requirements

1.2.1 Cardamoms with capsules

The cardamoms shall be nearly ripe fruits of *Elettaria cardamomum* (L.) Maton in the form of capsules, which have been dried. The capsules shall be of colour ranging from light green to brown, cream and white; global in size or three-cornered having a ribbed appearance. The capsules may be clipped and their pedicels removed. The capsules shall be well formed and with sound cardamom seeds inside. The capsules may also be bleached. The cardamoms may be graded on the basis of colour, clipping, size, mass per litre, bleaching or otherwise, proportions of extraneous matter, or place of origin.

The cardamom seeds: Capsules may also be decorticated and the separated seeds packed for trade purposes. The cardamom seeds may be graded on the basis of mass per litre and proportions of extraneous matter.

Taste and aroma or flavour: The taste and aroma or flavour of cardamom capsules and seeds shall be characteristic and fresh. The material shall be free from foreign taste and aroma or flavour, including rancidity and mustiness.

Freedom form moulds, insects etc.: Cardamom capsules and seeds shall be free from visible moulds and insect infestation.

Mass per litre: The mass of cardamom capsules and seeds shall be prescribed in Table A1.1. The mass of cardamoms contained in 1 l shall be determined in accordance with the method given in 4 of IS: 1797–1973.

Moisture content: The moisture content in all grades of cardamom except the grade "Bleached" or "Half bleached" shall not exceed 10 per cent when determined in accordance with the method given in 10 of IS: 1797–1973. The moisture content in the grade bleached or half bleached shall not exceed 13 per cent when determined by the same method.

Table A1.1 Grades and specifications for cardamom

Grade designation	Trade name	Extraneous matter per cent by mass max	Empty and malformed capsules, per cent by count max	Unclipped capsules, per cent by count max	Immature and shrivelled capsules, per cent by weight max	Blacks and splits, per cent by count max	Size (DIA) of hole in mm of the sieve on which retained	Mass in g/l minimum	Colour	General characteristics
Alleppey green										
AGEB	Cardamom extra bold	Nil	2.0	Nil	2.0	Nil	7.0	435		Kiln dried, 3 cornered and having a ribbed appearance
AGB	Cardamom bold	Nil	2.0	Nil	2.0	Nil	6.0	415	Deep green	
AGS	Cardamom superior	Nil	3.0	Nil	5.0	Nil	5.0	385	green or	
AGS1	Shipment green 1	Nil	5.0	Nil	7.0	10.0	4.0	350	light green	
AGS2	Shipment green 2	Nil	5.0	Nil	7.0	12.0	4.0	320	Creamy	
AGL	Light green	Nil	–	Nil	–	15.0	3.5	260		
Coorg green										
CGEB	Coorg extra bold	Nil	Nil	Nil	Nil	Nil	8.0	450	Golden to light	Global shape skin ribbed or smooth; the pedicels seperated
CGB	Bold	Nil	2.0	Nil	3.0	Nil	7.5	435	creamy	
CG1	Superior	Nil	3.0	Nil	5.0	Nil	6.5	415	Creamy	
CG2	Superior Coorg green or Motta green	Nil	5.0	3.0	7.0	Nil	6.0	385	light greenish to	
CG3	Shipment	Nil	5.0	5.0	7.0	10.0	5.0	350	greenish or	
CG4	Light	Nil	–	–	–	15.0	3.5	280	browinish to brown	

Grade	Description								Color	Remarks
Bleached or half bleached										
BL 1		Nil	Nil	Nil	Nil	Nil	8.5	340	Pale, creamy or dull white	Fully developed capsules bleached, global or 3-cornered with skin ribbed or smooth
BL 2		Nil	Nil	Nil	Nil	Nil	7.0	340		
BL 3		Nil	Nil	Nil	Nil	Nil	5.0	300		
Bleached white										
BW 1	Mysore/Mangalore bleachable cardamom A. clipped	Nil	1.0	Nil	Nil	Nil	7.0	460	White, light green or light gray	Fully developed capsules suitable for bleaching
BW 2	Mysore/Mangalore bleachable cardamom A. unclipped	Nil	1.0	Nil	Nil	Nil	7.0	460		
BW 3	Mysore/Mangalore bleachable bulk cardamom: clipped	Nil	2.0	Nil	Nil	Nil	4.3	435		
BW 4	Mysore/Mangalore bleachable bulk cardamom: unclipped	Nil	2.0	Nil	Nil	Nil	4.3	435		
Mixed										
MEB	Mixed extra bold	—	2.0	—	2.0	Nil	7.0	435	—	Dried and mixed capsules of different varieties of *Elettaria cardamomum*
MB	Mixed bold	—	2.0	—	2.0	Nil	6.0	415	—	
MS	Mixed superior	—	3.0	—	5.0	Nil	5.0	385	—	
MS 1	Mixed shipment I	—	5.0	—	7.0	10.0	4.0	350	—	
MS 2	Mixed shipment II	—	5.0	—	7.0	12.0	4.0	320	—	
ML	Mixed light	—	—	—	—	15.0	3.5	260	—	

Volatile oil content: The volatile oil content of different grades of cardamom capsules shall be not less than 3.5 per cent on capsule basis and in the case of seed grades it shall not be less than 4.0 per cent on the seed basis when determined in accordance with the method given in 15 of IS: 1797–1973.

Extraneous matter: The proportion of calyx pieces, stalk bits and other extraneous matter in cardamom in capsules and seeds shall be not more than 5 per cent and 2 per cent by mass, respectively, when determined in accordance with the method given in 5 of IS: 1797–1973.

Empty and malformed capsules: The proportion of immature and shrivelled capsules, which are not fully developed, shall be not more than 7 per cent by mass and shall be determined after separating them in accordance with the method given in 5 of IS: 1797–1973. Percentage of immature and shrivelled capsules shall be not more than the limit prescribed for each grade in Table A1.1.

Blacks and splits: Blacks include capsules having visible blackish to black colour and splits include those capsules which are open at corners for more than half the length. The proportion of blacks and splits shall be not more than 15 per cent by count. For this purpose, 100 capsules shall be taken from the sample and the number of blacks and splits separated and counted.

Light seeds: The proportion of light seeds in cardamom seeds shall be not more than 5 per cent by mass, when separated in accordance with the method given in 5 of IS: 1979–1973. Light seeds shall include seeds which are brown or red in colour broken or immature and shrivelled.

Grades: The cardamoms in capsules and seeds may also be graded before packing. There shall be 25 grades of cardamom with capsules and 3 of cardamom seeds. The designations of the grades and their requirements are given in Table 1.1 for cardamom with capsules and in Table A1.2 for cardamom seeds.

1.3   Packing and marking

1.3.1   Packing

Capsules: Cardamoms with capsules shall be packed in clean, sound and dry tinplate containers or wooden cases suitably lined with polyethylene or waterproof paper or kraft paper or in new jute bags lined with polyethylene or waterproof paper.

Table A1.2 Grade designations of cardamom seeds and their requirements

Grade designation	Trade name	Extraneous matter, per cent by mass max	Light seeds, per cent by mass max	Mass in g/l min	General characteristics
CS1	Prime	0.5	3.0	675	Decorticated and dry seeds of any variety of *Elettaria cardamomum*
CS2	Shipment	1.0	5.0	660	
CS3	Brokens	2.0	–	–	

Seeds: Cardamom seeds shall be packed in clean and dry tinplate containers or wooden cases lined with polyethylene or waterproof paper or kraft paper.

1.3.2   *Marking*

The following particulars shall be marked or labeled on each container:

(a) Name of the material, and the trade name or brand name if any;
(b) Name and address of the manufacturer or the packer;
(c) Batch or code number;
(d) Net mass in metric units;
(e) Grade of the material (if graded);
(f) Country of origin.

1.4   Sampling

Representative samples of the material shall be drawn and tested for conformity to this specification as prescribed in 3 of IS: 1797–1973.

2   LARGE CARDAMOM – SPECIFICATION

2.1   Scope

This standard specifies the requirements and methods of sampling and test for large cardamom, in capsules and seeds of *Amomum subulatum* Roxb., and other related species.

2.2   Reference

The Indian standards listed below are necessary adjuncts to this standard.

IS No.	Title
1070 : 1977	Water of general laboratory use
1797 : 1985	Method of test for spices and condiments
13145 : 1991	Method of sampling for spices and condiments.

2.3   Requirements

2.3.1   *Description*

Large cardamom in capsules: Large cardamom are the dried, nearly ripe fruits of *Amomum subulatum* Roxb., and other related species. The capsules shall be of the colour ranging from brown to pink, ovoid and more or less triangular shaped having ribbed appearance. The capsules may be clipped and their pedicels removed. The capsules shall be well formed with sound seeds inside.

Large cardamom seeds: The cardamom capsules may also be decorticated and the seeds separately packed for trade purpose.

2.3.2   Odour and Taste

The odour and taste of large cardamom capsules and seeds shall be characteristic and fresh. They shall be free from foreign odour and taste.

2.3.3   Freedom from insects, moulds

Large cardamom capsules and seeds shall be free from living insects and mould and shall be practically free from dead insects and rodent contamination visible to the naked eye with such magnification as may be necessary. If the magnification exceeds 10X, this fact shall be stated in the test report.

Extraneous matter: Large cardamom capsules and seeds shall be free from visible dirt or dust. The proportion of pieces of calyx and stalk and other extraneous matter shall not be more than 5 per cent when determined by the method specified in 4 of IS 1797 : 1985.

2.3.4   Empty and shrivelled capsules

The proportion of immature and shrivelled capsules shall not be more than 7 per cent when determined after separating them in accordance with the method given in 4 of IS 1797 : 1985.

2.3.5   Light seeds

The proportion of light seeds in large cardamom seeds shall not be more than 5 per cent when determined in accordance with the method given in 4 of IS 1797 : 1985.

2.3.6   Mass per litre

The mass of large cardamom capsules and seeds contained in 1 l shall be determined in accordance with the method given in 3 of IS 1797 : 1985.

2.3.7   Chemical requirements

The large cardamom capsules and seeds shall also comply with the requirements given in Table A1.1.

Notes

1   The determination of moisture content and total ash shall be made on the whole capsules.
2   The determination of volatile oil shall be made on the seeds obtained after separation of the skin and decortication.

2.4 Packing

Cardamom capsules shall be packed in clean, sound and dry tinplate containers or in suitably lined wooden cases or in suitably lined new jute bags. Lining materials may be for example, water proof paper, kraft paper or plastic material.

Cardamom seeds shall be packed in clean and dry tinplate containers, or wooden cases suitably lined, for example, with water proof paper, kraft paper or plastic material of food grade quality.

2.5 Marking

The following particulars shall be marked or labelled on each container:

(a) Name of the material, trade name or brand name, if any;
(b) Name and address of the manufacturer or packer;
(c) Batch or code number;
(d) Net mass;
(e) Year of harvest.

2.6 Sampling

Representative samples of large cardamom in capsules and seeds shall be drawn by the method prescribed in IS 13145 : 1991.

2.7 Tests

Tests shall be carried our in accordance with 3.4, 3.6, 3.7, 3.8, 3.9 and col. 4 of Table 3.

2.8 Quality of reagents

Unless specified otherwise, pure chemicals and distilled water shall be employed in tests.

Note

'Pure chemicals' mean chemicals that do not contain impurities which affect the results and analysis.

Table A1.3 Chemical requirements

Sl. No.	Characteristics	Requirement	Methods of Test, Ref. to Cl of IS 1797 : 1985
1	Moisture, content percent (m/m), *Max*	12.0	9
2	Volatile oil content, (ml/100 g) on dry basis, *Min*	1.0	15

Appendix 2
Selected recipes

RECIPES WITH CARDAMOM

The use of spices and herbs has evolved into an art in the Indian cuisine, as there is no other people who uses so much of spices in their food. For the Western mind it is often difficult to fathom the deep involvement and bond between the Indian people and spices and herbs. Though cardamom is not as commonly used as that of black pepper, chillies, ginger or coriander, it is nevertheless an essential ingredient in many meat and vegetable dishes, in sweets as well as in spice blends. An effort is made here by the editors to present an assortment of recipes where cardamom is used as one of the ingredients.

LAMB BIRIYANI

Ingredients

LAMB

- 1 kg lamb chops
- 1½ tablespoon ground coriander
- 1 teaspoon ground paprika
- ¼ teaspoon ground cumin, aniseed turmeric, cinnamon, cardamom, cloves
- 2 teaspoons grated fresh ginger
- 2 cloves garlic, crushed
- 1 teaspoon fresh lemon juice
- 3 tablespoons oil
- 4 large onions, finely chopped
- 3 tablespoons ground almonds
- 6 tablespoons plain yoghurt

RICE

- 4 cups long grain rice, washed and drained
- 3 tablespoons ghee
- 2 onions, finely sliced
- 1 teaspoon grated fresh ginger
- 1 clove garlic crushed
- 6 cups (1½ l) water
- ¼ teaspoon each ground turmeric, cinnamon, cardamom, cloves and 1 bay leaf
- 4 tablespoons evaporated milk
- Salt to taste

Method

To prepare lamb: Remove excess fat from chops, place lamb in a bowl. In another bowl, mix coriander, paprika, cumin, aniseed, turmeric, cinnamon, cardamom, cloves, ginger, garlic, vinegar and lemon juice. Add to lamb and mix well. Allow to stand for 15 min. Heat oil in a large saucepan, sauté onions until golden brown. Add marinated meat. Cover saucepan and simmer with no added water for 40 min. Remove from heat, add almonds and yoghurt. Stir to prevent sticking. Cook, covered, until liquid has dried up and oil rises. Remove saucepan from heat and put aside.

To prepare rice: In large saucepan, melt ghee, sauté onions until golden brown. Add ginger and garlic and sauté for 2 min. Add rice, water, turmeric, cinnamon, cardamom, cloves and bay leaf. Mix well. Boil, covered, until rice is three-quarters cooked, then reduce heat to very low. Add evaporated milk, and continue to cook until remaining liquid is absorbed. Stir rice with a fork, remove from heat. Transfer half the rice to a serving dish or casserole. Spread half the curry lamb over it. Make another layer of rice, and top dish with remaining lamb. Garnish with fried onion rings and cashew nuts.

BANNU KABAB

Ingredients

- Chicken breasts – 12 pieces (boneless)
- Ginger paste – 40 g
- Garlic paste – 40 g
- Lemon juice – 45 ml
- White pepper powder – 5 g
- Gram flour – 60 g
- Bread crumbs – 80 g
- Ginger – 20 g
- Coriander – 20 g
- Cardamom powder – 2 g
- Egg yolks – 3
- Oil – 10 ml
- Salt – To taste

Method

Clean each breast piece and cut into two. Mix ginger paste, garlic paste, lemon juice, salt and pepper and rub the chicken pieces with this mixture. Keep aside. Clean and chop the ginger and coriander. Heat butter and oil in a pan. Add gram flour and cook till golden brown. Add the marinated chicken and sauté. Add bread crumbs and mix well. Skewer each chicken piece horizontally and cook for about 7–8 min in a moderately hot tandoor. Coat the chicken pieces with the egg yolks and cook further till the coating turns golden brown. Remove and sprinkle cardamom powder and coriander. Serve immediately.

SPICED MUTTON

Ingredients

- 2 lb lean minced mutton
- 2 medium sized tomatoes
- 4 medium sized onions
- $\frac{1}{2}$ teaspoon pepper
- $\frac{1}{2}$ teaspoon ground cardamom
- 1 teaspoon salt
- 1 small jar unflavoured curd.

- 4 oz, butter, $\frac{1}{2}$ teaspoon clove powder
- 1 tablespoon dry ginger powder
- 1 tablespoon turmeric
- $2\frac{1}{2}$ teaspoon coriander
- 1 clove garlic
- Coriander leaves to garnish.

Method

Melt the butter in a heavy saucepan. Put in the minced onions and cook gently until the onions are browned. Add the minced mutton to which the turmeric, cardamom, ginger, coriander and pepper have been added. Stir and cook for 7–8 min. Add the cloves, cinnamon, curd, salt and the crushed garlic. Peel the tomatoes and stir into the meat mixture. Bring to boil. As soon as it boils, turn heat down, cover the dish, and simmer for 20 min. Serve on a bed of hot rice and garnish with a little chopped coriander leaves.

CHICKEN LIVER DELIGHT

Ingredients

- $\frac{3}{4}$ lb chicken liver
- 1 tablespoon flour
- Salt and black pepper
- 2 teaspoons Indian curry powder
- $\frac{1}{4}$ teaspoon ground cardamom
- 1 oz butter
- 2 small onions
- 4 oz mushrooms
- Coriander leaves
- $1\frac{1}{2}$ lb potatoes.

Method

Peel potatoes, cut into small pieces and boil in salt water. Slice the liver, wash, dry and dust all over with seasoned flour and about $\frac{1}{4}$ teaspoon Indian curry powder. Melt the butter, add the remaining curry powder and cardamom and when it is really hot, fry the liver till it is lightly brown all over. Peel and slice the onions and the mushrooms. Chop the coriander leaves and fry with the liver. Sieve the cooked potatoes and arrange them to form a border around the serving dish. Remove liver from the pan and dish up in the centre of the potato border.

MADRASI PULAO

Ingredients

- 1 cup rice
- $\frac{1}{4}$ cup butter

- $\frac{1}{4}$ cup chicken stock base
- 6 cloves
- 1 stick (3") cinnamon
- 3 teaspoons ground cardamom
- $\frac{1}{2}$ teaspoon salt
- 2 cups water
- $\frac{1}{2}$ cup raisins
- $\frac{1}{4}$ cup toasted almonds, sliced thinly.

Method

Fry rice in butter in a heavy saucepan until it is golden brown. Blend in chicken stock base. Add cloves, cinnamon stick, cardamom powder and salt. Stir in water. Cover and simmer over a low heat for 25–30 min. Remove cinnamon stick and the whole cloves. Rinse and drain raisins. Stir raisins and almonds into rice. Serve at once.

SUGAR PLUM COOKIES

Ingredients

- $\frac{3}{4}$ cup butter
- $\frac{1}{3}$ cup sifted powdered sugar
- 1 teaspoon lemon peel
- $\frac{1}{2}$ teaspoon ground cardamom
- $\frac{1}{3}$ teaspoon salt
- 1 tablespoon cold water
- 2 cups sifted flour
- $\frac{1}{3}$ cup toasted almonds, chopped
- Sifted powdered sugar.

Method

Cream butter and powdered sugar together till light and fluffy. Blend in lemon peel, cardamom, salt, vanilla and cold water. Gradually add sifted flour, blending well after each addition. Stir in chopped almonds. Roll dough into small balls and place on a lightly greased baking sheet. Bake high in the oven at 400 °F for 10–12 min, or until lightly brown. Roll at once in sifted, powdered sugar and cool. Roll in sifted, powdered sugar again just before serving.

CARDAMOM CHOCOLATE CREAMS

Ingredients

- 1 bar (4 oz) sweet chocolate
- $\frac{3}{4}$ cup whipping cream

- $^3/_4$ cup milk
- 1 tablespoon sugar
- $^1/_4$ teaspoon ground cardamom
- $^1/_4$ teaspoon salt
- 3 eggs
- 1 oz grated semi-sweet chocolate.

Method

Combine chocolate, whipping cream, milk, sugar, ground cardamom and salt in a small saucepan. Heat and stir constantly until chocolate is melted and mixture blended. Beat eggs well and then pour warm chocolate mixture over them. Turn into 6 baking cups. Place cups in a pot of water. Bake in a moderately slow oven (325 °F) for an hour. Place cups on a wire rack to cool. Sprinkle grated semi-sweet chocolate over pudding, allowing it to melt. Chill 6–8 h before serving.

BANANA COCONUT TOFFEE

Ingredients

- Banana, boiled and mashed after removing threads and seeds – 5 cups
- Milk – 2 cups
- Grated coconut – 2 cups
- Melted ghee – 1 cup
- Sugar – 5 cups
- Water – 2 cups
- Cardamom powder – 1 teaspoon.

Method

Grind the grated coconut coarsely and stir it in a hot skillet to remove moisture. Pour ghee and stir till it is light brown. Prepare sugar syrup. Add mashed banana mixed in milk. When it boils add coconut gratings. As it thickens to two third consistency add cardamom powder and pour into a greased plate. Cut into desired shapes while still hot.

SWEET CASHEW

Ingredients

- Cashewnut dried in the sun and fried in ghee (the nuts should not change colour) – 4 cups
- Sugar – 2 cups™
- Water – 1 cup
- Butter – $^1/_2$ teaspoon

- Lime juice – $\frac{1}{2}$ teaspoon
- Salt – a pinch (2 g)
- Cardamom powder – $\frac{1}{2}$ teaspoon
- Pepper powder – $\frac{1}{2}$ teaspoon
- Gingely seeds (husked white) – 1 dessert spoon.

Method

Melt sugar and water over fire and strain. Keep over fire and as it thickens add butter. When it melts add lime juice, salt, cardamom powder and pepper powder. Remove at once from fire and add fried gingely seeds. Beat the syrup well and as the colour changes add the cashew nuts. Keep over low fire until the nuts get coated with sugar and separate. Let the frosted nuts remain in the vessel till it cools and then store in glass jars.

SEMOLINA BALLS

Ingredients

- Bombay rava (Semolina) – 2 cups
- Melted ghee – 3 dessert spoons
- Cashewnut chopped – 2 dessert spoons
- Sugar – 2 cups
- Raisins – 1 dessert spoon
- Cardamom powder – 1 teaspoon
- Hot milk – $\frac{1}{2}$ cup.

Method

Fry and drain chopped cashewnut and raisins in hot ghee. Add semolina to the remaining ghee and fry over low fire. Immediately add sugar and cardamom powder and take from fire. Sprinkle hot milk and mix all the ingredients together and shape into balls.

WHEAT BURFI

Ingredients

- Wheat flour – 1 cup
- Sugar – 2 cups
- Water – 1 cup
- Melted ghee – 1$\frac{1}{2}$ cup
- Cardamom powder – $\frac{1}{2}$ teaspoon.

Method

Prepare sugar syrup of one-thread stage. As soon as the syrup is ready take it off from fire. Keep one cup of ghee on fire and when it is hot lower the heat and fry the flour in it till flavour comes out. Meanwhile keep the remaining ghee on a low fire and keep it hot. When the flour is fried remove it for 5 min and pour the sugar syrup into it. Return to fire and keep stirring constantly. Add the hot ghee gradually. After a few minutes of stirring, ghee comes out of the mixture. Add cardamom powder and quickly transfer into a flat plate. Cut the desired shape while it is still hot.

SWEET SAFFRON RICE

Ingredients

- Saffron strands – $\frac{1}{2}$ teaspoon
- Hot milk – 2 tablespoon
- Rice – $\frac{3}{4}$ cup (150 g)
- Ghee – 45 g
- Cardamom capsules – 3
- Stick cinnamon – 1 inch
- Cold milk – 2/3 cup (170 m)
- Sugar – 4 tablespoons
- Thick cream – 2 tablespoons
- Flaked almond (roasted) – 30 g
- Walnut (shelled) – 30 g
- Raisins – $\frac{1}{4}$ cup (30 g).

Method

Soak saffron strands in the hot milk. Cook rice in boiling water for 5 min, then drain. Melt ghee in a saucepan, sauté rice, cardamom and cinnamon for 3 min. Add cold milk and sugar, cook over low heat, covered, until the rice is fully cooked. Add cream and saffron milk, cook for 1 min. Spoon into serving dish, sprinkle with almonds, walnuts and raisins.

KERALA CHICKEN CURRY

Ingredients

- Chicken – $1\frac{1}{2}$ nos. (1.4 kg)
- Red chilli whole – 20 g
- Coriander leaves – 15 g
- Turmeric – 5 g
- Ginger and garlic – 15 g
- Sauf, cardamom, cinnamon, clove and bay leaf – 4 g
- Grated coconut – 1 coconut
- Red onion (small) – 25 g
- Roast and grind all the above ingredients to a fine paste
- Oil – 100 g

- Chopped onion – 150 g
- Chopped tomato – 60 g
- Curry leaves – 1 spring
- Salt – To taste.

Method

Clean, remove skin, wash and cut each chicken into 8 pieces (total 12 pieces) and keep aside.

Heat oil in saucepan or hundi. Fry chopped onions till brown. Add curry leaves and tomatoes and sauté for few minutes. Add ground masala, salt and stir and cook the masala well. Add ½ cup of water and cook again for 3 min. Now add chicken, stir and cook for 10–15 min. Adjust seasoning, remove from fire and serve hot with Kerala rice, appam, Kerala paratha etc.

VEGETABLE SAFFRANI KHORMA

Ingredients

- Carrot – 100 g
- Beans – 100 g
- Cauliflower – 100 g
- Potato – 100 g
- Green peas – 100 g
- Onion – 200 g
- Green chillies – 6
- Ginger and garlic paste – 5 g
- Oil – 50 g
- Cashewnut – 100 g
- Khus-khus – 50 g
- Cardamom, cinnamon, cloves, bay leaves – 5 g
- Saffron – ¼ g
- Milk – ¼ cup
- Salt – To taste.

Method

Peel and cut vegetable into cubes. Boil it separately and drain. Peel, boil and puree the onion. Soak khus-khus and cashewnuts in warm water for 1½ h. Strain and grind to a smooth paste. Chop green chilli, make ginger-garlic paste. Soak saffron in a little warm milk.

Heat oil, put whole garam masala, stir for a few seconds. Add onion paste and stir till it becomes transparent. Then add ginger and garlic paste and chopped green chillies. Stir for a few seconds and when it gives the cooked flavour, add khus-khus and cashewnut paste. Stir and cook till the paste is cooked when the oil separates out. Now add the boiled vegetables and cook for some more time. Finish by adding the saffron and adjust the seasoning.

CARDAMOM SOUFFLE

Ingredients

- Sugar – 200 g
- Eggs – 4
- Milk – 200 ml
- Fresh Cream – 200 ml
- Gelatine – 10 g
- Cardamom powder – 1 teaspoon.

Method

Separate egg yolks in a basin and whisk with ½ the sugar. Pour hot milk in that and whisk thoroughly and cool. Melt gelatine, cool and add into the mixture. Whip the cream and keep aside. Beat the egg whites with remaining sugar till stiff and slowly fold the cream, the egg whites and cardamom powder with the mixture. Make it cool and pour into the moulds. Keep in refrigerator for 1 h. Demould and serve.

FRUIT BUNS

Ingredients

- Hot roll mix – 1 packet
- Sweetened fruits juice – ¾ cup
- Ground cardamom – ½ teaspoon
- Ground mace – ¼ teaspoon
- Sugar – 3 tablespoons
- Glace fruit – 1 cup
- Oil – 1 tablespoon.

Method

Remove yeast package from hot roll mix. Heat fruit juice to lukewarm. Turn into a large bowl and sprinkle in yeast and stir till dissolved. Blend in cardamom, mace and sugar. Add flour mixture and stir until blended. Work in mixed glace fruits. Form dough into a ball. Spread oil over the surface and turn in bowl to oil lightly on all sides. Cover and leave for about an hour in a warm place till it has increased to about twice the original size. Punch down, cover and let rest for a further 5 min. Oil hands lightly and form into a dozen small balls. Place on a lightly greased baking tin. Cover and leave for a further 35 min for them to rise in bulk. They should double in size again in this time. Bake in a oven (400 °F) 25–30 min or until rich golden brown. Turn out into wire racks to cool.

MURGI–DAHI–KALIYA

Chicken cooked in yoghurt and spices, served with rice or bread.

Ingredients

- Fleshy chicken, about 2 kgs, after cleaning – 2
- Sour yoghurt/curds – 4 cups
- Salt – $2\frac{1}{2}$ teaspoons
- Medium onion – 1
- Asafoetida – 1 teaspoon
- Clarified butter/ghee – 1 cup
- Milk – $\frac{1}{2}$ cup
- Almonds – 25

Spices

- Black pepper corns – 10
- Dry ginger – 8 cm
- Cardamom – 6
- Large cardamom – 6
- Cinnamon sticks – 6
- Dry red chillies – 5
- Bay leaves – 2
- Dry coriander seeds – $1\frac{1}{2}$ teaspoons
- Saffron – $\frac{1}{2}$ teaspoon.

Cut chickens into large pieces. Grind ginger into fine paste. Grind cumin, dry red chillies and dry coriander seeds together. Powder asafoetida and half of listed quantity of cardamoms, cinnamon and black pepper. Finely chop onion. Blanche and slice almonds, and fry to light golden colour.

Heat ghee in wide, heavy-bottom pan. Season with asafoetida, bay leaves and remaining whole spices. Stir and add chopped onion. Fry till onion turns golden brown. Add ground ginger and chicken.

Churn yoghurt and put into pan. Add salt. Mix well. Cover pan with fitting lid. Allow to simmer for 20 min on slow fire. Add ground chillies, cumin and coriander seeds. Stir well and cover.

To avoid ingredients from sticking to bottom of pan, stir after every few minutes with a flat spoon.

After all liquid has evaporated and chicken pieces are almost tender, fry well, till ghee oozes out. Add $2\frac{1}{2}$ cups hot water. Sprinkle powdered spices, and bring to a boil. Remove.

If desired, steam some potatoes. Peel and cut into large pieces, and deep fry. Add to chicken, but increase quantity of hot water proportionately.

Serve in a deep dish. Garnish with fried almonds.

STUFFED GREEN CAPSICUM

Ingredients

- Capsicum – 450 g (8 medium size)
- Oil – 100 g
- Salt – To taste
- Onion – 150 g
- Minced mutton – 500 g
- Green chillies – 4–5
- Turmeric powder – 2 g
- Chilli powder – 5 g
- Coriander powder – 5 g
- Cinnamon – Roasted and powdered
- Cardamom – 5 g
- Cloves – 3–4
- Ginger garlic juice – 5 ml
- Coriander leaves – ¼ bunch.

Method

Chop onions, green chillies, coriander leaves and ginger. Slice the top of the capsicum, remove the seeds, apply salt and keep it upside down. Heat oil add mustard seeds, chopped onion, green chillies, ginger and powdered garam masala (powdered spice mixed well).

Add minced mutton, cook well, check for seasoning and add lime juice. Stuff this mixture in capsicum and cover the top with maida batter. Heat remaining oil and cook the stuffed capsicum till it is soft.

FRUIT CHAT

A sweet and sour tropical salad with a dash of spice.

Ingredients

- Apple – 1
- Papaya – ¼ kg
- Pineapple – ¼ kg
- Grapes, black – 100 g
- Green – 100 g
- Sweet lime/orange segments – 150 g
- Lime – 1
- Chat masala powder (Black salt, Mango powder, pepper powder, cloves, cardamom, chilli powder, ajwain) – 1 teaspoon
- Salt – To taste
- Coriander leaves – 5 g.

Method

Peel and cut the fruits into dices. Mix with chat masala. Add lime juice, check the seasoning and sprinkle finely chopped coriander leaves on top. Serve cold.

CARDAMOM CHOCOLATE CREAMS

Ingredients

- Sweet chocolate – 1 bar (4 oz)
- Whipping cream – $^3/_4$ cup
- Milk – $^3/_4$ cup
- Sugar – 1 tablespoon
- Ground cardamom – $^1/_4$ teaspoon
- Salt – $^1/_4$ teaspoon
- Eggs – 3
- Grated semi-sweet chocolate – 1 oz.

Method

Combine chocolate, whipping cream, milk, sugar, ground cardamom and salt in a small saucepan. Heat and stir constantly until chocolate is melted and mixture blended. Beat eggs well and then pour warm chocolate mixture over them. Turn into 6 baking cups. Place cups in a pot of water. Bake in a moderately slow oven (325 °F) for an hour. Place cups on a wire rack to cool. Sprinkle grated semi-sweet chocolate over pudding, allowing it to melt. Chill 6–8 h before serving.

MUSHROOM MASALA

Ingredients

- Mushrooms shredded – 250 g
- Green chillies – 2
- Big onions – 2
- Ginger shredded – 1 teaspoon
- Garlic cut into thin pieces – 3 flakes
- Clove – 4
- Cardamom – 2
- Cinnamon – 2 pieces
- Chilli powder – 1 teaspoon
- Pepper powder – $^1/_2$ teaspoon
- Turmeric powder – $^1/_2$ teaspoon.

Method

Fry the onions, ginger, green chillies, curry leaves in oil, add the mushrooms, masala paste and water. Close the vessel with lid. Cook till done for 15 min and serve hot.

MUSHROOM CUTLET

Ingredients

- Mushroom – 250 g
- Potato – 200 g
- Big onions – 4
- Ginger shredded – 1 teaspoon
- Green chillies – 4
- Curry leaves – 6
- Chilli powder – 100 g
- Green peas – 100 g
- Turmeric powder – 1 teaspoon
- Masala powder (cinnamon, clove, cardamom 1:2:1) – 2 teaspoons.

Method

Cook the shredded mushrooms with turmeric powder, chilli powder, coriander powder, masala and salt. Fry the green chillies, onions, curry leaves and add to the cooked mushrooms. Mash the cooked potatoes and peas and add to the above. Make into small balls and dip in beaten egg white. Immerse in powdered bread crumbs and deep fry in oil till golden brown.

TANDOORI SQUAB (OR QUAIL)

Ingredients

4 squabs or 8 quails, ready to cook

Method

Clarified butter for basting radishes and lime juice for garnish.
The marinade

- Clarified butter – 3 tablespoons
- Plain yoghurt – $\frac{1}{4}$ cup
- Lemon juice – $1\frac{1}{2}$ tablespoons
- Prune butter – 1 tablespoon
- 2 large cloves of garlic peeled – $\frac{1}{2}$ oz
- $1\frac{3}{4}$ inch piece fresh ginger peeled and cut into chunks – $\frac{1}{2}$ oz
- Lemon zest – $\frac{1}{2}$ teaspoon
- Saffron threads, crushed – $\frac{1}{2}$ teaspoon
- Ground cardamom – 1 teaspoon
- Cayenne pepper – $\frac{1}{4}$ teaspoon
- Black pepper – $\frac{1}{2}$ teaspoon
- Salt – To taste.

1   With poultry shears, cut the backbones or the squabs. If you are using quails, split them lengthwise. Pull the skin off the squabs or quail, using paper towel to get a better grip. Place the squabs or quail, breast side up, on the work surface. Press down with the heel of your palm to break the breastbone and flatten. Using a fork make diagonal slashes, a half-inch-deep and one-inch-apart, on the meat, along the grain. Make a slit with a paring knife in the meat between the thigh and the breast and secure the ends of the leg through the slit. Place the squabs or quails in a shallow dish.

2   Process all the ingredients of the marinade in the blender or a food processor until they are liquefied. Pour and marinade over the squabs or quail. Turn and toss to coat thoroughly. Cover and marinate at room temperature for 15 min, or 4 h in the refrigerator (maximum, overnight). Take the squabs or quail from the refrigerator at least 30 min before cooking.

3   Light and prepare a covered charcoal grill until white ash covers the coals and the heat subdues to a moderately hot level. Scatter presoaked grapevine cuttings over the coals. Lightly brush the racks with oil.

4   Coat the squabs or quail lightly with clarified butter and place them breast side down, on rack. Barbecue, covered, with the vents open, turning three times, for 17 min (15 min for quail) or until the meat is cooked medium rare (the meat will be pink near the joints). Transfer the squabs or quail to a heated platter and surround with radishes. Serve sprinkled generously with lime juice.

TANDOORI CHICKEN

Ingredients

4 Cornish hens, or poussin (very young chickens), or 2 small ($2\frac{1}{2}$ pound) broilers
$1\frac{1}{2}$ cups plain yoghurt.

Method

Oil for basting.
Slices of tomato and cucumber for garnish.
The marinade

- Olive oil – $1\frac{1}{2}$ cup
- Vinegar – $\frac{1}{4}$ cup
- $\frac{1}{3}$-inch piece fresh ginger peeled and cut into chunks – 2 oz
- 8 large cloves of garlic peeled – 2 oz
- Ground cumin – 2 tablespoons
- Ground coriander – 2 tablespoons
- Ground cardamom – 2 teaspoons
- Ground clove – 1/2teaspoon
- Cayenne pepper – 1–2 teaspoons
- Black pepper – 1 teaspoon
- Salt – $2\frac{1}{2}$ teaspoons
- Red food colouring – 1 tablespoon
- Yellow food colouring – 2 tablespoons.

1   With poultry shears, cut the backbones of the hens. Place the hens, breast side up, on the kitchen counter. Press down on the breast bone with the heel of your palm to flatten. If you are using broiler chickens, split them lengthwise into halves.

2   Pull the skin off the hens, using paper towel to get a better grip. Prick the hens all over with a fork or a thin skewer. Make diagonal slashes, a half inch deep and one inch apart, on the meat, along the grain. Make a slit with a paring knife in the meat between the thigh and the breast. Secure the ends of the legs through the slit. Put the hens in a large bowl and set aside.

3   Process all the ingredients of the marinade in a blender or food processor until they are liquefied. Pour the marinade over the hens. Mix thoroughly, turning several times to coat the pieces evenly with the marinade. Cover and refrigerate overnight (a maximum of 2 days) or marinate at room temperature for at least 1 h (preferably four).

4   Take the hens from the refrigerator at least 1 h (maximum, 5 h) before cooking.

5   Light and prepare a covered charcoal grill until white ash covers the coals and the heat subdues to a moderately hot level. Scatter presoaked grapevine cuttings over the coal. Lightly brush racks with oil.

6   Coat the hens lightly with oil and place them, breast side down, on the rack. Barbecue, covered with the vents open, turning three or four times, without basting, for about 25 min – until the juices run clear when pierced with a knife at the joint. Serve the hens garnished with slices of tomato and cucumber.

YOGHURT FISH

Ingredients

- Fish pieces 1" thick and 3" long slices – 4 kg
- Whole cumin seeds – 4 teaspoons
- Cardamom – 4
- Coriander powder – 1 tablespoon
- Pepper corns – 4
- Chilli powder – 2 teaspoons
- Mango powder – 1 teaspoon
- Yoghurt – 250 g
- Oil – 1 tablespoon
- Finely chopped fresh Coriander (optional) – 1 tablespoon.

Grind all the above to a fine powder.

Method

Beat yoghurt. When smooth, mix in spices. Heat oil in a pan, gradually add the yoghurt mixture a tablespoon at a time. Place the fish pieces in the same pan and cook gently, stirring occasionally until the fish is cooked and flakes easily (about 10 min). Add salt to taste, garnish with finely chopped fresh coriander.

RECIPE FOR GARAM MASALA

Ingredients

- Cumin seeds – $\frac{1}{2}$ oz
- Cardamom – 2 oz
- Black pepper – $\frac{1}{4}$ oz
- Cinnamon – $\frac{1}{4}$ oz
- Cloves buds (optional) – $^{1}/_{4}$ oz.

SPICES FRUIT

Ingredients

- 1 lb canned apricot halves
- 1 lb canned pin apple-mango mixture
- 1 lb canned papaya chunks
- 1 thinly sliced lemon with skin on
- $\frac{1}{2}$ cup sugar
- 3 tablespoons ground cardamom
- 6 clove buds
- 1 teaspoon cinnamon or one 3" stick.

Method

Drain fruit reserving all syrup. Combine drained appricots, pineapple-mango mixture, papaya chunks, lemon slices and quartered orange slices in a china bowl. Turn reserved fruit syrup into a saucepan. Stir in sugar, cardamom and cinnamon and cloves. If cinnamon is in stick form, break it in small pieces and add it. Simmer 25 min. Remove from heat and cool, pour over fruit. Cover and refrigerate overnight.

MUGHAL SHAHI KORMA

Ingredients

- 750 g lean mutton, cut into small pieces
- Juice of 1 lemon
- 1 teaspoon saffron
- 1 cup boiling water
- 50 g fresh ginger, scrubbed but not peeled
- 100 g cashew nut
- 4 green chillies, stalks removed
- Seeds from 20 cardamom fruits
- 10 cloves
- 2 teaspoon ground coriander seeds
- 1 teaspoon cumin seeds

- 1 teaspoon ground cinnamon
- 300 ml plain unsweetened curd
- 50 g ghee
- 2 medium size onions, peeled and thinly sliced
- 2 teaspoon salt
- 4 cloves garlic, peeled and thinly sliced
- 1½ teaspoon freshly ground pepper
- 300 ml thick cream
- 50 g blanched almonds
- Fresh coriander to garnish.

Method

Rub the pieces of mutton with lemon, and put to one side. Immerse saffron in boiling water and allow to stand for 10 min to release the flavour and colour. Put ginger, cashew nuts, chillies, cardamom seeds, cloves, ground coriander and seeds, cumin seeds, cinnamon and curd in a liquidiser and blend to a smooth paste. Draw in any excess lemon juice from the mutton into the liquidiser and mix in with the paste. Heat ghee in a heavy-based saucepan and fry the mutton gently. Remove the mutton and keep aside. Fry onions and garlic in ghee until soft, and then add the liquidised spice mixture and bring to boil. Next add saffron and its water, stir in well and add mutton. Sprinkle in salt and pepper, and bring to boil. Cover and simmer until the mutton is tender. Reduce heat, and stir in cream and almonds. Cook gently for a further 2–3 min but do not allow to boil. Turn out into a heated serving dish. Garnish with coriander leaves.

MOCK EGG CURRY

Ingredients (For eggs)

- 250 g paneer
- 100 g turdal
- ½ teaspoon turmeric
- 1 tablespoon cornflour
- 1 onion
- 2 cloves garlic
- 2 cups nutrinugget
- 1 cup curd
- ¼ teaspoon nutmeg
- 6 cardamoms
- 1 green chilli
- 4 tablespoons chopped coriander leaves
- 4 tablespoons flour
- 1 cup water
- Salt to taste
- Oil for frying.

Method

Soak the Nutrinuggets in yoghurt (curd) for 2 h. Soak the turdal for 2 h and grind to a smooth paste with salt and turmeric. Prepare small balls (size of an egg yolk) out of the turdal paste. Steam in a cooker till firm. Cool.

Peel and slice onion and garlic. Sauté in little oil with green chilli till soft. Grind it with nuggets, nutmeg, cardamom and salt. Knead it to a hard paste.

Mix paneer with little salt and knead to a smooth dough, sprinkle cornflour to bind. Prepare an outer covering for the dal yolk out of the paneer. Shape into size of a real egg. Roll each egg in the ground nutrinugget paste to form and outer layer about $\frac{1}{2}$ cm thick.

Prepare a thin batter with flour and water. Dip each egg in batter. Deep fry in oil till golden brown. Keep aside.

Ingredients (For the curry)

- 3 onions
- 4 large tomatoes
- 8 cardamoms
- 2 cloves
- 1 stick cinnamon
- 1 piece ginger
- 3 cloves garlic
- 1 teaspoon chilli powder
- $\frac{1}{2}$ teaspoon poppy seeds
- 1 tablespoon coriander powder
- $\frac{1}{2}$ teaspoon turmeric
- 4 tablespoons chopped coriander
- 4 tablespoons fresh cream
- a pinch of saffron
- Salt to taste
- Oil for frying.

Method

Soak saffron in a teaspoon of hot milk. Chop tomatoes. Boil in a tablespoon of water. Cool, mash and sieve. Peel and grate onions. Grind cardamoms, cloves, cinnamon, ginger, garlic, popy seeds and coriander powder to a fine paste.

Heat oil. Sauté the grated onions, add the ground paste, turmeric and chilli powder. Fry well till oil separates. Add the tomato puree, salt and slowly slide the 'fried eggs' in the curry. Simmer till thick on low heat. Pour the soaked saffron and cream. Garnish with coriander leaves. Serve hot.

GRILLED STEAKS

Ingredients (For steaks)

- 500 g minced meat
- $\frac{1}{2}$ teaspoon papaya paste

- 1 level tablespoon ginger-garlic paste
- 2 big onions
- 6 green chillies
- 10 cashewnuts
- 2 teaspoon poppy seeds
- 3 cm cinnamon
- 8 cloves
- 10 cardamoms
- Juice of 2 limes
- 1 bunch coriander leaves
- 1 bunch mint leaves
- 1 tablespoon fried garam flour to bind the steaks
- Salt to taste
- Oil for frying.

Method

Apply papaya paste and ginger-garlic paste to the minced meat and marinate for ½ h. Roast onions, green chillies, cashewnuts and poppy seeds and grind to a paste. Add the onion paste to the minced meat, along with powdered cinnamon, cardamoms and cloves, and the rest of the ingredients. Divide the mixture into eight balls. Flatten the balls to form steaks. Heat oil in a pan and fry the steaks on both sides for 3 min.

Ingredients (For garnishing)

- 100 g grated cheese
- 1 egg yolk
- 4 tablespoons whipped cream
- 8 greased tomatoes
- Salt and pepper to taste
- Little chopped parsley
- Boiled peas and fried potatoes (optional).

Method

Mix grated cheese, egg yolk, cream, salt and pepper to a creamy mixture and spread on the steaks. Line the dish with tomatoes and place under the grill till the cheese melts and gets a light brown colour. Sprinkle with chopped parsley and serve with peas and fried potatoes.

CARDAMOM BUTTER COOKIES

Ingredients

- ½ lb soft butter or margarine
- 1 cup flour

- ½ teaspoon ground cardamom
- ½ cup sugar
- ½ cup corn starch.

Method

Beat the butter until it is light and creamy. Mix flour, cardamom, sugar and starch and add to the butter mixture. Continue to cream till it is well mixed. Wrap in foil and chill thoroughly. Shape the mixture into two inch balls. Flatten with fork. Place two inches apart on ungreased baking tins. Bake at 325 °F for about 20 min or until delicately browned. Decorate with jelly or cream as desired.

ELAICHI MURGH (CARDAMOM-SCENTED BRAISED CHICKEN WITH NECTARINES)

Ingredients

- 24 cardamom pods for 3–3½ pound dressed chicken
- 1/3 cup minced onion
- ¾ stick (6 tablespoons) unsalted butter
- 1 tablespoon peeled and minced ginger
- 2 teaspoon sweet paprika
- ½ teaspoon ground cinnamon
- ½ teaspoon freshly ground pepper
- ¼ teaspoon ground cloves
- ¼ teaspoon freshly grated nutmeg
- 1¼ pounds (about 4) nectarines
- 1½ teaspoon sugar.

Method

In a mortar with a pestle or between sheets of wax paper with a rolling pin pulverize the cardamom pods. In a large heavy flameproof casserole large enough to hold the chicken sauté the onion in three tablespoons of the butter over moderately high heat, stirring, until it is golden, add the ginger and sauté the mixture, stirring, for 2 min. Remove the casserole from the heat and stir in the paprika, the cardamom, the cinnamon, the pepper, the cloves, and the nutmeg. In a food processor or blender blend the mixture with half the nectarines, peeled, pitted and chopped, pulsing, until it is a coarse puree, adding water if necessary to obtain a thick sauce-like consistency, and season the puree with salt.

Add two tablespoons of the remaining butter to the casserole and heat it over moderately high heat until it is hot. Add the chicken on its side and sear it, swirling the casserole gently, for 2 min. Turn the chicken and sear the other side, swirling the casserole, for 2 min. Turn the chicken breast side up, turn off the heat, and let the chicken cook in the residual heat for 1 min. Add the nectarine mixture, bring it to a boil, and simmer the chicken, covered tightly with foil and the lid, for 20 min. Turn the chicken on its side and simmer it, covered tightly, for 6 min. Turn the chicken on its other side and simmer it, covered tightly, for 6 min more. Turn the chicken breast side up, turn off the heat, and let the chicken stand, covered, for 30 min.

In a heavy skillet sauté the remaining nectarines peeled, pitted, and cut into $\frac{1}{2}$-inch slices, in the remaining one tablespoon butter over moderately high heat, for 1 min sprinkle them with the sugar, and sauté them, turning them, until they are glazed. Transfer the chicken to a cutting board, and quarter the chicken. Arrange the chicken on a heated platter, pour some of the sauce over it, and garnish the platter with the nectarine slices. Serve the remaining sauce separately.

Index

9 780415 284936